THE OXFORD ENGINEERING SCIENCE SERIES

GENERAL EDITORS

J. M. BRADY, C. E. BRENNEN, A. L. CULLEN,
T. V. JONES, J. VAN BLADEL,
L. C. WOODS, C. P. WROTH

THE OXFORD ENGINEERING SCIENCE SERIES

1   D. R. RHODES: *Synthesis of planar antenna sources*
2   L. C. WOODS: *Thermodynamics of fluid systems*
3   R. N. FRANKLIN: *Plasma phenomena in gas discharges*
4   G. A. BIRD: *Molecular gas dynamics*
5   H.-G. UNGER: *Planar optical waveguides and fibres*
6   J. HEYMAN: *Equilibrium of shell structures*
7   K. H. HUNT: *Kinematic geometry of mechanisms*
8   D. S. JONES: *Methods in electromagnetic wave propagation* (Two volumes)
9   W. D. MARSH: *Economics of electric utility power generation*
10 P. HAGEDORN: *Non-linear oscillators* (Second edition)
11 R. HILL: *Mathematical theory of plasticity*
12 D. J. DAWE: *Matrix and finite element displacement analysis of structures*
13 N. W. MURRAY: *Introduction to the theory of thin-walled structures*
14 R. I. TANNER: *Engineering rheology*
15 M. F. KANNINEN and C. H. POPELAR: *Advanced fracture mechanics*
16 R. H. T. BATES and M. J. McDONNELL: *Image restoration and reconstruction*
18 K. HUSEYIN: *Multiple-parameter stability theory and its application*
19 R. N. BRACEWELL: *The Hartley transform*
20 J. WESSON: *Tokamaks*
21 P. B. WHALLEY: *Boiling, condensation, and gas–liquid flow*
22 C. SAMSON, M. Le BORGNE, and B. ESPIAU: *Robot control: the task function approach*
23 H. J. RAMM: *Fluid dynamics for the study of transonic flow*
24 R. R. A. SYMS: *Practical volume holography*
25 W. D. McCOMB: *The physics of fluid turbulence*
26 Z. P. BAZANT and L. CEDOLIN: *Stability of structures: principles of elastic, inelastic, and damage theories*
27 J. D. THORNTON: *Science and practice of liquid–liquid extraction* (Two volumes)
28 J. VAN BLADEL: *Singular electromagnetic fields and sources*

Robot Control
The Task Function Approach

CLAUDE SAMSON
Director of Research, INRIA
Valbonne, France

MICHEL Le BORGNE
Teacher, IFSIC and Researcher, IRISA
University of Rennes, France

and

BERNARD ESPIAU
Director of ISIA and Professor at ENSMP
Valbonne, France

CLARENDON PRESS · OXFORD
1991

Oxford University Press, Walton Street, Oxford OX2 6DP

Oxford New York Toronto
Delhi Bombay Calcutta Madras Karachi
Petaling Jaya Singapore Hong Kong Tokyo
Nairobi Dar es Salaam Cape Town
Melbourne Auckland

and associated companies in
Berlin Ibadan

Oxford is a trade mark of Oxford University Press

Published in the United States
by Oxford University Press, New York

© Claude Samson, Michel Le Borgne, and Bernard Espiau, 1990

All rights reserved. No part of this publication may be reproduced,
stored in a retrieval system, or transmitted, in any form or by any means,
electronic, mechanical, photocopying, recording, or otherwise, without
the prior permission of Oxford University Press

British Library Cataloguing in Publication Data
Samson, Claude
Robot control.
1. Industrial robots. Manipulators
I. Title II. Le Borgne, Michel III. Espiau, Bernard
629.8
ISBN 0-19-853805-7

Library of Congress Cataloging in Publication Data
Samson, Claude.
Robot control: the task function approach/Claude Samson,
Michel Le Borgne, and Bernard Espiau.
p. cm.—(The Oxford engineering science series; 22)
Includes bibliographical references
Includes index.
1. Robots—Control systems. 2. Manipulators (Mechanism) I. Le Borgne, Michel.
II. Espiau, Bernard. III. Title. IV. Series.
TJ211.35.S26 1990 629.8'92—dc20 90-36440
ISBN 0-19-853805-7

Typeset by Macmillan (India) Ltd., Bangalore 25
Printed in Great Britain by
Bookcraft (Bath) Ltd., Midsomer Norton, Avon

ANDERSONIAN LIBRARY

3 1. MAY 91

UNIVERSITY OF STRATHCLYDE

FOREWORD

*Michael Brady, Professor of Information Engineering,
University of Oxford*

In industry, robot arms whirl remorselessly and repeatedly to spot-weld automobile bodies, and smoothly execute flowing trajectories as they spray paint on metal panels. They seem so fast, powerful, and flawless, yet their crippling limitations are embarrassing and numerous. For example, most robot arms, despite appearances, are weaklings. An arm weighing 100 kg may not be capable of manipulating stably a load of more than 2 kg! Few arms, even in laboratories, can support a significant fraction of their own weight. In part, this is because conventional designs of critical subparts of arms, particularly the three-roll wrist, produce structures that are weak. Equally, however, it is because heavy loads are too much for most controllers to cope with. Again, despite appearances, most arms move at glacial speeds compared with what might be expected from the joint actuators. Again, the main culprit turns out to be the arm controller. Why?

Part of the answer is to be found in the complex dynamics of a robot arm. The equations of motion that govern each degree of freedom involve torques and motions inherited from adjoining links. The result for the conventional six degree-of-freedom serial-link robot arm (the easiest case) is a set of six coupled second-order differential equations. The inertial parameters of the arm vary throughout the workspace as rapidly as the arm configuration changes. This renders ineffective most schemes for plant identification in adaptive control. Singularities are numerous, the reach and velocity constraints severe, and the effective manipulability of the arm, as measured by expressions involving the Jacobian of the change from world coordinates to joint coordinates, depends crucially on configuration. Add to this the fact that, in reality, a robot joint is made non-linear by friction, by backlash in the gearing, and by quantization in the computer's representation of position and velocity. Finally, throw in the uncertainty of position and force from sensors that are inherently noisy. The result is a plant that is considerably more complex to control than those for which automatic control theory developed.

Perhaps reflecting these observations, although automatic control has been the subject of intense study and has seen many exciting developments, its techniques have not much influenced the development of practical robot control systems, which continue, outside the laboratory, to be dominated by classical PD controllers with high gains that serve as low-pass filters. Perhaps investigators and implementers were not familiar with the techniques of modern control theory. Perhaps also, those control theories do not extend to the complex, rapidly changing plants that are robot arms. There has for some time been a need for synthesis between theory and practice in robot control: a unified concept of robot control rooted in the techniques of modern control theory that takes account of real robot dynamics. To achieve such a synthesis for all robots in all applications would be a major contribution. This book proposes a particularly novel and interesting approach to such a synthesis. Of course there is much to do, as the authors point out in the Preface. For example, time is treated as continuous, not sampled as it would be for computer control, and there is no mention of robot programming, though the robot derives its usefulness from being programmed to work in a variety of tasks that are not completely determined. Again, robots do not consist only of serial link robot arms. Four-bar linkages, multi-fingered hands, and legged systems feature closed kinematic chains. Nevertheless, the scope of the authors' treatment of robot positioning, and the synthesis that they achieve for position control, is an exciting beginning.

PREFACE

A complete approach to the problem of controlling robot manipulators needs to bring together three scientific branches: computer science, mechanics, and automatic control.

Computer science issues extend over several levels in which the degree of abstraction varies considerably. In a modern robotic system, we may for instance find at the highest level an advanced programming structure which includes facilities for *off-line* (planning) and *on-line* (execution control) reasoning, as well as intelligent and ergonomic man–machine communication tools. Another issue is the effective implementation of the control techniques and algorithms. This may involve distributed architectures and multiprocessor structures which communicate with each other through buses, networks, or more complex connections. Usually, the real-time problems directly related to the dynamic characteristics of the controlled system are located at the lowest level of those structures.

While robotics has always closely followed the progress of computer science techniques (hardware and software), the situation is somewhat different for the control techniques themselves. This clearly appears when comparing the evolution of automatic control theory in the last thirty years and the studies or applications in robot control reported in the literature to date.

Automatic control, as a scientific discipline of its own, started with the *frequency domain* and *transfer function* approaches. Then linear system theory became important with the growing success of *state-space* representations. With it, structural questions arose, which led to a first approach to *controllability* and *observability* concepts. In parallel, *feedback control* techniques were developed: pole shifting, and linear quadratic regulation with or without using state observers. More recently, *geometric* approaches for linear time-invariant systems were proposed. Finally, techniques emanating from non-linear system theory were applied to control design. *Feedback linearization* is an example of those. Let us not forget *adaptive control* and *robust control* techniques which have been studied in parallel. Among the latter, let us mention those based on singular perturbation theory (or multiple time-scale systems) and H^∞ techniques.

It is interesting to compare the above evolution, resulting from progress at the conceptual level, with the history of robot control, the evolution of which was mainly driven by application requirements. In the 1960s, the first robots were simply viewed as a stacking of independent actuator/load modules, with electric motors controlled in a common way, i.e. by regulating the speed of rotation. The desired rotation velocity, being the input of the controller, was soon to be considered as a true control variable. The notion of *velocity control* in robotics probably comes from there. It was necessary however to take into account the fact that controlling the position of a robot is often more important than controlling its velocity. As a consequence, some position-controlled loop had to be introduced and combined with the velocity control philosophy. This was achieved by making the *desired velocity* depend on the robot's position error, the ideal position of the robot being expressed either in the *joint space* or in the user's Cartesian *task space*. A much addressed problem then was the generation of a 'trajectory' in this latter space, while taking into consideration various constraints due to the nature of the task and to the actuators. Control in the Cartesian space also increased the emphasis on the problems of reachable space and of *kinematic singularities*. The case of redundant systems (usually when the number of joints is greater than six) led to many variations on the theme of *generalized inverses* of Jacobian matrices. In order to improve the overall control performance, it appeared that it was no longer possible to neglect dynamic coupling phenomena (Coriolis and centrifugal forces) occurring during fast movements, especially in the case of *direct-drive* robots for which there is no mechanical system of reduction between the joints and the actuators. This led to explicitly modelling the robot's dynamic equations by using *Lagrange* or *Newton–Euler* formulations. From there it soon appeared, assuming that the derived mathematical model was perfect, that it was quite simple to design a non-linear control input which transforms the initial non-linear system into a decoupled time-invariant linear one composed of pure double integrators. Classical linear feedback loops, designed in the adequate work space (joint space or Cartesian space) could then be applied to this last system. In the robotics literature, this linearization technique, which has been later and independently generalized in non-linear control theory, received the names of *computed torque method, inverse dynamics technique*, and *dynamic control*. Practical implementation of this technique stirred the problem of on-line computations of several terms present in both the robot dynamic equations and the control expression, such as the robot inertia matrix and extraneous forces (Coriolis, centrifugal, gravity, and frictional forces) which are directly compensated. This has justified many studies aimed at minimizing the number of arithmetic operations needed to compute these terms. More recently, the search for distributed (parallel or pipeline) algorithms has taken over from these studies. Along with the trend to better model the robotic system in order to improve the performance of control, *adaptive control* techniques appeared. The underlying idea in these techniques is to use on-line identification of the system to adapt a predefined set of control parameters. The study of *robustness* oriented techniques was in turn motivated by the fact that models never reflect the real system precisely, and additionally all elements in the system are not equally worth modelling. The best known of them were derived from variable structure systems (*sliding mode control*) based on classical switching methods in non-linear systems. In parallel, it appeared more and more clearly that the distinctive characteristic of a robot lay in its ability to interact with its environment and that new generations of robot controllers would have to incorporate sensor-based control techniques. Methods based on classical mechanical models were proposed to describe the interactions between the robot and its environment: representation of a task through the formalism of linkages, mechanics of gripping, modelling using mechanical impedance·concepts, ... The inadequacy of the standard control schemes supported by the existing industrial robot controllers was also revealed, with other difficult problems, when trying to design control strategies based on the measurements of these interactions through dedicated sensors. Despite this, methods built on classical robot concepts were developed for controlling contact forces. Roughly, the basic idea consisted of splitting the space into two subspaces, one being 'force-controlled' and the second 'motion-controlled', by analogy with the kinematics of contacts (*hybrid control*).

Finally, recent technological developments (in space, undersea or nuclear applications) and the never ending search for performance improvement led to the integration of various sources of mechanical flexibility into robot models. Despite initial studies, many of which are oriented toward adaptive control solutions, and encouraging preliminary results, the control of flexible robots still remains a largely unsolved problem.

From this short presentation of the history of robot control we may make two observations. Firstly, the attempts to utilize the results of 'automatic control' in industrial robot controllers have been rather limited, although a robot manipulator can be considered as a particular kind of dynamic system. Secondly, past investigators have studied many robot control issues from their own specific perspectives, without trying to develop a general approach.

Observing this widespread evolution, which may not be the best approach in the development of advanced industrial controllers, we have tried to define a *unified control concept* for rigid robots. Thus, the central contribution of this book is to show that for rigid robots with complete torque control vector the following topics can be treated within a *unified control concept*:

(1) the specification of tasks;

(2) the notion of redundancy;

(3) the use of dynamic information obtained from exteroceptive sensors;

(4) the analysis of a *wide class of control schemes*.

However, not all aspects of the global problem of robot control are covered in the book and it is important to clearly delimit the framework of the proposed analysis. We are mainly interested in the control of robot manipulators from the point of view of automatic control, e.g. special attention is paid to the problems of *stability* and *robustness*. The design and analysis of *feedback loops* are the central issues. So, despite their importance, programming questions and off-line trajectory generation are not treated. Obviously, higher level problems, such as planning or robot vision, also fall outside the scope of this book for the same reasons. The proposed analysis is also based on a certain number of assumptions; for example the *rigidity* of the robot (links and transmissions), the full dimension of the control vector (as many actuators as there are links) and the measurement of all joint variables and joint velocities. Most of the analysis is derived in *continuous-time* and the effects of *discretization* of the control, *computational delays* and *mechanical elasticity* are only partially treated.

The assumption of rigidity is needed for ensuring that exact finite-dimensional models with correct control dimensions can be used effectively. In fact, the treatment of truly flexible structures would require modelling and control tools quite different from that used here for rigid robots. However, certain flexibilities in the system, which can be considered as external to the robot, are allowed because they do not invalidate the assumptions of the analysis; for example, elastic contacts between the robot and its environment. The choice of performing the analysis in continuous-time results first from the obvious fact that it is a necessary step before coming to discretization problems. However, the question is more complex than simply going from a continuous-time theory to a discrete-time implementation of the control. For instance, truly difficult problems of instability may result from asynchronism and communication delays in distributed robotic systems. In a non-linear (and even linear) context, general powerful theoretical tools for analysing these questions do not exist. For the sake of integrity, the discrete-time control issues are considered beyond the bounds of the book and their discussions are restricted to some simple and illustrative examples.

The book is structured as follows.

Chapter 0 is a brief review of the mathematical tools used in the following chapters. It includes basic concepts on control theory, differentiable manifolds and calculus. Less classical is a global version of the Implicit Function Theorem.

Chapter 1 aims to study the fundamental configuration spaces in robotics. Here, for example, may be found an analysis of the basic problems of representing orientations, focusing on the geometric aspects. This problem is indeed difficult to approach, and is unavoidable owing to the topology of the space of rotations. The reader will then find a study of the space tangent to the space of displacements. Using the formalism of screws, a calculus on the space of displacements, later applied to the analysis of sensor characteristics (Chapter 7), is derived.

Chapter 2 is devoted to the derivation of basic robot models. From a control point of view, the state-space model is the core of the description of the dynamic behaviour of a system. For this reason, the dynamic model is first established, from both Lagrange and Newton–Euler formulations. As robot modelling is not the basic topic of the book, we briefly recall the notions of direct and inverse kinematics and present the main techniques for the computation of the variational models.

Chapter 3 is dedicated to the mathematical characterization of the fundamental concept of a robot task. The basic idea behind the approach developed in this book is that many robot tasks may be reduced to a problem of *positioning*, and that the control problem may generally be stated as the regulation of a certain vector function, called the *task function*, which characterizes the task. The variables of this function are the joint position and the time index. The explicit dependence on time allows for a possible evolution of the goals to be taken into account; for example the motion of an object to be tracked. Once given the task function, the *ideal trajectory* is defined as the trajectory of the robot which keeps the task function equal to the null vector during the duration of the task. The important concept of ρ-admissibility of the task is introduced to characterize the conditioning of the task with respect to the control problem. It also provides a way of evaluating the intrinsic difficulty of realizing the task and of designing a (good) control. The notion of *task singularity* arises on this occasion from a new point of view, more general than the classical one of kinematic singularity.

In Chapter 4, an important special case of robot task is considered, when the system is *redundant*. After a formal description of the notion of redundancy, some definitions are proposed. Then the treatment of redundancy is immersed in a classical problem of constrained minimization. An original feature of this treatment is that it is entirely performed at the task definition level, and not at the control level as is frequently done. It is also shown that the classical differential motion approach of redundancy is contained in this treatment. Some of the results derived in this chapter are applied in Chapter 7 for the design of sensor-based task functions, a specific case being hybrid force/position control problems.

Once the work of defining and analysing a task has been done, and once a well-conditioned task function characterizing this task has been chosen by the user, the control problem itself remains to be solved, i.e. the problem of designing a control vector liable to effectively perform the regulation of the task function. This problem is addressed in Chapter 5. After a brief study of the classical control techniques of d.c. motors aimed at demonstrating some basic characteristics of robot control, a general non-linear proportional-derivative (P.D.) control scheme is proposed. It is shown that the main existing control techniques, ranging from the simplest linear P.D. controls to adaptive techniques and non-linear decoupling, are particular cases of this scheme and that they can be derived from it by making modelling assumptions and choosing an underlying minimization method (Newton or gradient).

A stability analysis of this general control scheme is presented in Chapter 6 where some fundamental stability conditions are derived. The most important one is a positivity condition which relates the so-called *task Jacobian matrix* (i.e. the Jacobian of the task function) to a *model* of this matrix, chosen by the control designer and used in the control expression. The analysis, based on robustness considerations, also provides conditions bearing upon the size of the (possibly non-linear) control gains. Finally, in order to meet the practical implementation problems to some extent, a brief analysis of discretization and delay effects is presented.

Chapter 7 is devoted to the design and analysis of task functions involving the use of exteroceptive sensors like proximity sensors, local range finders and force sensors. A preliminary analysis points out the existence of an important feature, the *interaction screw*, which contains essential information about the interaction between a sensor and its environment. For non-contact sensors, a methodological tool for the design of sensor-based tasks is then proposed: it is based on an analogy with the classical analysis of linkages between rigid bodies. The design of the complete task function is performed by using the 'redundancy' approach of Chapter 4, which leads to the concept of hybrid tasks. A last section focuses on some specific problems related to the use of force sensors.

In general, we have tried to maintain a didactic character to the overall book. This is why we have attempted to illustrate some concepts which are not always self-evident with examples or starting from simple situations that can easily be analysed. We have also tried to examine the practical consequences of the results presented. To this end, in most of the chapters, commonly encountered cases and design choices are described and discussed.

We hope that this book will be useful to educators and students as well as to the designers of advanced robot controllers, or to users wishing to understand the specific problems of robot control in depth.

Valbonne and Rennes C. S.
February 1990 M. Le B.
  B. E.

ACKNOWLEDGEMENTS

The authors wish to express their gratitude to all those who helped them in preparing this book, in particular:

- Laurent Baratchart, Directeur de Recherche at INRIA Sophia-Antipolis, for his contribution to the proof of Theorem 6.1;
- Brian Armstrong, from the University of Wisconsin (USA), for having read the first version of the book and for his useful comments;
- Nejat Olgac, Professor at the University of Connecticut (USA), for his careful and detailed reading of the book and for the many insightful corrections that he suggested about the form and contents.

The authors are also appreciative of the Oxford University Press editors for their remarkable work, and of Professor Michael Brady for writing the book's foreword.

AUTHORS

Claude Samson, INRIA, 2004 Route des Lucioles, Sophia Antipolis, 06565 Valbonne, Cedex, France
(Also at GRECO CNRS, 'SARTA', ENSIEG-BP 46, 38402 Saint-Martin d'Hères, France)

Michel Le Borgne, IRISA, Campus de Beaulieu, 35042 Rennes, Cedex, France

Bernard Espiau, INRIA, 2004 Route des Lucioles, Sophia Antipolis, 06565 Valbonne, Cedex, France

CONTENTS

0 MATHEMATICAL PRELIMINARIES — 1

Introduction — 1
0.1 Linear algebra and linear control theory — 1
0.2 Non-linearities — 8
0.3 Bibliographic note — 17

1 FUNDAMENTAL CONFIGURATION SPACES — 18

Introduction — 18
1.1 Notation, basis, and frames — 18
1.2 Attitudes of rigid bodies — 21
1.3 Screws — 41
1.4 Bibliographic note — 54
Appendix A1.1 Proof of proposition P1.6 — 55

2 BASIC MODELS OF ROBOTS — 56

Introduction — 56
2.1 The dynamic model — 56
2.2 Direct and inverse kinematics — 62
2.3 The differential motion model — 71
2.4 Bibliographic note — 75

3 THE TASK FUNCTION CONCEPT — 77

Outline — 77
3.1 An informal approach — 77
3.2 ρ-admissible functions and tasks — 85
3.3 Non-admissibility and task singularities — 92
3.4 Admissibility of usual robotic tasks — 103
3.5 Bibliographic note — 108

CONTENTS

4 TASK REDUNDANCY — 109

- 4.1 Locally redundant systems — 109
- 4.2 Redundant tasks — 117
- 4.3 Examples of secondary tasks — 142
- 4.4 Monitoring of the transient phase of the secondary cost minimization — 160
- 4.5 Bibliographic note — 162
- Appendix A4.1: Proofs of lemmas of Chapter 4 — 163

5 CONTROL: A GENERAL APPROACH — 168

- 5.1 General background — 168
- 5.2 A general control structure — 182
- 5.3 Application to prominent control schemes — 196
- 5.4 Bibliographic note — 216

6 STABILITY ANALYSIS — 218

- 6.1 Preliminary results — 218
- 6.2 A general stability theorem — 224
- 6.3 Principles of application to robot control — 229
- 6.4 Effects of measurement errors and disturbances — 240
- 6.5 Effects of discretization and measurement delays — 245
- 6.6 Use of the positivity condition of $J_T J_q^{-1}$ — 248
- 6.7 Bibliographic note — 253
- Appendix A6.1 Proofs of lemmas of Chapter 6 — 254
- Appendix A6.2 Proof of Theorem 6.1 — 258

7 SENSOR-BASED TASKS — 265

- Introduction — 265
- 7.1 Modelling of the interactions with the environment — 266
- 7.2 Simplified models of interactions: a tool for the user — 277
- 7.3 Immersion of sensor-based tasks in the task redundancy framework — 293
- 7.4 Summary — 308
- 7.5 Examples — 309
- 7.6 Tasks based on the use of force sensors — 324
- 7.7 Concluding remarks — 346
- 7.8 Bibliographic note — 347

CONTENTS

Appendix A7.1 A model of optical proximity sensors ... 348
Appendix A7.2 Combination of generalized springs ... 350

REFERENCES ... 352

INDEX ... 361

0
MATHEMATICAL PRELIMINARIES

INTRODUCTION

For the reader's convenience, we shall quickly review some mathematical concepts used in the book. As we do not intend to present a comprehensive treatise on control, we restrict our presentation to material which we feel is not common knowledge in the robotics community. However, a few sections of the book use mathematical concepts going beyond those explained in this preliminary chapter. These sections will be indicated by an asterisk (*). As the material in this first chapter is fairly standard, few proofs are offered; the bibliographic note at the end of the chapter provides access to detailed developments. We assume the reader is familiar with linear algebra, matrices, elementary calculus and topology, and linear control theory. Some basic definitions in these fields will be recalled when necessary to set the notation used in the book.

0.1 LINEAR ALGEBRA AND LINEAR CONTROL THEORY

0.1.1 Matrices

Matrices will be used to represent linear maps. Usually we need not distinguish sharply between a matrix A and the associated linear map. Thus if A is an $n \times m$ matrix, the kernel or null space of A is a subspace defined by

$$\text{Ker}(A) = N(A) = \{x \mid x \in \mathbf{R}^m \ \& \ Ax = 0\} \tag{0.1.1}$$

while the image or range of A is the subspace

$$\text{Im}(A) = R(A) = \{y \mid y \in \mathbf{R}^n;\ \ y = Ax \text{ with } x \in \mathbf{R}^m\}. \tag{0.1.2}$$

Given an $n \times n$ matrix A and a norm $\|\cdot\|$ defined on $\mathbf{R}^n$, we define the associated norm of A as

$$\|A\| = \sup_{x \in \mathbf{R}^n \to 0} \frac{\|Ax\|}{\|x\|} = \sup_{\|x\| = 1} \|Ax\|. \tag{0.1.3}$$

It is well known that the equalities above define a norm on the space $M_n(\mathbf{R})$ of $n \times n$ real matrices with the essential properties

$$\|Ax\| \leq \|A\| \, \|x\|, \qquad (0.1.4)$$

$$\|AB\| \leq \|A\| \, \|B\|. \qquad (0.1.5)$$

This type of matrix norm associated with a norm on the vector space can be related to the entries or the eigenvalues of the matrix. Recall that the spectral radius of a matrix B is the largest module $\rho(B)$ of the eigenvalues. In the usual case of the Euclidean norm, we have the following result.

Proposition P0.1 *If*

$$\|x\| = \left(\sum_{i=1}^n x_i^2\right)^{1/2}$$

is the Euclidean norm on $\mathbf{R}^n$, then the associated matrix norm on $M_n(\mathbf{R})$ is $\|A\| = (\rho(A^T A))^{1/2}$.

0.1.2 Positive matrices and non-degenerate matrices

A concept used in control analysis is that of positive matrix.

Definition D0.1 *An $n \times n$ matrix A is positive if for every vector x in $\mathbf{R}^n$, $\langle Ax, x \rangle = x^T A x \geq 0$, where $\langle , \rangle$ denotes the usual scalar product in $\mathbf{R}^n$.*

This property does not imply that the entries of A are positive nor that A is symmetric. Thus our definition of a positive matrix is different from that often used in numerical analysis for example, where a positive matrix is a matrix with positive entries. The concept of positive matrix can be geometrically interpreted in dimension 2 or 3. A matrix A is positive if for every x the angle between Ax and x is less than $\pi/2$. We speak of a strictly positive matrix whenever $x^T A x > 0$ for any vector x different from the null vector.

Proposition P0.2 *If a matrix A is positive (respectively strictly positive), then the eigenvalues of A have positive (respectively strictly positive) real parts. The converse is not true.*

The proof is immediate by computing $x^T A x$ with x being first the real part and then the imaginary part of an eigenvector associated with the eigenvalue $a + ib$, ($i^2 = -1$). Unfortunately this property is not sufficient, as shown by the following example:

$$A = \begin{bmatrix} 1 & 3 \\ 0 & 1 \end{bmatrix}.$$

The only eigenvalue is 1 but, for example, $(-1 \;\; 1)A(-1 \;\; 1)^T = -1$.

0.1 LINEAR ALGEBRA AND LINEAR CONTROL THEORY

It is also easy to verify that a matrix A is positive if and only if its symmetric part $\frac{1}{2}(A + A^T)$ is positive.

Recall now that a matrix A is non-degenerate if, for all y, $y^T A x = 0$ implies $x = 0$. Given a matrix A, the mapping $y \to \langle Ax, y \rangle$ is a linear form associated with each vector x. Let us denote by $\langle Ax, . \rangle$ this element of the dual of $\mathbf{R}^n$. When the matrix A is non-degenerate we get the following important property.

Proposition P0.3 *If A is an $n \times n$ non-degenerate matrix, the mapping $x \to \langle Ax, . \rangle$ is an isomorphism from $\mathbf{R}^n$ onto its dual.*

This is an immediate consequence of the definition. Another immediate consequence is that a matrix is non-degenerate if and only if it is regular.

0.1.3 Generalized inverses

Our modelling of robot control problems will use a *state space* and an *output space*, in the traditional way. In the case of redundant robots, or more generally of redundant tasks as we shall see in Chapter 4, the dimensions of the state space and of the output space are different. So we shall have to consider one-sided inverses of rectangular matrices. This section is devoted to that question.

Let A be an $m \times n$ real matrix.

Definition D0.2 A^g *is a generalized inverse of A if and only if*

$$A A^g A = A. \tag{0.1.6}$$

Definition D0.3 A^r *is a reflexive generalized inverse of A if and only if it is a generalized inverse and*

$$A^r A A^r = A^r. \tag{0.1.7}$$

Proposition P0.4 A^+ *is the unique pseudoinverse of A if and only if it is a reflexive generalized inverse, and*

$$(A^+ A)^T = A^+ A, \tag{0.1.8}$$

$$(A A^+)^T = A A^+. \tag{0.1.9}$$

Since $\dim(A) = m \times n$, A can be considered as the matrix of a linear mapping from $\mathbf{R}^n$ to $\mathbf{R}^m$. Given two norms, n_1 defined on $\mathbf{R}^n$ and n_2 defined on $\mathbf{R}^m$, and given a vector y of $\mathbf{R}^m$, consider the following problem. Find a vector $\hat{x}$ in $\mathbf{R}^n$ with:

$$\left. \begin{array}{c} \| A\hat{x} - y \|_{n_2} = \min_x \| Ax - y \|_{n_2} \\ \hat{x} \text{ is of minimum norm } n_1 \text{ among the solutions} \end{array} \right\}.$$

The proposition indicates that the matrix A^+ above corresponds to the solution of the problem in the sense that

$$\hat{x} = A^+ y. \quad (0.1.10)$$

The pseudoinverse A^+ is unique once the norms n_1 and n_2 are chosen.

Proposition P0.5 *If A is of full rank $m < n$ and the norms are the Euclidean norms, the pseudoinverse of A is given by:*

$$A^+ = A^T(AA^T)^{-1}. \quad (0.1.11)$$

0.1.4 Particular subspaces and projection operators

If $R(A)$ denotes the range of A and $N(A)$ the null space (or kernel) of A, we have the basic decomposition of $\mathbf{R}^n$:

$$\mathbf{R}^n = N(A) \oplus R(A^g A). \quad (0.1.12)$$

Further, if A^g is the pseudoinverse A^+, $R(A^+ A)$ is the orthogonal complement to $\mathbf{R}^n$ of $N(A)$, and $N(A^+)$ is the orthogonal complement to $\mathbf{R}^m$ of $R(A)$.

A general projection operator onto $N(A)$ is

$$P^B_{N(A)} = I - B(AB)^{-1}A \quad (0.1.13)$$

where B is an $n \times m$ matrix such that AB is regular (i.e. with rank m). When using A^+, we may define the following *orthogonal* projection operators:

$$P_{N(A)} = I - A^+ A \quad (0.1.14)$$

$$P_{R(A^+)} = A^+ A. \quad (0.1.15)$$

Of course, all these projection operators have the property of being idempotent (i.e. $P^2 = P$).

0.1.5 Stationary linear systems: state-space representation

A state-space representation of a controlled stationary linear system is given by

$$\left. \begin{array}{l} \dot{X} = AX + BU \\ Y = CX \end{array} \right\} \quad (0.1.16)$$

where $X \in \mathbf{R}^n$ is the state of the system, $U \in \mathbf{R}^m$ a vector of inputs, and $Y \in \mathbf{R}^p$ a vector of outputs. A, B, and C are constant matrices with suitable dimensions. The dot over X stands for time derivative. This system is called (CS).

0.1 LINEAR ALGEBRA AND LINEAR CONTROL THEORY

A system with no input; called (S):

$$\left.\begin{array}{l}\dot{X} = AX \\ Y = CX\end{array}\right\} \qquad (0.1.17)$$

is stable if for any t_0 and ε, there exists an η such that

$$\|X(t_0)\| < \eta \text{ implies } \|X(t)\| < \varepsilon, \text{ for all } t > t_0.$$

Such a system is asymptotically stable if it is stable and $\lim_{t \to \infty} \|X(t)\| = 0$.

Proposition P0.6 *System (S) is asymptotically stable if and only if all the eigenvalues of A have strictly negative real parts.*

It is a trivial remark that in this case $\lim_{t \to \infty} \|Y(t)\| = 0$ also.

For a system with inputs like (CS), asymptotic stability is no longer a sufficient concept. The state of the system may never reach zero if the inputs remain non-null. The correct notion is now of bounded input bounded output (BIBO) stability.

Definition D0.4 *The system (CS) is BIBO stable if for every bounded input $U(t)$ the output $Y(t)$ remains bounded.*

It is evidently obvious to say that the output may remain bounded while $X(t) \to \infty$. If the state-space representation is a minimal representation (i.e. controllable and observable) of the input–output system, this cannot happen. More precisely, we have the following.

Proposition P0.7 *If system (CS) is controllable and observable, it is BIBO stable if and only if all the eigenvalues of A have strictly negative real parts.*

0.1.5.1 Practical stability tests

Two criteria are traditionally used to check the stability of a linear time-invariant system. The best known, the *Nyquist* criterion, is adequate when graphic representation of a transfer function is used. The *Routh* criterion is more interesting with regard to the above definitions, as it works on the characteristic polynomial. Let us write

$$D(\lambda) = \det(A - \lambda I) = a_0 \lambda^n + \ldots + a_n = 0, \qquad a_0 > 0. \qquad (0.1.18)$$

A first necessary condition for having $\text{Re}(\lambda_i) < 0$ for any $i = 1, \ldots, n$ is that

$$a_i > 0, \qquad i = 1, 2, \ldots, n. \qquad (0.1.19)$$

The second step starts from the so-called triangular 'Routh table':

$$\left.\begin{array}{cccc} a_0 & a_2 & a_4 & \ldots \\ a_1 & a_3 & a_5 & \ldots \\ b_1 & b_2 & b_3 & \ldots \\ c_1 & c_2 & c_3 & \ldots \end{array}\right\} \quad (0.1.20)$$

the components of which are computed in the following way:

$$b_1 = -\frac{1}{a_1}\det\begin{bmatrix} a_0 & a_2 \\ a_1 & a_3 \end{bmatrix}; \quad b_2 = -\frac{1}{a_1}\det\begin{bmatrix} a_0 & a_4 \\ a_1 & a_5 \end{bmatrix};$$

$$b_3 = -\frac{1}{a_1}\det\begin{bmatrix} a_0 & a_6 \\ a_1 & a_7 \end{bmatrix}\ldots \quad (0.1.21)$$

Then, by shifting down the computations in the table, and using the same rule,

$$c_1 = -\frac{1}{b_1}\det\begin{bmatrix} a_1 & a_3 \\ b_1 & b_2 \end{bmatrix}; \quad c_2 = -\frac{1}{b_1}\det\begin{bmatrix} a_1 & a_5 \\ b_1 & b_3 \end{bmatrix};$$

$$c_3 = -\frac{1}{b_1}\det\begin{bmatrix} a_1 & a_7 \\ b_1 & b_4 \end{bmatrix}\ldots \quad (0.1.22)$$

The total number of rows is $(n+1)$. The basic Routh result is that the number of roots of $D(\lambda)$ with positive real parts is equal to the number of sign changes in the first column of the Routh table. As a_0 is chosen positive, a corollary is that *the system is asymptotically stable if and only if all the elements of the first column are positive.*

0.1.5.2 Discrete-time case

We consider here only the case of a linear stationary system, called DS:

$$x(k+1) = Ax(k). \quad (0.1.23)$$

Briefly, we have the following.

Theorem 0.1 *System* (DS) *is asymptotically stable (with the same definition as in the continuous-time case) if and only if all the eigenvalues of A lie inside the unit circle.*

A practical stability test is Jury's criterion, which is itself a simplified version of the Schur–Cohn criterion. Let us consider the characteristic equation of A:

$$H(\lambda) = \sum_{i=0}^{n} a_i \lambda^i, \quad \text{with } a_n > 0. \quad (0.1.24)$$

0.1 LINEAR ALGEBRA AND LINEAR CONTROL THEORY

The Jury table has the following structure:

$$\left.\begin{matrix} a_0 & a_1 & a_2 & \cdots & a_n \\ a_n & a_{n-1} & a_{n-2} & \cdots & a_0 \\ b_0 & b_1 & b_2 & \cdots & b_{n-1} \\ b_{n-1} & b_{n-2} & b_{n-3} & \cdots & b_0 \\ c_0 & c_1 & \cdots & c_{n-2} & \\ c_{n-2} & c_{n-3} & \cdots & c_0 & \\ \cdots \\ y_0 & y_1 & y_2 & y_3 \\ y_3 & y_2 & y_1 & y_0 \\ z_0 & z_1 & z_2. \end{matrix}\right\} \quad (0.1.25)$$

The elements of the Jury table are constructed in the following way:

$$b_k = \det\begin{bmatrix} a_0 & a_{n-k} \\ a_n & a_k \end{bmatrix}; \quad c_k = \det\begin{bmatrix} b_0 & b_{n-1-k} \\ b_{n-1} & b_k \end{bmatrix};$$

$$d_k = \det\begin{bmatrix} c_0 & c_{n-2-k} \\ c_{n-2} & c_k \end{bmatrix} \cdots$$

$$z_0 = \det\begin{bmatrix} y_0 & y_3 \\ y_3 & y_0 \end{bmatrix}; \quad z_1 = \det\begin{bmatrix} y_0 & y_2 \\ y_3 & y_1 \end{bmatrix}; \quad z_2 = \det\begin{bmatrix} y_0 & y_1 \\ y_3 & y_2 \end{bmatrix}. \quad (0.1.26)$$

The necessary and sufficient conditions for having all the roots of H inside the unit circle are

$$H(1) > 0; \quad H(-1) \begin{cases} > 0 & n \text{ even} \\ < 0 & n \text{ odd}, \end{cases} \quad (0.1.27)$$

and

$$\left.\begin{matrix} |a_0| < a_n \\ |b_0| > |b_{n-1}| \\ |c_0| > |c_{n-2}| \\ |d_0| > |d_{n-3}| \\ \cdots \\ \cdots \\ |y_0| > |y_3| \\ |z_0| > |z_2|. \end{matrix}\right\} \quad (0.1.28)$$

This leads for example to the following particular expressions.

For $n = 3$:

$$\left.\begin{aligned} a_0 + a_1 + a_2 + a_3 &> 0 \\ -a_0 + a_1 - a_2 + a_3 &> 0 \\ a_3 - |a_0| &> 0 \\ a_0 a_2 - a_1 a_3 - a_0^2 + a_3^2 &> 0; \end{aligned}\right\} \quad (0.1.29)$$

for $n = 4$:

$$\left.\begin{aligned} a_0 + a_1 + a_2 + a_3 + a_4 &> 0 \\ -a_0 + a_1 - a_2 + a_3 - a_4 &< 0 \\ a_4^2 - a_0^2 - |a_0 a_3 - a_1 a_4| &> 0 \\ (a_0 - a_4)^2(a_0 - a_2 + a_4) + (a_1 - a_3)(a_0 a_3 - a_1 a_4) &> 0. \end{aligned}\right\} \quad (0.1.30)$$

0.2 NON-LINEARITIES

0.2.1 Differentiable manifolds

Even in robotics, the state spaces or output spaces used in the modelling of control problems may be spaces more complex than the usual $\mathbf{R}^n$. As a consequence we shall need some differential geometry.

A C^∞ (respectively C^k) differentiable manifold M of dimension n is a metric space with the following properties:

(1) if $p \in M$, then there exist some neighbourhood U of p and a homeomorphism x from U onto an open subset of $\mathbf{R}^n$;

(2) for every (x, U) and (y, V) defined as above, the maps

$$y \circ x^{-1}: x(U \cap V) \longrightarrow y(U \cap V) \qquad (0.2.1)$$

$$x \circ y^{-1}: y(U \cap V) \longrightarrow x(U \cap V) \qquad (0.2.2)$$

are C^∞ (respectively C^k) when defined.

A couple (x, U) is called a *chart* or a *local coordinate system* on M. Two charts (x, U) and (y, V) are compatible if they have property (2). A set of compatible charts $\{(x_i, U_i); i \in I\}$ such that $M = \bigcup_{i \in I} U_i$ is an *atlas* on M.

The concept of a chart appears quite frequently in mechanics when one describes a mechanical system with a set of parameters $(q_1, \ldots, q_n)$, often called generalized coordinates. This notion is essential in the Lagrangian formalism. In general, mechanical equations are established in the domain of the chart so defined and one does not bother about changing these local coordinates. We will see in the next chapter that, in fact, different local coordinates must be used when dealing with attitudes of rigid bodies.

0.2 NON-LINEARITIES

Various concepts from differential geometry will be used at times throughout this book. All these concepts are classical, and can be found in numerous introductory books on differential geometry. Let us summarize the most important of them.

The interesting class of mappings between differential manifolds is the class of differentiable mappings. Let M and N be two differential manifolds with dimensions m and n respectively. A map $f: M \to N$ is differentiable if for every $p \in M$, for every chart (x, U) around p, and for every chart (y, V) around $f(p)$, the map $y \circ f \circ x^{-1}$ is C^∞ (respectively C^k), from $\mathbf{R}^m$ to $\mathbf{R}^n$. A diffeomorphism is a bijective differentiable map with a differentiable inverse. Diffeomorphisms are isomorphisms for differential manifolds.

A more delicate notion arising naturally with differential manifolds is the notion of tangent space or tangent bundle. There are different equivalent ways of building the tangent space of a manifold. We will concentrate on only two different methods for building the tangent space at a point p, also named the fibre over p, and denoted by TM_p. The interested reader is referred to basic books in differential geometry for explanations of how these fibres are 'glued' together to form a new differential manifold: the tangent bundle.

Our purpose is to generalize the familiar notion of a tangent plane at a point p of, say, a sphere. One method is suggested if we notice that the tangent plane at a point p to a sphere can be seen as the vector space of all the tangent vectors at p to all the differentiable curves passing through p. Of course two differentiable curves may have the same tangent vector at p.

Following this idea, let M be a differential manifold with $p \in M$. A differentiable curve or path on M is a differentiable map defined on some interval I in $\mathbf{R}$ with image in M.

Consider the differentiable curves $c: (-\varepsilon, \varepsilon) \to M$, each defined on some interval around 0 with $c(0) = p$. If (x, U) is a chart around p, we define the relation

$$c_1 \underset{p}{\approx} c_2 \qquad (0.2.3)$$

if $x \circ c_1$ and $x \circ c_2$ mapping $\mathbf{R}$ to $\mathbf{R}^n$ have the same derivative $(x \circ c)'$ at 0.

It is easy to check that this relation is an equivalence relation, independent of the choice of the chart (x, U). It remains to turn the quotient space into a vector space. For this, if c is a curve with $c(0) = p$ let us denote by $[c]_p$ the class of c. With (x, U) as above we define

$$\lambda_1 [c_1]_p + \lambda_2 [c_2]_p = [c]_p \qquad (0.2.4)$$

where

$$c(t) = x^{-1}[x(p) + t\{\lambda_1 (x \circ c_1)'|_{t=0} + \lambda_2 (x \circ c_2)'|_{t=0}\}]. \qquad (0.2.5)$$

Although this way of building the fibre over p seems very complicated, it is in fact natural, as shown in Fig. 0.1.

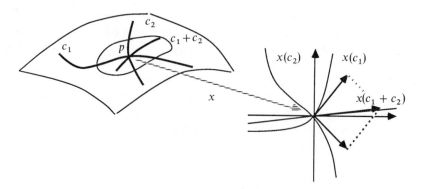

Fig. 0.1 A method of constructing the tangent space at p

The other way to construct the tangent space at p is valid only for manifolds which can be considered as submanifolds of some $\mathbf{R}^n$. This will be practically always the case. The delicate point is to define what a submanifold of $\mathbf{R}^n$ is.

Definition D0.5 *A subset M of $\mathbf{R}^n$ is a submanifold at $p \in M$ with dimension $m \leq n$ if there exists an open neighbourhood W of p in $\mathbf{R}^n$ and a diffeomorphism θ from W onto an open subset of $\mathbf{R}^n$ such that $\theta(W \cap M)$ is an open subset of an $\mathbf{R}^n$ subvector space F_p with dimension m.*
If M is a submanifold at every point p, M is a submanifold of $\mathbf{R}^n$.

The charts of M are then the couples (x, U) with $U = W \cap M$ and $x = \theta|_U$ the restriction of θ to U. The mapping $\phi = \theta^{-1}|_{x(U)}$ is a local parametrization of M (Fig. 0.2). The fibre over p is then easy to define: θ^{-1} is a diffeomorphism from $\theta(W)$ onto W. The tangent space at p is

$$(p, F_p) \approx (p, D\theta^{-1}|_{\theta(p)}(F_p)). \tag{0.2.6}$$

The notion of local parametrization is important in robot control. We will see in the next chapter that, for example, Euler angles are local parametrizations for the set of rotations.

The last thing which remains to be done is to define the diffferential of a differentiable map between two manifolds. Note that if $f: M \to N$ is a differentiable map and $c:(-\varepsilon, \varepsilon) \to M$ is a differentiable curve with $c(0) = p \in M$, then $f \circ c$ is a differentiable curve on N such that $f \circ c(0) = f(p)$. Denoting by $Df|_p$ the differential of f at p, it is natural to define

$$Df|_p([c]_p) = [f \circ c]_{f(p)}. \tag{0.2.7}$$

In this way we obtain a linear map from TM_p to $TN_{f(p)}$. The proof needs only repetitive use of the chain rule.

0.2 NON-LINEARITIES

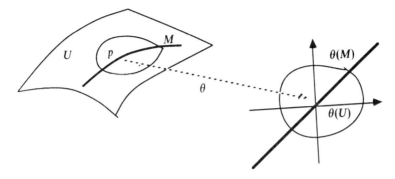

Fig. 0.2 A local submanifold at p

The rank of a differentiable mapping at a point p is simply the rank of the linear mapping $Df|_p$.

As a last point about manifolds, we would like to give the following useful criterion to recognize a submanifold of $\mathbf{R}^n$:

Proposition P0.8 *Let M be a manifold and $f: \mathbf{R}^n \to M$ be a differentiable map. Let p be a point of M and q be a point of $f^{-1}(p)$. If $Df|_q$ is onto, then $f^{-1}(p)$ is a submanifold of $\mathbf{R}^n$ at q with dimension $\dim(\operatorname{Ker} Df|_q)$.*

This proposition is a direct consequence of the Implicit Function Theorem. The proof is postponed to the next section.

As an application, consider the mapping: $f(x, y, z) = x^2 + y^2 + z^2 - 1$. We get: $Df|_{(x, y, z)} = [2x \; 2y \; 2z]$, thus $f^{-1}(0)$ is a submanifold of $\mathbf{R}^n$! More generally the mapping $f(x_1, \ldots, x_n) = x_1^2 + \cdots + x_n^2 - 1$ from $\mathbf{R}^n$ into $\mathbf{R}$ defines a submanifold of $\mathbf{R}^n$, the sphere $S^{n-1} = f^{-1}(\{0\})$. The dimension of S^{n-1} is $n - 1$.

0.2.2 Calculus and implicit function theorems

Let U and V be two open sets and $f: U \to V$ an onto mapping. We say that f is a diffeomorphism of class C^r if there exists a map $g: V \to U$ of class C^r such that $g \circ f$ is the identity on U. As U and V can be considered as submanifolds of $\mathbf{R}^n$, this definition is just a rephrasing of the general definition given in the previous section.

A whole group of theorems is known in calculus under the generic name of Implicit Function Theorems. They can all be derived from the following, called the Inverse Function Theorem.

Theorem 0.2 (Inverse Function Theorem) *Let $f: U \to \mathbf{R}^m$ be of class $C^r, r \geq 1$. U is an open set in $\mathbf{R}^m$. If $Df|_{p:} : \mathbf{R}^m \to \mathbf{R}^m$ is an isomorphism, then f is a local* diffeomorphism at p; that is, there exist neighbourhoods V included in U, and W included in $\mathbf{R}^m$, such that $f: U \to W$ is a diffeomorphism.

The following is the classical Implicit Function Theorem.

Theorem 0.3 (Implicit Function Theorem) *Let U be an open set of $\mathbf{R}^m \times \mathbf{R}^n$ and $f: U \to \mathbf{R}^n$ a C^r map, $r \geq 1$. Let $z_0 = (x_0, y_0) \in U$ and $c = f(z_0)$. Suppose that the partial derivative with respect to the second variable, $D_2 f(z_0): \mathbf{R}^n \to \mathbf{R}^n$, is an isomorphism. Then there exist open sets V included in $\mathbf{R}^m$ containing x_0 and W included in U containing z_0 such that, for each $x \in V$, there exists a unique $\xi(x) \in \mathbf{R}^n$ with $(x, \xi(x)) \in W$ and $f(x, \xi(x)) = c$. The map $\xi: V \to \mathbf{R}^n$, defined in this way, is of class C^r and its derivative is given by $d\xi(x) = [D_2 f(x, \xi(x))]^{-1} \circ D_1 f(x, \xi(x))$.*

These two basic theorems lead to other classical results such as the Constant Rank Theorem.

Theorem 0.4 (Constant Rank Theorem) *Let M and N be two manifolds with dimension n and m respectively. If $f: M \to N$ has rank k at p, then there is some coordinate system (x, U) around p and some coordinate system (y, V) around $f(p)$ with $y \circ f \circ x^{-1}$ in the form*

$$y \circ f \circ x^{-1}(a^1, \ldots, a^n) = (a^1, \ldots, a^k, \psi^{k+1}(a), \ldots, \psi^m(a)).$$

If f has rank k in a neighbourhood of p, then there are coordinate systems (x, U) and (y, V) such that:

$$y \circ f \circ x^{-1}(a^1, \ldots, a^n) = (a^1, \ldots, a^k, 0, \ldots, 0).$$

This theorem is also a consequence of the Inverse Function Theorem. The proof of Proposition P0.8 follows easily from this theorem.

Proof of Proposition P0.8 Let $\dim(M) = m$. Since $Df|_q$ is onto, we have $m \leq n$. Now by the Constant Rank Theorem there exists a chart (x, U) around q and a chart (y, V) around $f(q) = p$ such that

$$y \circ f \circ x^{-1}(a^1, \ldots, a^n) = (a^1, \ldots, a^m). \tag{0.2.8}$$

Let $(a_0^1, \ldots, a_0^n) = x(q)$. Then for all points $q' \in U \cap f^{-1}(p)$ we have, if $x(q) = (a^1, \ldots, a^n)$,

$$\begin{aligned} y \circ f \circ x^{-1}(a^1, \ldots, a^n) &= (a^1, \ldots, a^m) \\ &= y \circ f(q) \\ &= (a_0^1, \ldots, a_0^n) \end{aligned} \tag{0.2.9}$$

so that the first m components of $x(q')$ are constant. A slight modification of the chart x by a translation gives a chart x' such that $x'(U \cap f^{-1}(p))$ is an open subset of an $\mathbf{R}^n$ vector subspace with dimension $n - m$.

0.2.3 Global versions of Implicit Function Theorems

As we shall see in Chapter 3, it is important for control purposes to determine whether a function is a diffeomorphism between two precise sets, and not just a local diffeomorphism. The following theorem provides a means of checking whether a function is a global diffeomorphism. The theorem is an adaptation of a well-known theorem on Banach spaces.

Theorem 0.5 *Let Ω be an open set of $\mathbf{R}^n$ and $\Phi: \Omega \to \mathbf{R}^n$ a continuously differentiable mapping (respectively C^k). Let A be a subset of Ω and B a subset of $\mathbf{R}^n$ such that:*

(1a) *A is a closed set of $\mathbf{R}^n$;*
(1b) *A is a non-empty arcwise connected component of $\Phi^{-1}(B)$;*
(1c) *For each x in A, $D\Phi|_x$ is invertible and $\|[D\Phi|_x]^{-1}\| \leq K < \infty$;*

(2) *B is arcwise and simply connected.*
 Then Φ is a (respectively C^k) diffeomorphism from A onto B.

The proof will depend on the following lemma.

Lemma 0.1 *On the same hypothesis as the theorem, if Δ is the square $0 \leq s \leq 1$, $0 \leq t \leq 1$, and $F(s, t)$ satisfies the following conditions:*

(1) *$F(s, t): \Delta \to B$;*
(2) *$F(s, t)$ is continuous in (s, t) and for every s, $F(s, t)$ is differentiable in t;*
(3) *$F(s, t)$ has fixed endpoints, i.e. there exist y_0, y_1 in B such that for all $0 \leq s \leq 1$, $F(s, 0) = y_0$ and $F(s, 1) = y_1$;*
(4) *There exists $x_0 \in A$ such that $\Phi(x_0) = y_0$.*

Then there exists a function $G(s, t)$ from Δ to A which also satisfies (2) and in addition $\Phi[G(s, t)] = F(s, t)$ for all $(s, t) \in \Delta$.

Proof. From the local inverse theorem and since Ω is open in $\mathbf{R}^n$, there exist open (in $\mathbf{R}^n$) neighbourhoods U of x_0 and V of y_0 such that Φ is a diffeomorphism from U onto V. Then, for sufficiently small $\varepsilon \leq 1$, $F([0, 1] \times [0, \varepsilon])$ is included in U and it is possible to define a mapping $G_\varepsilon: [0, 1] \times [0, \varepsilon] \to U$ as $G_\varepsilon(s, t) = [\Phi|_U]^{-1} \circ F(s, t)$.

We then have the following properties:

1. $G_\varepsilon([0, 1] \times [0, \varepsilon])$ is included in A.

Since $G_\varepsilon(s, t)$ is continuous, $G_\varepsilon([0, 1] \times [0, \varepsilon])$ is arcwise connected and $A \cap G_\varepsilon([0, 1] \times [0, \varepsilon])$ is non-empty; so $A \cup G_\varepsilon([0, 1] \times [0, \varepsilon])$ is arcwise connected, included in $\Phi^{-1}(B)$ and contains A. By property (1b) of Theorem 0.5, $A \cup G_\varepsilon([0, 1] \times [0, \varepsilon])$ is included in A.

2. If we consider the family of functions $G_\varepsilon(s, t): [0, 1] \times [0, \varepsilon] \to A$ such that

$$F = \Phi \circ G_\varepsilon \quad \text{on} \quad [0, 1] \times [0, \varepsilon] \qquad (0.2.10)$$

this family is non-empty and two such functions G_ε and $G_{\varepsilon'}$ coincide on $[0, 1] \times [0, \varepsilon]$ by the local inverse theorem.

Let a be the upper bound of the ε's. We can now define $G: [0, 1] \times [0, a[\to A$ by $G(s, t) = G_\varepsilon(s, t)$ if $t \in [0, \varepsilon]$.

3. For all s, $G(s, a)$ is defined.

$G(s, t)$ must be differentiable in t, for $F(s, t)$ satisfies (2) of Lemma 0.1, and by the local inverse theorem. By the chain rule, we have for all $0 \leq s \leq 1$

$$\Phi'[G(s, t)]G'(s, t) = F'(s, t) \qquad (0.2.11)$$

where the prime denotes differentiation with respect to t. Hence

$$G'(s, t) = [\Phi'(G(s, t))]^{-1} F'(s, t) \qquad (0.2.12)$$

and

$$\|G'(s, t)\| \leq \|[\Phi'(G(s, t))]^{-1}\| \, \|F'(s, t)\|. \qquad (0.2.13)$$

Then

$$\|G'(s, t)\| \leq K \|F'(s, t)\| \leq M_s. \qquad (0.2.14)$$

Now integrating with respect to t between t_0 and t_1 we get

$$\|G(s, t_1) - G(s, t_0)\| \leq M_s |t_1 - t_0|, \qquad (0.2.15)$$

a Lipschitz condition for $t \to G(s, t)$. Therefore, since A is complete as a closed subset of $\mathbf{R}^n$ (property (1a) of Theorem 0.5), $\lim_{t \to a^-} G(s, t)$ exists for each s and $G(s, t)$ can be defined at $t = a$.

4. Finally, $a = 1$.

If this were not the case, consider the continuous path $s \to G(s, a)$. For each $s (0 \leq s \leq 1)$ we can choose U_s an open neighbourhood of $G(s, a)$ and V_s an open neighbourhood of $\Phi(G(s, a))$ such that $\Phi|_{U_s}$ is a diffeomorphism. But the path considered is compact, therefore there exists a finite covering of the curve $G(s, a)$ with neighbourhoods $U_{s_i}, i = 1, \ldots, n$. In each of these neighbourhoods we can define the function $G(s, t)$ for $0 \leq t \leq a + \varepsilon_i$ by the local inverse theorem. So $G(s, t)$ is defined for the rectangle $[0, 1] \times$

$[0, a + \min_i \varepsilon_i]$, contradicting the fact that a was the largest of such numbers.

Proof of the theorem Since A is non-empty, there exists an x_0 in A such that $\Phi(x_0) = y_0$ belongs to B. Let y be another point of B. Property (2) of Lemma 0.1 implies the existence of a differential path $F(t): [0, 1] \to B$ such that $F(0) = y_0$ and $F(1) = y$. As a particular case of the lemma, there exists a curve $G(t)$, $0 \le t \le 1$ in A such that $\Phi(G(t)) = F(t)$. Then $\Phi(G(1)) = y$ and Φ is onto.

Suppose now that there are two points x_1 and x_2 in A such that $\Phi(x_1) = \Phi(x_2) = y$. From property (1b) of Theorem 0.5 there exist C^1 curves f_1 and f_2 from x_0 to x_1 and x_2 respectively. Then both image curves $\Phi \circ f_1$ and $\Phi \circ f_2$ will join y_0 and y. As B is simply connected, there exists a function $F(s, t)$ from Δ to B such that $F(0, t) = \Phi \circ f_1(t)$ and $F(1, t) = \Phi \circ f_2(t)$ and $F(s, 0) = y_0$, $F(s, 1) = y$. By the argument of the lemma, we find a function $G(s, t)$ from Δ to A continuous and such that $\Phi(G(s, t)) = F(s, t)$ with $G(0, t) = f_1(t)$ and $G(1, t) = f_2(t)$. But then the continuous curve $G(s, 1)$ with endpoints x_1 and x_2 is mapped by Φ onto y. This contradicts the local inverse theorem and therefore Φ is one-to-one. Φ^{-1} is obviously a diffeomorphism, (respectively C^k).

It is possible to derive from this global inverse theorem, global versions of the Implicit Function Theorem and of the Constant Rank Theorem. However, the global result depends strongly on the topological properties (1a), (1b), and (2) of Theorem 0.5, which are rather clumsy to formulate. For example, a global Constant Rank Theorem requires, as a sufficient condition, that a precise minor of the Jacobian matrix of the mapping should be invertible on a set A with properties similar to those in the theorem.

0.2.4 Differential equations and stability

0.2.4.1 Existence theorem

Another tool from differential geometry is the basic theory of differential equations. Let M be an n-dimensional manifold and Ω an open subset of $M \times \mathbf{R}$ such that $M \times \{t_0\}$ is included in Ω. A vector field of class C^r on M is a C^r map which associates a vector $X(p, t)$ of the tangent space TM_p with each point $p \in M$. An integral curve of a vector field X through a point p is a mapping $\alpha: I \to M$, where I is an open interval containing t_0, $\alpha(t_0) = p$ and $\alpha'(t) = X(\alpha(t), t)$ for all t in I. We then say that (α, I) is a solution of the differential equation $dx/dt = X(x, t)$ with initial condition $\alpha(t_0) = p$. If (α_1, I_1) and (α_2, I_2) are two solutions, we define a partial order on the set of all the solutions by

$$(\alpha_1, I_1) < (\alpha_2, I_2) \quad \text{iff } I_1 \text{ is included in } I_2 \text{ \& } \alpha_2|_{I_1} = \alpha_1. \quad (0.2.16)$$

Theorem 0.6 (Existence and uniqueness of solutions) *Let X be a vector field of class C^r, on a manifold M and $p \in M$. Then there exists a unique maximal solution (α, I) of the differential equation associated with X.*

When a control feedback has been defined for a control problem, the closed-loop equation of the system is an ordinary differential equation. In the non-linear case (which is the usual case in robotics), solutions of the equation do not necessarily exist for all t. The following theorem is useful for proving the global existence of solutions.

Theorem 0.7 *Let I be an interval of $\mathbf{R}$ and A a compact subset of M. Let X be a C^∞ vector field and $(p_0, t_0) \in A \times I$. If for any integral curve (α, J) defined for all $t \in J$ and such that $\alpha(t_0) = p_0$, we have $\alpha(J)$ included in A, then there exists an integral curve $(\bar{\alpha}, I)$ defined on the whole I with $\bar{\alpha}(t_0) = p_0$.*

Roughly speaking, if it is possible to bound *a priori* a solution (i.e. without knowing if such a solution exists or not) of the differential equation ($\alpha(J)$ included in A), then such a solution does exist.

Proof Let $]a, b[$ be an open interval such that $(\alpha,]a, b[)$ is a maximal solution of the differential equation with $\alpha(t_0) = p$. The hypothesis implies that $\alpha(]a, b[)$ is included in the compact set A. If $\{t_k\}_{k \in N}$ is a sequence of elements in $]a, b[$ converging to b, $\alpha(t_k)$ is a sequence of points in A. Extracting a convergent subsequence, it is easy to show that $\alpha(b)$ can be defined as the limit of this sequence. If $]a, b[$ is strictly included in I, it is then possible to find a solution defined on $]a, b + \varepsilon[$ for some $\varepsilon > 0$ so that $(\alpha,]a, b[)$ is not maximal. This contradiction implies $]a, b[= I$.

0.2.4.2 Stability

Concerning the behaviour of the solutions of a differential equation, stability is a central issue in control theory. In the non-linear case, definitions of stability are numerous, reflecting the richness of the subject. Given a differential equation $dx/dt = X(x, t)$, where the vector field $X(x, t)$ depends on time, we shall say that $p \in M$ is an equilibrium point if $X(p, t) = 0$ for all t.

Definition D0.6 *An equilibrium point p_0 is stable (in the sense of Lyapunov) if, for every $\varepsilon > 0$ and every t_0, there exists a real number $\delta(t_0) > 0$ such that $\|p_0 - p\| < \delta(t_0)$ implies $\|\alpha(t) - p\| < \varepsilon$ for all $t > t_0$ with α a solution of the differential equation such that $\alpha(t_0) = p$.*

The equilibrium point is uniformly stable if $\delta(t_0)$ can be chosen independent from t_0.

In control applications, stability is often not a satisfactory property and asymptotic stability is preferred.

Definition D0.7 *An equilibrium point p_0 is (uniformly) asymptotically stable if it is (uniformly) stable and $\lim_{t \to \infty} \alpha(t) = p_0$ in the definition above.*

For linear autonomous systems, asymptotic stability of the origin implies the existence of a positive constant $k(p)$ such that $\|\alpha(t)\| \leq e^{-k(p)t}$. Thus the rate of convergence of $\alpha(t)$ to the equilibrium point is exponential. This warrants the following definition.

Definition D0.8 *An equilibrium point p_0 is exponentially asymptotically stable if it is asymptotically stable and if there exists a positive constant $k(p, t_0)$ and a T_1 such that $\|\alpha(t) - p_0\| \leq e^{-k(p, t_0)t}$ for all $t \geq T_1$.*

The dependence of k on the initial point p and initial time t_0 gives rise to various definitions of uniform exponential stability when independence is required.

0.3 BIBLIOGRAPHIC NOTE

The different theorems on calculus recalled in this chapter are classical except for the global versions of some of them, which seem to be less well known. All the material in calculus and differential geometry is developed in the classical treatises of Birkhoff and Rota (1989), Dieudonne (1968), and Spivak (1970). A Banach space version of the global inverse theorem can be found in Schwartz (1969).

For readers who are not familiar with control theory and in particular the state space, output function approach, excellent introductions are (among others) Astrom (1984), Kailath (1980), and Faurre (1984), or, at a more abstract level, Wonham (1985).

1
FUNDAMENTAL CONFIGURATION SPACES

INTRODUCTION

This chapter presents basic mathematical tools used throughout the book for modelling, control synthesis, and control analysis.

The central concept of the chapter is that of *configuration space*. Roughly speaking, the configuration space of a mechanical system is a set such that each element represents a possible location of the mechanical system. For example, the location of a rigid polygon is located by three parameters (two for the translation, one for the orientation). For a particle, the configuration space is simply $\mathbf{R}^3$, each point representing a possible position of the particle. For a system consisting of n particles, we need $(\mathbf{R}^3)^n$ to describe the 'position' of the system. If the positions of the n particles are subject to constraints, then a submanifold of $(\mathbf{R}^3)^n$ may be a suitable configuration space. More generally the configuration space of a mechanical system which can be described with a finite number of parameters is a differentiable manifold M. In this chapter we examine the configuration space for *attitudes* and for *locations* of rigid bodies.

An essential notion in mechanics is that of velocity. The tangent bundle TM to the configuration space is a good candidate to describe the set of all possible velocities. The tangent space TM_p at a point p is the set of all possible velocities when the mechanical system is in a position corresponding to p.

The other fundamental mechanical idea is the concept of forces. As we shall see, it is possible to model the action of forces locally as a linear form defined on the tangent space TM_p. With this approach, modelling of constraints put on the displacements of a rigid body is natural and straightforward.

Control objectives need also a 'good' parametrization of the output space. In robotics, output functions are often more or less linked to locations and attitudes of rigid bodies, so we also examine this problem in this chapter.

1.1 NOTATION, BASIS, AND FRAMES

The modelling and study of control problems in robotics imply the use of many frames, matrices, and vectors, so it is necessary to establish notational conventions that will make equations easier to write and simpler to read.

1.1 NOTATION, BASIS, AND FRAMES

A frame F_i is defined by its origin O_i and a basis $\{x_i, y_i, z_i\}$ in the vector space $\mathbf{R}^3$. We shall denote such a frame by F_i or (O_i, x_i, y_i, z_i). All bases considered will be orthonormal bases for the usual inner product in $\mathbf{R}^3$ with direct orientation given by the canonical basis.

If v is a vector in $\mathbf{R}^3$, we shall denote by $[v]_i$ the 3×1 matrix composed of the coordinates of v relative to the basis $\{x_i, y_i, z_i\}$.

If M is a point, $[M]_i$ will represent the 4×1 matrix of the homogeneous coordinates of M relative to the frame (O_i, x_i, y_i, z_i). That is,

$$[M]_i = \begin{bmatrix} [O_i M]_i \\ 1 \end{bmatrix} \tag{1.1.1}$$

where $O_i M$ is the vector defined by the two ordered points O_i and M.

1.1.1 Change-of-frames formulae

If $F_i(O_i, x_i, y_i, z_i)$ and $F_j(O_j, x_j, y_j, z_j)$ are two frames, R_{ij} will denote the 3×3 matrix whose columns are the respective coordinates of x_j, y_j, z_j in the basis $\{x_i, y_i, z_i\}$. With the notation defined above,

$$R_{ij} = \begin{bmatrix} [x_j]_i & [y_j]_i & [z_j]_i \end{bmatrix}. \tag{1.1.2}$$

R_{ij} is a rotation matrix and the following relations are well known:

$$(R_{ij})^{-1} = R_{ji} = (R_{ij})^T \tag{1.1.3}$$

$$[v]_i = R_{ij}[v]_j.$$

In order to obtain analogous formulae for point coordinates, let us introduce the homogeneous matrix

$$\bar{R}_{ij} = \begin{bmatrix} R_{ij} & [O_i O_j]_i \\ 0\ 0\ 0 & 1 \end{bmatrix}. \tag{1.1.4}$$

The following change-of-coordinates and inversion formulas are straightforward:

$$(\bar{R}_{ij})^{-1} = \bar{R}_{ji} = \begin{bmatrix} (R_{ij})^T & -R_{ij}^T[O_i O_j]_i \\ 0\ 0\ 0 & 1 \end{bmatrix}, \tag{1.1.5}$$

$$[M]_i = \bar{R}_{ij}[M]_j.$$

Now if $F_k(O_k, x_k, y_k, z_k)$ is a third frame, the following relationships are useful for dealing with successive changes of frame:

$$\bar{R}_{ik} = \bar{R}_{ij}\bar{R}_{jk} \tag{1.1.6}$$

$$R_{ik} = R_{ij}R_{jk}.$$

Given a reference frame F_0, it is possible to associate with each frame F_i a homogeneous matrix $\bar{R}_{0i}$. Reciprocally, each homogeneous matrix $\bar{R}$ defines a unique frame F. From this remark it is possible to consider the set of frames to be identical (isomorphic) to the set of homogeneous matrices. The isomorphism depends on the choice of the reference frame, but the change-of-frame formula establishes a link between two representations.

Another equivalent point of view is to consider a frame F_i as the image of the reference frame F_0 through the mapping $\bar{r}_{0i}$ with matrix $\bar{R}_{0i}$. This mapping is an orientation-preserving isometry and it is well known that the set of all orientation-preserving isometries in three dimensions is a group isomorphic to the group of homogeneous 4×4 matrices $\bar{R}_{ij}$ introduced above. This group is often known as the Special Euclidean group, denoted by SE_3.

From all these considerations, it appears that SE_3 can be considered as the configuration space for frames or equivalently for rigid bodies. We shall denote by $\bar{r}$ an element of this group with associated homogeneous matrix $\bar{R}$.

Definition D1.1 *The location of a frame* F *relative to a reference frame* F_0 *is the unique element* $\bar{r}$ *in* SE_3 *such that* $F = \bar{r}(F_0)$.

1.1.2 Frames in robotics

Dealing as it does with moving bodies, moving objects, and moving cameras, robotics implies the use of many frames. It is commonly observed that the choice of the frames has consequences for the complexity of the equations appearing in robot control problems.

For the present, let us consider the problem of associating a frame with each link of the manipulator. From a control-theory point of view, it is often enough to know that a frame is associated with each link and to know the transformation matrix from one frame to another. In practice it eases the work to have a rule for choosing the frames, in particular if the rule leads to frames giving simple equations. A popular rule used in robotics is the Denavit–Hartenberg convention. More recently, a modified Denavit–Hartenberg notation has been introduced to avoid the disadvantages of the traditional form for closed kinematic chains in particular. The conventions are as follows in the case of an open kinematic chain (Fig. 1.1).

1. The manipulator is composed of $n + 1$ links and n joints. Links are $L_0, L_1, \ldots, L_n$, where L_0 is the base of the manipulator and L_n the link of the end-effector. The ith joint connects link L_{i-1} to link L_i.

2. A frame $F_i(O_i, x_i, y_i, z_i)$ is associated with L_i. The joint is either a pure rotation around an axis, or a pure translation along an axis.

1.2 ATTITUDES OF RIGID BODIES

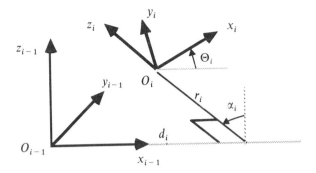

Fig. 1.1 The modified Denavit–Hartenberg parameters

3. z_i is the axis of the ith joint.
4. x_i is the axis orthogonal to z_i and z_{i+1}.

The position of F_i with respect to F_{i-1} is defined by four parameters:

(1) α_i is the angle between z_{i-1} and z_i around x_{i-1};
(2) d_i is the distance between O_{i-1} and z_i measured along x_{i-1};
(3) Θ_i is the angle between x_{i-1} and x_i around z_i;
(4) r_i is the distance between x_{i-1} and O_i along z_i.

If the ith joint is a rotation, the joint variable is Θ_i. If it is a translation, the joint variable is r_i. The matrix of the transformation from F_{i-1} to F_i is

$$\begin{bmatrix} \cos(\Theta_i) & -\sin(\Theta_i) & 0 & d_i \\ \cos(\alpha_i)\sin(\Theta_i) & \cos(\alpha_i)\cos(\Theta_i) & -\sin(\alpha_i) & -r_i\sin(\alpha_i) \\ \sin(\alpha_i)\sin(\Theta_i) & \sin(\alpha_i)\cos(\Theta_i) & \cos(\alpha_i) & r_i\cos(\alpha_i) \\ 0 & 0 & 0 & 1 \end{bmatrix}. \qquad (1.1.7)$$

Remark. If the $(i-1)$th joint and the ith joint have parallel axes, the axis x_{i-1} is not uniquely defined.

1.2 ATTITUDES OF RIGID BODIES

1.2.1 Attitudes

As seen in the previous section, the location of a frame F_j relative to another frame F_i is completely described by the homogeneous matrix $\bar{R}_{ij}$. Now if we consider a rigid body such as a link, the end-effector, or a tool tied to the manipulator, it is possible to associate with this body a frame such that every point of the body has invariant coordinates in that frame. The location of the rigid body is then defined by the homogeneous matrix giving the position of this frame relative to a reference one.

In this section we shall be concerned only with the rotation matrix R_{ij}, defined as follows.

Definition D1.2 *If the location of a rigid body (frame) relative to a reference frame is given by a homogeneous matrix $\bar{R}_{ij}$, the associated rotation matrix R_{ij} (or equivalently the rotation) is called the attitude of the rigid body (frame).*

As we have already noted with frames, there is a one-to-one correspondence between attitudes of rigid bodies (or of the associated basis), and rotations or rotation matrices as soon as a reference basis is chosen. It is also easy from (1.1.6) to find the new correspondence if the reference basis is changed. This allows us to consider the configuration space for *attitudes* of rigid bodies to be the group of rotations isomorphic to the group of usual rotation matrices on $\mathbf{R}^3$. Mathematicians often denote this by SO_3, and we shall not make any distinction between a rotation r and its associated matrix R.

1.2.1.1 The differential structure on SO_3

It is well known that rotation matrices are 3×3 matrices R such that

$$RR^T = Id, \quad \det(R) = +1. \tag{1.2.1}$$

The set of all 3×3 matrices R may be considered as $\mathbf{R}^9$, while the set of symmetric 3×3 matrices may be considered as $\mathbf{R}^6$. The map $\phi(A) = AA^T$ defined from $\mathbf{R}^9$ into $\mathbf{R}^6$ is a differentiable map. It is straightforward to verify that its rank is 6 at every matrix A such that $AA^T = Id$. From Proposition P0.8 we know that the set $O(3)$ of matrices A such that $AA^T = Id$ is a submanifold of the set of 3×3 matrices. Its dimension is $9 - 6 = 3$.

Now if R is an element of $O(3)$ with $\det(R) = 1$, let (x, U) be a chart around this element. By narrowing, if necessary, the open set U, and since the determinant map is continuous, it is possible to suppose that $\det(A) = 1$ for all A in U. In this way we obtain an atlas for SO_3. The compatibility relations are still satisfied after narrowing some U. This shows that SO_3 is a three-dimensional submanifold of $\mathbf{R}^9$.

In order to describe the tangent space at a point R, let us consider a differentiable path $t \to R(t)$ on SO_3. Since for all t

$$R(t)R^T(t) = Id, \tag{1.2.2}$$

1.2 ATTITUDES OF RIGID BODIES

differentiating with respect to t gives

$$\frac{dR}{dt} R^T + R \left(\frac{dR^T}{dt} \right) = 0. \qquad (1.2.3)$$

This last equation shows that $(dR/dt) R^T$ is a skew symmetric matrix.

Definition D1.3 *If*

$$v = \begin{pmatrix} v_1 \\ v_2 \\ v_3 \end{pmatrix}$$

is a vector, the associated skew symmetric matrix is

$$AS(v) = \tilde{v} = \begin{bmatrix} 0 & -v_3 & v_2 \\ v_3 & 0 & -v_1 \\ -v_2 & v_1 & 0 \end{bmatrix}. \qquad (1.2.4)$$

The mapping AS defines a one-to-one correspondence between three-dimensional vectors and 3×3 skew symmetric matrices.

If we consider a vector $\omega(t)$ such that $\tilde{\omega}(t) = (dR/dt) R^T$, we get

$$\frac{dR}{dt} = \tilde{\omega} R(t). \qquad (1.2.5)$$

This equation shows that the tangent bundle over SO_3 is isomorphic to $\mathbf{R}^3 \times SO_3$ by the mapping

$$(v, R) \longrightarrow \tilde{v} R, \qquad (1.2.6)$$

the fibre over each rotation R being isomorphic to the vector space of 3×3 skew symmetric matrices.

1.2.1.2 Parametrizations

Although matrices are convenient for computation, they do not exhibit two important elements of a rotation: the axis and the angle. Moreover, it is in general impossible for a human being to deduce the nature of a rotation from a simple examination of the matrix. On the other hand it is not a natural operation to discover the matrix of a rotation from a more usual representation. Last but not least, local parametrizations are necessary for control purposes and the nine entries of a rotation matrix are not independent parameters.

Since SO_3 is a three-dimensional differentiable manifold, it is possible to find a diffeomorphism from an open set of $\mathbf{R}^3$ to an open set of SO_3. It would be ideal to have a diffeomorphism from a subset of $\mathbf{R}^3$ to the whole SO_3. This would imply that small changes in attitude would result in small changes in the parameters representing the rotation, with the converse also true. The interest of such a diffeomorphism would be to map every vector field on SO_3 onto a vector field on the more manageable space $\mathbf{R}^3$ with many properties unchanged. Unfortunately such an ideal set of parameters does not exist, as the topology of SO_3 is quite different from that of $\mathbf{R}^3$. In particular, SO_3 is *compact* (i.e. closed and bounded) and $\mathbf{R}^3$ is not.

The practical consequence of these topological problems is that every differentiable map from an open subset of $\mathbf{R}^3$ onto SO_3 has *singular* points. At these points the differential of the map is not invertible so, in the neighbourhood of a singular point, small changes in the rotation result in large changes in the parameters. Robot wrists with three degrees of freedom illustrate this fact. They necessarily have singular configurations.

Despite this drawback, sets of three parameters are used in practice because they are closer to common experience than matrices are. As they are diffeomorphisms from an open set of $\mathbf{R}^3$ to an open set of SO_3, we call them charts or chart-type representations. (They are charts in the meaning of differentiable manifold theory.)

1.2.2 Chart representations: Euler angles

A general method of finding a chart representation is to define a set of Euler angles. Let F and F_0 be two frames with the same origin. An axis Δ of the reference frame F_0 and an axis δ of the frame F are chosen. The intersection of the two planes orthogonal to Δ and δ is a line Ou (when defined). The three Euler angles associated with this choice are (Fig. 1.2):

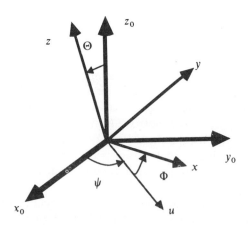

Fig. 1.2 The classical Euler angles

1.2 ATTITUDES OF RIGID BODIES

(1) Ψ, giving the position of the line Ou in the plane orthogonal to Δ;

(2) Θ, defining the position of δ in a plane orthogonal to Ou;

(3) Φ, the angle between Ou and an axis linked to F and contained in the plane orthogonal to δ.

The axis Ou is not defined if Δ and δ coincide. Then the Euler angles are not defined, and this is precisely the configuration where singularity occurs.

The image of the three parameters (Ψ, Θ, Φ) is the rotation obtained as the product of three elementary rotations r_Ψ, r_Θ, r_Φ:

(1) r_Ψ has Δ as axis and Ψ as angle;

(2) r_Θ has Ou as axis and Θ as angle;

(3) r_Φ has δ as axis and Φ as angle.

Another equivalent point of view is to consider that the location of the frame F with respect to the frame F_0 is defined by two intermediate frames.

$$F_1 = (O, u, v, \Delta) \quad \text{and} \quad F_2 = (O, u, w, \delta)$$

the axes v and w being defined once u, δ and Δ are given.
The change-of-frame formula applies, and gives:

$$r_{FF_0} = r_{FF_2} \circ r_{F_2 F_1} \circ r_{F_1 F_0} = r_\Phi \circ r_\Theta \circ r_\Psi.$$

The most popular choices are $\Delta = (O, z_0)$ and $\delta = (O, z)$ which gives the classical Euler angles (Fig. 1.2) and $\Delta = (O, z_0)$, $\delta = (O, x)$ which gives the roll, pitch, and yaw angles, schematized for an aircraft in Fig. 1.3.

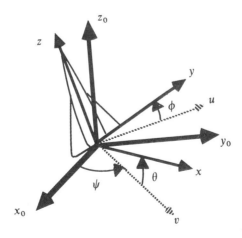

Fig. 1.3 The roll, pitch, and yaw angles

1 FUNDAMENTAL CONFIGURATION SPACES

The classical Euler angles are not defined if the two axes (O, z_0) and (O, z) coincide. In particular, this chart is not suitable for representing small attitude errors. In other words, classical Euler angles define a chart with open domain $SO_3 - \{\text{Rot}|Oz_0\}$ where $\{\text{Rot}|Oz_0\}$ is the set of rotations with axis $\Delta = Oz_0$.

The roll, pitch, yaw (RPY) representation is defined by the choice of the axis mentioned above:

(1) angle $(x_0, v) = \psi$ is the yaw angle;

(2) angle $(v, x) = \Theta$ is the pitch angle;

(3) angle $(u, y) = \Phi$ is the roll angle.

The rotation matrix giving the position of the basis (x, y, z) relative to the basis (x_0, y_0, z_0) is then obtained from (1.1.6):

$$\begin{bmatrix} \cos(\Theta)\cos(\psi) & -\sin(\psi)\cos(\Theta) & -\sin(\Theta) \\ \cos(\Phi)\sin(\psi)-\sin(\Theta)\cos(\psi)\sin(\Phi) & \cos(\Phi)\cos(\psi)+\sin(\Theta)\sin(\Phi)\sin(\psi) & -\sin(\Theta)\sin(\psi) \\ \sin(\psi)\sin(\Phi)+\sin(\Theta)\cos(\psi)\cos(\Phi) & \sin(\Phi)\cos(\psi)-\sin(\Theta)\sin(\psi)\cos(\Phi) & \cos(\psi)\cos(\Theta) \end{bmatrix}$$
(1.2.7)

which shows that it is impossible to determine Φ and ψ if $\Theta = \pi/2$. In this case there is an infinity of couples (Φ, ψ) corresponding to one rotation matrix. The closed subset $NRPY$ of rotations with matrices

$$\begin{pmatrix} 0 & \cos(\alpha) & -\sin(\alpha) \\ 0 & \sin(\alpha) & \cos(\alpha) \\ 1 & 0 & 0 \end{pmatrix}$$
(1.2.8)

cannot be suitably represented. Since the domain of this chart contains the identity matrix, it is a suitable parametrization for attitude errors. The main drawback of Euler angles is the difficulty of computing the parameters corresponding to the product of two rotations from the parameters representing the two rotations.

1.2.3 Quaternions

1.2.3.1 Definition and elementary properties

There are several ways of introducing quaternions. As mathematical objects they can be considered as couples (α, β) where α is an element of $\mathbf{R}$ and β is an element of $\mathbf{R}^3$. So, basically, the set of quaternions is $\mathbf{R}^4 = \mathbf{R} \times \mathbf{R}^3$. The sum

1.2 ATTITUDES OF RIGID BODIES

of two quaternions is the usual componentwise sum on $\mathbf{R}^4$:

$$(\alpha_1, \beta_1) + (\alpha_2, \beta_2) = (\alpha_1 + \alpha_2, \beta_1 + \beta_2). \tag{1.2.9}$$

Quaternions also inherit the vector space structure on $\mathbf{R}^4$.

For every quaternion $\lambda = (\alpha, \beta)$ there is a unique decomposition

$$\lambda = (\alpha, 0) + (0, \beta) \tag{1.2.10}$$

as the sum of an element belonging to $\mathbf{R} \times \{0\}$ and an element of $\{0\} \times \mathbf{R}^3$. We shall make no distinction between $\mathbf{R}$ and $\mathbf{R} \times \{0\}$ and between $\mathbf{R}^3$ and $\{0\} \times \mathbf{R}^3$. With an obvious abuse of notation, we can write every quaternion as the sum of a real number and a vector:

$$\lambda = \alpha + \beta. \tag{1.2.11}$$

α is called the real part of λ and denoted by $R(\lambda)$, while β is called the pure part of λ and denoted by $P(\lambda)$. A quaternion λ is a scalar if $P(\lambda) = 0$ and it is a pure quaternion or a vector if $R(\lambda) = 0$.

The operation that gives particular properties to the set of quaternions is the product. If $\lambda_1 = \alpha_1 + \beta_1$ and $\lambda_2 = \alpha_2 + \beta_2$ are two quaternions, their product is defined as:

$$\left. \begin{array}{l} R(\lambda_1 \lambda_2) = \alpha_1 \alpha_2 - \langle \beta_1, \beta_2 \rangle \\ P(\lambda_1 \lambda_2) = \beta_1 \times \beta_2 + \alpha_1 \beta_2 + \alpha_2 \beta_1 \end{array} \right\}, \tag{1.2.12}$$

where $\times$ denotes the usual cross product; or equivalently

$$\lambda_1 \lambda_2 = [\alpha_1 \alpha_2 - \langle \beta_1, \beta_2 \rangle] + [\beta_1 \times \beta_2 + \alpha_1 \beta_2 + \alpha_2 \beta_1]. \tag{1.2.13}$$

It is straightforward although tedious to check that this product is associative and distributive with respect to the sum. There is a unit element, the scalar 1, and every quaternion different from 0 has an inverse. It is important to notice that this product is *not commutative*.

Let us denote by $\mathbf{H}$ (in honour of Hamilton) the set $\mathbf{R}^4$, together with the sum and product just defined. $\mathbf{H}$ is a division ring.

1.2.3.2 Basis, conjugate, and norm

Let $\{1, i, j, k\}$ be the canonical basis of $\mathbf{R}^4$ where 1 stands for $(1, 0, 0, 0)$, i stands for $(0, 1, 0, 0)$ and so on. The following properties are straightforward.

Proposition P1.1

$$\left. \begin{array}{l} 1\,1 = 1; \quad 1\,i = i; \quad 1\,j = j; \quad 1\,k = k, \\ i^2 = j^2 = k^2 = -1, \\ ij = -ji = k; \quad jk = -kj = i; \quad ki = -ik = j. \end{array} \right\} \tag{1.2.14}$$

1 FUNDAMENTAL CONFIGURATION SPACES

It is possible to define the conjugate of a quaternion, as for complex numbers:

$$\bar{\lambda} = R(\lambda) - P(\lambda). \tag{1.2.15}$$

Proposition P1.2

$$\begin{aligned} \bar{\bar{\lambda}} &= \lambda; \\ \lambda &= \bar{\lambda} \Leftrightarrow \lambda \in \mathbf{R}, \\ \lambda &= -\bar{\lambda} \Leftrightarrow \lambda \in \mathbf{R}^3, \\ \overline{\lambda + \mu} &= \bar{\lambda} + \bar{\mu}, \\ \overline{\lambda \mu} &= -\bar{\mu}\bar{\lambda}. \end{aligned} \tag{1.2.16}$$

(*The non-commutativity also appears in this last formula.*)

Computations are straightforward. The usual inner product on $\mathbf{R}^4$ can be expressed as

$$\langle \lambda, \mu \rangle = \tfrac{1}{2}(\bar{\lambda}\mu + \bar{\mu}\lambda). \tag{1.2.17}$$

The corresponding norm is $\|\lambda\| = \sqrt{\lambda\bar{\lambda}} = \sqrt{\alpha^2 + \|\beta\|^2}$.

Proposition P1.3 *For every quaternion $\lambda \neq 0$,*

$$\lambda^{-1} = \frac{\bar{\lambda}}{\|\lambda\|^2}. \tag{1.2.18}$$

For every pair of quaternions λ and μ,

$$\|\lambda\mu\| = \|\lambda\| \, \|\mu\|. \tag{1.2.19}$$

As a consequence of this last property, the subset of quaternions with norm equal to 1 with multiplication as its law of composition forms a group. This non-commutative group of unitary quaternions can also be considered as the three-dimensional sphere S^3 endowed with a group structure. This group will be the basic tool for representing a rotation as a quaternion. Unitary quaternions are sometimes called 'Euler parameters'.

When restricted to $\mathbf{R}^3$ considered as the subset of pure quaternions, the product and inner product have some nice properties.

Proposition P1.4 *If v_1 and v_2 are vectors in $\mathbf{R}^3$, then*

$$\langle v_1, v_2 \rangle = \frac{-1}{2}(v_1 v_2 + v_2 v_1). \tag{1.2.20}$$

1.2 ATTITUDES OF RIGID BODIES

When restricted to $\mathbf{R}^3$ the commutator of two quaternions, $1/2[v_1, v_2]$, is the usual cross product,

$$v_1 \times v_2 = \tfrac{1}{2}(v_1 v_2 - v_2 v_1) = \tfrac{1}{2}[v_1, v_2]. \tag{1.2.21}$$

1.2.3.3 Rotations and quaternions

Basic geometric transformations of the plane $\mathbf{R}^2$ can be expressed in terms of operations on complex numbers. Since the set of quaternions contains the space $\mathbf{R}^3$, it is hoped that some analogous property holds for geometric transformations on $\mathbf{R}^3$. Let us examine the problem for rotations.

It is possible to associate with each unitary quaternion a mapping ρ_λ from $\mathbf{H}$ onto itself by

$$\left. \begin{array}{c} \mathbf{H} \longrightarrow \mathbf{H} \\ \mu \longrightarrow \rho_\lambda(\mu) = \lambda \mu \lambda^{-1}. \end{array} \right\} \tag{1.2.22}$$

The following theorem summarizes the properties of this mapping.

Theorem 1.1

1. *The set of pure quaternions $\mathbf{R}^3$ is invariant under the mapping ρ_λ, and ρ_λ is a rotation when restricted to $\mathbf{R}^3$.*

2. *$\rho_\lambda = \rho_\mu$ is equivalent to $\lambda = \varepsilon\mu$ ($\varepsilon = \pm 1$).*

3. *For every rotation $r \in SO_3$ there exists a unitary quaternion $\lambda \in S^3$ such that $r = \rho_\lambda$.*

This result is a classical one. Let us merely sketch the proof.

For point (1) it is enough to use the easily proved property that the product of a pure quaternion with itself is a negative real number. Then, if v is in $\mathbf{R}^3$ we have

$$(\lambda v \lambda^{-1})^2 = \lambda v \lambda^{-1} \lambda v \lambda^{-1} = \lambda v^2 \lambda^{-1} = v^2, \tag{1.2.23}$$

since a quaternion and a real number commute.

A direct computation using Propositions P1.3 and P1.4 shows that ρ_λ is an isometry. Since it is also a linear map, it remains to show that it is a positive isometry. But

$$\rho_\lambda(i) \times \rho_\lambda(j) = \tfrac{1}{2}[\rho_\lambda(i)\rho_\lambda(j) - \rho_\lambda(j)\rho_\lambda(i)]$$
$$= \tfrac{1}{2}[\lambda(ij - ji)\lambda^{-1}]$$
$$= \rho_\lambda(k), \tag{1.2.24}$$

so ρ_λ maps a direct base onto a direct base.

Suppose now that $\rho_\lambda = \rho_\mu$. Then for every $v \in \mathbf{R}^3$,

$$\lambda v \lambda^{-1} = \mu v \mu^{-1}, \tag{1.2.25}$$

which implies
$$(\mu^{-1}\lambda)v = v(\mu^{-1}\lambda) \tag{1.2.26}$$
for all v in $\mathbf{R}^3$. Since the only elements of $\mathbf{H}$ which commute with every element of $\mathbf{R}^3$ are the real numbers, $\mu^{-1}\lambda$ belongs to $\mathbf{R}$. It is then easy to show that it is 1 or -1.

We have obtained a mapping π from the sphere S^3 to the rotation group SO_3, defined by $\pi(\lambda) = \rho_\lambda$. Recall that S^3 together with the quaternion product is a group. The following proposition is a direct consequence of the theorem.

Proposition P1.5 *The map π is a continuous group homomorphism and is onto. Its kernel is $\{\pm 1\}$.*

The practical interest of this proposition is that every rotation can be represented by a quaternion; moreover, the product (composition) of two rotations is represented by the product of the corresponding quaternions. From a topological point of view, SO_3 is the image of a compact set S^3 through a continuous mapping π, so it is also a compact set. Unfortunately, for every rotation r there are two quaternions λ and $-\lambda$ such that $\pi(\lambda) = \pi(-\lambda) = r$. Worse, there is no means of choosing between them. This can be stated more precisely as follows.

Proposition P1.6 (see Appendix A1.1) *There exists no section $f: SO_3 \to S^3$ of the map π (i.e such that $\pi \circ f = Id_{SO_3}$) which is either continuous, or a group homomorphism.*

In other words, there is no canonical representation of a rotation by a quaternion. The situation is very similar to that encountered when one wants to define the square root or the logarithm of a complex number. It is well known that trying to find a continuous square root or a continuous logarithm along a closed path around zero leads to assigning two different values to the starting point. In fact, the proof of the proposition uses this property of complex numbers. The proof is held over to Appendix A1.1.

1.2.3.4 Axis and angle of a rotation represented by a quaternion

When working with rotations, it is often useful to know the axis and the angle of a rotation.

Proposition P1.7 *Let $r = \pi(\lambda)$ be a rotation represented by the quaternion λ. Then $P(\lambda)$ is a vector along the axis of the rotation, and $2\arccos(R(\lambda))$ the angle of rotation, with an orientation induced by $P(\lambda)$.*

1.2 ATTITUDES OF RIGID BODIES

Remark. The vector space $\mathbf{R}^3$ is given an orientation by the basis $\{i, j, k\}$. Now there exists a unique orientation of the plane orthogonal to the one-dimensional subspace generated by $P(\lambda)$ such that, if $\{v_1, v_2\}$ is a direct basis for this orientation, $\{v_1, v_2, P(\lambda)\}$ is a direct basis in $\mathbf{R}^3$. This orientation is called the orientation induced by $P(\lambda)$.

Proof of the proposition Let $\lambda = \alpha + \beta$ with $\alpha \in \mathbf{R}$ and $\beta \in \mathbf{R}^3$. An easy computation shows that $\lambda \beta \lambda^{-1} = \beta$, so β is a vector of the axis of the rotation. If u is a unitary vector orthogonal to β and Θ is the angle of the rotation, we get the following relationships:

$$\cos(\Theta) = \langle u, r(u) \rangle = -\tfrac{1}{2}(u\lambda u \lambda^{-1} + \lambda u \lambda^{-1} u)$$
$$= \alpha^2 - \|\beta\|^2 \tag{1.2.27}$$

$$u \times r(u) = \frac{\beta}{\|\beta\|} \sin(\Theta) = \tfrac{1}{2}[u, \lambda u \lambda^{-1}] \tag{1.2.28}$$

which gives
$$\cos(\Theta) = \alpha^2 - \|\beta\|^2, \tag{1.2.29}$$
$$\sin(\Theta) = 2\alpha \|\beta\|. \tag{1.2.30}$$

Since $\alpha^2 + \|\beta\|^2 = 1$ it is then possible to find Φ such that
$$\alpha = \cos(\Phi), \quad \|\beta\| = \sin(\Phi), \quad 0 \leq \Phi \leq \pi. \tag{1.2.31}$$

The above equations imply $\Theta = 2\Phi$ and $\Phi = \arccos(\alpha)$.

As a consequence of this last proposition, if a rotation is defined by a unitary vector u of its axis and its angle Θ (with the orientation induced by u) a quaternion λ representing this rotation is:

$$\lambda = \cos(\Theta/2) + \sin(\Theta/2)u. \tag{1.2.32}$$

1.2.3.5 Quaternion representation and rotation matrix

Given a rotation r such that $r = \pi((\lambda_0, \lambda_1, \lambda_2, \lambda_3))$, it may be useful to obtain the matrix R of r relative to the basis $\{i, j, k\}$. The computation is straightforward:

$$\begin{bmatrix} \lambda_0^2 + \lambda_1^2 - \lambda_2^2 - \lambda_3^2 & 2(\lambda_1\lambda_2 - \lambda_3\lambda_0) & 2(\lambda_0\lambda_2 + \lambda_1\lambda_3) \\ 2(\lambda_0\lambda_3 + \lambda_1\lambda_2) & \lambda_0^2 + \lambda_2^2 - \lambda_1^2 - \lambda_3^2 & 2(\lambda_2\lambda_3 - \lambda_0\lambda_1) \\ 2(\lambda_1\lambda_3 - \lambda_0\lambda_2) & 2(\lambda_0\lambda_1 + \lambda_2\lambda_3) & \lambda_0^2 + \lambda_3^2 - \lambda_1^2 - \lambda_2^2 \end{bmatrix}. \tag{1.2.33}$$

A brief examination of the above matrix leads to another useful expression of the rotation matrix:

$$(2\lambda_0^2 - 1)Id_3 + 2P(\lambda)P(\lambda)^T + 2\lambda_0 AS(P(\lambda)). \tag{1.2.34}$$

Conversely, if a rotation r is specified by its matrix R relative to the basis $\{i, j, k\}$, a quaternion λ representing r can be found from the well-known relationships:

$$\operatorname{tr}(R) = 1 + 2\cos(\Theta) \tag{1.2.35}$$

$$\tfrac{1}{2}(R - R^{\mathrm{T}}) = AS(\sin(\Theta)u), \tag{1.2.36}$$

where $\operatorname{tr}(R)$ represents the trace of the matrix R, u is a unitary vector of the axis of the rotation and Θ is the angle of the rotation for the orientation induced by u. Then we get:

$$\lambda_0 = \tfrac{1}{2}\sqrt{(1 + \operatorname{tr}(R))}, \tag{1.2.37}$$

$$AS\left(\begin{bmatrix}\lambda_1 \\ \lambda_2 \\ \lambda_3\end{bmatrix}\right) = \frac{1}{2\sqrt{(1 + \operatorname{tr}(R))}}(R - R^{\mathrm{T}}). \tag{1.2.38}$$

Remarks.

1. If $\operatorname{tr}(R) = -1$, then $\Theta = \pi$ and $(R - R^{\mathrm{T}}) = 0$. The last equation is indeterminate. Let $R = [c_1, c_2, c_3]$, where c_i represents the ith column of the matrix R. It is then easy to check that the pure part of the quaternion we are looking for is a non-null vector among $\{i + c_1, j + c_2, k + c_3\}$ divided by its norm. Such a vector always exists.

2. If $\operatorname{tr}(R) \neq -1$, it is possible to compute the pure part of the quaternion using

$$\begin{bmatrix}\lambda_1 \\ \lambda_2 \\ \lambda_3\end{bmatrix} = \frac{1}{\sqrt{1 + \operatorname{tr}(R)}}(i \times c_1 + j \times c_2 + k \times c_3). \tag{1.2.39}$$

1.2.3.6 Summary and supplementary results

Let us summarize the relationships between the representations of a rotation by quaternions, by axis and angle, and by rotation matrix.

From a quaternion representation $\lambda = \alpha + \beta = (\lambda_0, \lambda_1, \lambda_2, \lambda_3)$ we obtain the rotation matrix

$$\begin{bmatrix} \lambda_0^2 + \lambda_1^2 - \lambda_2^2 - \lambda_3^2 & 2(\lambda_1\lambda_2 - \lambda_3\lambda_0) & 2(\lambda_0\lambda_2 + \lambda_1\lambda_3) \\ 2(\lambda_0\lambda_3 + \lambda_1\lambda_2) & \lambda_0^2 + \lambda_2^2 - \lambda_1^2 - \lambda_3^2 & 2(\lambda_2\lambda_3 - \lambda_0\lambda_1) \\ 2(\lambda_1\lambda_3 - \lambda_0\lambda_2) & 2(\lambda_0\lambda_1 + \lambda_2\lambda_3) & \lambda_0^2 + \lambda_3^2 - \lambda_1^2 - \lambda_2^2 \end{bmatrix},$$

$$\tag{1.2.40}$$

the axis $u = \beta/\|\beta\|$ and the angle $\Theta = 2\arccos(\alpha)$.

From a matrix representation we get a quaternion representation $\lambda = \alpha + \beta$ with

$$\left.\begin{array}{l} \alpha = \tfrac{1}{2}\sqrt{1 + \operatorname{tr}(R)} \\ AS(\beta) = \dfrac{1}{2\sqrt{1 + \operatorname{tr}(R)}} (R - R^T). \end{array}\right\} \quad (1.2.41)$$

The axis and angle can be obtained from the relation

$$\tfrac{1}{2}(R - R^T) = AS((\sin(\Theta))u). \quad (1.2.42)$$

Given the axis represented by the unitary vector u and the angle Θ we can derive a quaternion representation

$$\lambda = \cos(\Theta/2) + u \sin(\Theta/2) \quad (1.2.43)$$

and the matrix of the rotation

$$R = \cos(\Theta) Id_3 + \sin(\Theta)\tilde{u} + (1 - \cos(\Theta))uu^T. \quad (1.2.44)$$

1.2.4 Geometry of the attitude space

It is usual in control theory to work with output errors since it is convenient to transform many control problems into regulation problems. For the control of the orientation of a body one is thus led to define an attitude error. This problem motivates the following study of the geometry of SO_3.

1.2.4.1 The exponential map

Recall that the exponential of a square matrix is defined as

$$\exp(M) = Id + \frac{M}{1!} + \frac{M^2}{2!} + \cdots + \frac{M^n}{n!} + \cdots \quad (1.2.45)$$

Now if u is a unitary vector and Θ a real number, $\exp(\Theta\tilde{u})$ is a rotation matrix. The vector u is a unitary vector on the axis and Θ is the angle of the rotation measured with the orientation induced by u. (The proof is easy by using an orthonormal basis $\{v_1, v_2, u\}$ and straightforward computation.) Conversely, there exists for every rotation r with matrix R a unitary vector u and a real number Θ such that $R = \exp(\Theta\tilde{u})$. This representation is not unique as the mapping $\Theta \to \exp(\Theta\tilde{u})$ is periodic, with period 2π.

Nevertheless, for every rotation r with angle not equal to π or zero, it is possible to find a unique couple (u, Θ) such that $0 < \Theta < \pi$ and $R = \exp(\Theta\tilde{u})$.

1.2.4.2 Geodesics on SO_3^*

Distances on a differential manifold such as SO_3 are usually measured in two stages: first, the length of a differentiable path is defined, and then shortest paths are searched for. The distance between two points is then defined as the length of the shortest path(s) joining the two points.

In measuring the length of differential paths we need to define a metric on the tangent space to SO_3. We have seen in section 1.2.1 that the tangent vectors to SO_3 at a point R (denoted by $TSO_3|_r$) can be considered as elements of $\mathbf{R}^3$ through the mapping $v \to \tilde{v}R$. So it is natural to choose as inner product on $TSO_3|_r$ the one induced by the canonical inner product on $\mathbf{R}^3$:

$$\langle \tilde{v}_1 R, \tilde{v}_2 R \rangle_{TSO_3|_r} = \langle v_1, v_2 \rangle_{\mathbf{R}^3}. \tag{1.2.46}$$

We shall not go into details, but interested readers may check that this inner product defines a bi-invariant Riemannian metric on SO_3 (invariant with respect to the left and right translations).

Now let $c: t \to r(t)$ be a differentiable path on SO_3, defined on the interval $[t_0, t_1]$. The length of this path is defined as

$$l(c) = \int_{t_0}^{t_1} \left\| \frac{dr}{dt}(s) \right\| ds \tag{1.2.47}$$

where $\|\cdot\|$ is the norm on tangent vectors corresponding to the inner product defined above.

The search for shortest paths on Riemannian manifolds is a classical problem, and necessary conditions for a path to be of minimal length are classical results. A path satisfying these conditions is called a geodesic.

Fortunately, SO_3 is not only a differentiable manifold but also a Lie group. Geodesics on Riemannian Lie groups are well known. If we translate the theorem describing them into our particular case, it reads as follows.

Theorem 1.2 *Geodesics on SO_3 are paths of the type $t \to \exp(t\tilde{u})r$ where u is a unitary vector and r a rotation.*

What is the length of a geodesic g if $t \in [t_0, t_1]$? From

$$\frac{d}{dt}(\exp(t\tilde{u})) = \tilde{u}\exp(t\tilde{u}) \tag{1.2.48}$$

we get

$$l(g) = \int_{t_0}^{t_1} \|\tilde{u}\exp(s\tilde{u})\| \, ds = \int_{t_0}^{t_1} \|u\| \, ds = (t_1 - t_0) \tag{1.2.49}$$

since u is a unitary vector.

1.2.4.3 Distances on SO_3^*

Let r_1 and r_2 be two rotations. There always exists a geodesic starting in r_1 and ending in r_2. More precisely, if $r_2 r_1^{-1} = \exp(\Theta \tilde{u})$ then $r(t) = \exp(t\tilde{u}) r_1$ is a geodesic with $r(0) = r_1$ and $r(\Theta) = r_2$. This property leads to the following definition:

Definition D1.4 *The distance between two rotations r_1 and r_2 is the shortest length of the geodesics starting at r_1 and ending at r_2.*

If r_1 and r_2 are two rotations as above, the rotation $r_2 r_1^{-1}$ with matrix $R_2 R_1^{-1}$ can be represented as an exponential $R_2 R_1^{-1} = \exp(\Theta \tilde{u})$ with $0 \leq \Theta \leq \pi$ and $\|u\| = 1$. The length of the geodesic

$$r(t) = \exp(t\tilde{u}) r_1, \quad 0 \leq t \leq \Theta \qquad (1.2.50)$$

is then Θ. Since the mapping $\Theta \to \exp(\Theta \tilde{u})$ is periodic with period 2π, the above geodesic is a shortest path between r_1 and r_2. Moreover if $r_1 \neq r_2$ and $r_2 r_1^{-1}$ is not of angle π, there exists a unique shortest path from r_1 to r_2. If $r_2 r_1^{-1}$ is of angle π there are two shortest paths from r_1 to r_2.

Remark. The existence of two shortest paths from a rotation r_1 to a rotation r_2 with distance $\rho(r_1, r_2) = \pi$ is another fact which emphasizes the difference between the topology of SO_3 and the topology of a space such as $\mathbf{R}^3$ or $\mathbf{R}^n$. This difference explains why some non-linear techniques such as linearization fail if we want a control defined on the whole of SO_3. It is not surprising that the singularities encountered in linearization occur precisely when the rotations are of angle π.

Definition D1.5 *If $R_2 R_1^{-1} = \exp(\Theta \tilde{u})$; $\|u\| = 1$; $0 \leq \Theta \leq \pi$, then the vector u is called the direction of the shortest path from r_1 to r_2.*

If $r_2 r_1^{-1}$ is considered as an error, r_1 being the actual rotation and r_2 the desired rotation, we call the vector u with Θ and u as above, the *attitude error*.

The following theorem is obtained by combining results from sections 1.2.4.2 and 1.2.4.3.

Theorem 1.3 *If $r_2 r_1^{-1}$ is represented by a quaternion $(\lambda_0, \lambda_1, \lambda_2, \lambda_3)$ with $\lambda_0 \geq 0$, then:*

1. *If $\rho(r_1, r_2) \neq \pi$ the shortest path from r_1 to r_2 is of direction $u = 1/(\sqrt{1 - \lambda^2}) P(\lambda)$ and the distance between r_1 and r_2 is $\rho(r_1, r_2) = 2 \arccos(\lambda_0)$.*
2. *If $\rho(r_1, r_2) = \pi$ then $P(\lambda)$ is the direction of the shortest path.*

If $r_2 r_1^{-1}$ is represented by a matrix R and u is the direction of the shortest path from r_1 to r_2, $\Theta = \rho(r_1, r_2)$ then

$$\tfrac{1}{2}(R - R^T) = AS(\sin(\Theta)u). \tag{1.2.51}$$

Remark. The last expression has been proposed by several authors as a measure of an attitude error. The essential drawback of this expression is that it is null if $\Theta = \pi$, i.e. precisely when the error is maximum. Despite this, it is a convenient approximation if the distance between the two rotations remains small.

1.2.5 Parametrizations and differentials

1.2.5.1 Quaternions and rotation velocities

In the previous section we have seen how to represent a rotation by a quaternion. The aim of this section is to find the relation between rotation velocities and vectors tangent to S^3. This is done by computing the differential of the mapping $\pi: S^3 \to SO_3$.

Thus, let $t \to \lambda(t)$ be a differential path on the unit sphere S^3, defined on $]-\varepsilon, +\varepsilon[$ and such that $\lambda(0) = \lambda$. Then $t \to \pi(\lambda(t))$ is a differential path on SO_3. Let $r(t) = \pi(\lambda(t))$ and u be a constant vector in $\mathbf{R}^3$. From the expression (1.2.5) of the derivative of a rotation matrix we get

$$\frac{dR(t)u}{dt} = \omega(t) \times R(t)u \tag{1.2.52}$$

with $\omega(t)$ the velocity of $R(t)$ at time t. Considering vectors as quaternions and using Proposition P1.4 we get

$$\left.\frac{dR(t)u}{dt}\right|_{t=0} = \tfrac{1}{2}[\omega(0), R(0)u] = \tfrac{1}{2}[\omega(0), \lambda u \lambda^{-1}]. \tag{1.2.53}$$

On the other hand,

$$R(t)u = \lambda(t) u \lambda(t)^{-1}. \tag{1.2.54}$$

Differentiating both sides gives

$$\frac{dR(t)u}{dt} = \frac{d\lambda}{dt} u \lambda^{-1} + \lambda u \frac{d\lambda^{-1}}{dt} \tag{1.2.55}$$

From $\lambda \lambda^{-1} = 1$ we get

$$\frac{d\lambda^{-1}}{dt} = -\lambda^{-1} \frac{d\lambda}{dt} \lambda^{-1},$$

1.2 ATTITUDES OF RIGID BODIES

and using this last expression in (1.2.55) leads to

$$\frac{dR(t)u}{dt} = \left[\frac{d\lambda}{dt}\lambda^{-1}, \lambda u \lambda^{-1}\right]. \tag{1.2.56}$$

Let $\lambda = \alpha + \beta$ with $\alpha = R(\lambda)$ and $\beta = P(\lambda)$. Then

$$R\left(\frac{d\lambda}{dt}\lambda^{-1}\right) = \alpha\frac{d\alpha}{dt} + \beta\frac{d\beta}{dt}. \tag{1.2.57}$$

But $\lambda\bar{\lambda} = \alpha^2 + \langle \beta, \beta \rangle = 1$. Differentiating on both sides shows

$$R\left(\frac{d\lambda}{dt}\lambda^{-1}\right) = 0. \tag{1.2.58}$$

We have proved that $(d\lambda/dt)\lambda^{-1}$ is a pure quaternion, so it is a vector and for every vector u we have from (1.2.53) and (1.2.21):

$$\frac{d\lambda}{dt}\lambda^{-1} \times \lambda u \lambda^{-1} = \tfrac{1}{2}\omega \times \lambda u \lambda^{-1} \tag{1.2.59}$$

from which we derive the following relation between angular velocity and quaternion velocity.

Proposition P1.8 *If $t \to \lambda(t)$ is a differentiable path on S^3 and $r(t) = \pi(\lambda(t))$ is the corresponding path on SO_3, the angular velocity and the quaternion velocity are related by*

$$\omega = 2\frac{d\lambda}{dt}\lambda^{-1}; \quad \frac{d\lambda}{dt} = \tfrac{1}{2}\omega\lambda. \tag{1.2.60}$$

Remark. The product in the above expression is of course the product of quaternions. If λ is written using the basis $\{1, i, j, k\}$, the formula gives the coordinates of ω in the basis $\{i, j, k\}$ of $\mathbf{R}^3$.

It is a straightforward computation to obtain the matrix of the differential of π (in the canonical basis):

$$d\pi|_\lambda = 2\begin{bmatrix} -\lambda_1 & \lambda_0 & -\lambda_3 & \lambda_2 \\ -\lambda_2 & \lambda_3 & \lambda_0 & -\lambda_1 \\ -\lambda_3 & -\lambda_2 & \lambda_1 & \lambda_0 \end{bmatrix}. \tag{1.2.61}$$

It is easy to check that $d\pi|_\lambda$ is always of rank 3, showing that π is a local diffeomorphism (by the Local Inverse Theorem 0.2) from S^3 to SO_3. Of course, π is not a global diffeomorphism since it is not even injective.

The relation $d\lambda/dt = \frac{1}{2}\omega\lambda$ can also be given a matrix form:

$$\frac{d\lambda}{dt} = \frac{1}{2}\begin{bmatrix} -\lambda_1 & -\lambda_2 & -\lambda_3 \\ \lambda_0 & \lambda_3 & -\lambda_2 \\ -\lambda_3 & \lambda_0 & \lambda_1 \\ \lambda_2 & -\lambda_1 & \lambda_0 \end{bmatrix}\omega. \quad (1.2.62)$$

1.2.5.2 A local parametrization with quaternions

Being a local diffeomorphism is a nice property of the mapping π. In order to find a local parametrization of SO_3 around the identity, it suffices to find an open set containing the identity such that the restriction of π to this open set is bijective.

Consider the open half-sphere S^3_+ defined as:

$$S^3_+ = \{(\lambda_0, \lambda_1, \lambda_2, \lambda_3) \in S^3 \mid \lambda_0 > 0\}.$$

From Theorem 1.1, the restriction of π to S^3_+ is one-to-one. The image of the restriction of π to S^3_+ is the open subset of rotations with angle not equal to π. Let us denote by Ret the subset of rotations with angle π. Since $\pi_{|S^3_+}$ is a bijective local diffeomorphism from S^3_+ onto $\{SO_3-Ret\}$, it is a diffeomorphism.

We do not yet have a local parametrization because S^3_+ is not an open subset of $\mathbf{R}^3$. Let us just remark that the canonical projection from $\mathbf{R}^4$ onto $\mathbf{R}^3$ along the first component

$$(\lambda_0, \lambda_1, \lambda_2, \lambda_3) \longrightarrow (\lambda_1, \lambda_2, \lambda_3)$$

induces a diffeomorphism from the half-sphere S^3_+ onto the unit ball:

$$B(0, 1) = \{v \in \mathbf{R}^3 \mid \|v\| < 1\}.$$

If the rotation r is already represented by a quaternion λ with $R(\lambda) > 0$, then the local parametrization which is obtained maps the rotation r onto the vector $P(\lambda)$.

1.2.5.3 Mappings defined on the set of rotations

Many robotic control problems use functions depending on the location of one or several rigid bodies; think of an end-effector equipped with sensors, for example. The outputs of this kind of sensor are signals depending on the environment and on the location of the effector. In particular they depend on the attitude of the end-effector (cf. Chapter 7).

So, consider a vector function $f: SO_3 \to \mathbf{R}^p$. The image of a rotation r is denoted by $f(r)$. If the mapping f is differentiable (as a mapping between two differentiable manifolds), small variations of f are related to small variations of the attitude r by the differential of f:

$$df|_r : TSO_3|_r \longrightarrow T\mathbf{R}^p|_{f(r)}$$

Given a function $f(r, x, t)$ depending on many variables, where $r \in SO_3$, we shall denote by $\partial f/\partial r$ the differential of the partial function $r \to f(r, x, t)$.

1.2.5.4 Differentials of various parametrizations of rotations

So far, we have encountered different local parametrizations of rotations. To summarize and introduce some notation we have the following.

1. The classical Euler angles. These are not of interest in control because the domain of this chart does not contain the identity.
2. The roll, pitch, and yaw angles. Let us denote by $RPY(r)$ the vector consisting of the three angles associated with the rotation r. RPY is a local parametrization defined on $SO_3 - NRPY$.

As previously noted, this parametrization, although natural, is not convenient for computations.

3. The parametrization derived from the quaternion representation of rotations and denoted $Q(r)$:

$$Q(r) = P(\lambda) \Leftrightarrow \pi(\lambda) = r \quad \& \quad R(\lambda) > 0. \tag{1.2.63}$$

Q is a mapping from $SO_3 - Ret$ onto the unit sphere of $\mathbf{R}^3$.

4. The parametrization derived from the matrix representation

$$AS(u\sin(\Theta)) = \tfrac{1}{2}(R - R^T). \tag{1.2.64}$$

Let us denote by $PL(r)$ the vector $u\sin(\Theta)$. 'P' stands for parametrization while 'L' is reminiscent of logarithm, since for small rotation angles we have $\exp(AS(PL(r))) \approx R$. PL also has $\{SO_3 - Ret\}$ as domain.

When a parametrization is chosen, the differential of this parametrization is of constant use in control problems:

Proposition P1.9 *For an r belonging to the domain of the associated parametrization we have*

$$\frac{d(RPY)}{dr} = \begin{bmatrix} \sin(\Theta) & 0 & 1 \\ \sin(\Phi)\cos(\Theta) & \cos(\Phi) & 0 \\ \cos(\Phi)\cos(\Theta) & -\sin(\Phi) & 0 \end{bmatrix}^{-1}, \tag{1.2.65}$$

$$\frac{dQ}{dr} = \tfrac{1}{2}(\lambda_0 Id - AS(P(\lambda))) \tag{1.2.66}$$

$$\frac{d(PL)}{dr} = \tfrac{1}{2}(\mathrm{tr}(R)Id - R). \tag{1.2.67}$$

d$(RPY)/$dr is obtained from the decomposition of the angular velocity of frame **F** (with the notation of section 1.2.2):

$$\omega = \dot{\psi} z_0 + \dot{\Theta} u + \dot{\Phi} x.$$

d$Q/$dr is derived from the more general differential d$\lambda/$dr obtained in (1.2.60). d$(PL)/$dr can also be obtained from d$\lambda/$dr if one observes that:

$$PL(r) = u\sin(\Theta) = 2u\sin(\tfrac{1}{2}\Theta)\cos(\tfrac{1}{2}\Theta) = 2\lambda_0 P(\lambda) \qquad (1.2.68)$$

if $\lambda = (\lambda_0, \lambda_1, \lambda_2, \lambda_3)$ represents r. A straightforward computation leads to

$$\frac{d}{dr}(2\lambda_0 P(\lambda)) = -P(\lambda)P(\lambda)^T + \lambda_0^2 Id_3 - \lambda_0 AS(P(\lambda)) \qquad (1.2.69)$$

which can be written as:

$$\frac{d}{dr}(2\lambda_0 P(\lambda)) = (2\lambda_0^2 - \tfrac{1}{2})Id_3 - [P(\lambda)P(\lambda)^T + (\lambda_0^2 - \tfrac{1}{2})Id_3 + \lambda_0 AS(P(\lambda))].$$

$$(1.2.70)$$

Then using (1.2.34) and (1.2.41) we obtain the stated result.

As we shall see later, positivity of some matrices plays an important role in stability and robustness of control algorithms. Concerning parametrizations of rotations, we have the following result.

Proposition P1.10

$\dfrac{dQ}{dr}(r)$ *is a positive matrix for all r in its domain (i.e. with angle less than π).*

$\dfrac{d(PL)}{dr}(r)$ *is positive for all r with angle less than $\tfrac{1}{2}\pi$.*

Proof. The expression

$$\frac{dQ}{dr} = \tfrac{1}{2}(\lambda_0 Id - AS(P\lambda)) \qquad (1.2.71)$$

is the decomposition of d$Q/$dr as the sum of a symmetric matrix $\tfrac{1}{2}\lambda_0 Id$ and a skew-symmetric one. From the definition of Q, λ_0 is positive and thus d$Q/$dr is positive in its domain.

To prove the second assertion, let us consider the expression

$$\frac{d}{dr}(2\lambda_0 P(\lambda)) = -P(\lambda)P(\lambda)^T + \lambda_0^2 Id_3 - \lambda_0 AS(P(\lambda)). \qquad (1.2.72)$$

1.3 SCREWS

By the same argument as in the preceding proof, the symmetric part of $d(PL)/dr$ is

$$\lambda_0^2 Id - P(\lambda)P(\lambda)^T. \tag{1.2.73}$$

Now for every vector v we have

$$\begin{aligned}
v^T(\lambda_0^2 Id - P(\lambda)P(\lambda)^T)v &= \lambda_0^2 \|v\|^2 - \langle P(\lambda), v \rangle^2 \\
&\geq \|v\|^2 (\lambda_0^2 - \|P(\lambda)\|^2) \\
&= \|v\|^2 (2\lambda_0^2 - 1).
\end{aligned} \tag{1.2.74}$$

Thus a sufficient condition for $d(PL)/dt$ to be a positive matrix is $\lambda_0^2 \geq \frac{1}{2}$. Taking $v = P(\lambda)$ in the above equation shows that the condition is necessary.

1.3 SCREWS

1.3.1 The tangent space to the frame space

1.3.1.1 Frame velocities

Let us return to the problem of the position of a frame F_i relative to a frame F_0. Suppose now that the frame F_i is moving so that its position, described by the matrix

$$\bar{R}_{0i} = \begin{bmatrix} R_{0i}(t) & [O_0 O_i(t)]_i \\ 0 & 1 \end{bmatrix}, \tag{1.3.1}$$

is a function of time. Differentiating with respect to time gives

$$\frac{d\bar{R}_{0i}}{dt} = \begin{bmatrix} \tilde{\omega}(t) R_{0i}(t) & \dfrac{d[O_0 O_i(t)]_i}{dt} \\ 0 & 0 \end{bmatrix}. \tag{1.3.2}$$

This relation shows that an element of the tangent space to the frame space at $\bar{R}_{0i}$ can be considered as a vector (ω, v) in $\mathbf{R}^3 \times \mathbf{R}^3$ with ω as above and $v = d[O_0 O_i(t)]_i/dt$. The velocity of a frame is known when the velocity of a point and the rotational velocity of the frame are known.

1.3.1.2 Composition of frame velocities

Consider now three frames F_0, F_1, F_2 such that F_2 is moving relative to F_1 with velocity (ω_{21}, v_{21}) and F_1 is moving relative to F_0 with velocity (ω_1, v_1). Let us compute the velocity of F_2 relative to F_0. From equation (1.1.6) we get

$$\bar{R}_{02}(t) = \bar{R}_{01}(t)\bar{R}_{12}(t). \tag{1.3.3}$$

Differentiating with respect to t gives

$$\frac{d\bar{R}_{02}(t)}{dt} = \frac{d\bar{R}_{01}(t)}{dt}\bar{R}_{12} + \bar{R}_{01}\frac{d\bar{R}_{12}(t)}{dt} \qquad (1.3.4)$$

and in a more expanded form

$$\begin{bmatrix} \tilde{\omega}_2 \bar{R}_{02} & v_2 \\ 0 & 0 \end{bmatrix} = \begin{bmatrix} \tilde{\omega}_1 \bar{R}_{01} & v_1 \\ 0 & 0 \end{bmatrix} \begin{bmatrix} R_{12} & [O_1 O_2]_1 \\ 0 & 1 \end{bmatrix}$$
$$+ \begin{bmatrix} R_{01} & [O_0 O_1]_0 \\ 0 & 1 \end{bmatrix} \begin{bmatrix} \tilde{\omega}_{21} \bar{R}_{12} & v_{21} \\ 0 & 0 \end{bmatrix}, \qquad (1.3.5)$$

from which we obtain

$$[\tilde{\omega}_2]_0 R_{02} = [\tilde{\omega}_1]_0 R_{02} + [\tilde{\omega}_{12}]_0 R_{02} \qquad (1.3.6)$$

$$v_2 = [\tilde{\omega}_1]_0 R_{01}[O_1 O_2]_1 + v_1 + R_{01}[v_{21}]_1. \qquad (1.3.7)$$

Thus

$$\omega_2 = \omega_1 + \omega_{21} \qquad (1.3.8)$$

$$v_2 = v_1 + v_{21} + \omega_1 \times O_1 O_2. \qquad (1.3.9)$$

This formula is very familiar in mechanics. Now consider a moving frame $F_i(O_i, x_i, y_i, z_i)$ with velocity $(\omega_i, V(O_i))$ and a point M linked to the frame F_i (i.e. $d[O_i M]_i/dt = 0$). From the above formula we get

$$V(M) = V(O_i) + \omega_i \times O_i M. \qquad (1.3.10)$$

It is possible to define a vector field on $\mathbf{R}^3$ by associating with each point its velocity as if it were a point linked to the frame F_i. From the above, this vector field is well defined if its value is known at one point and if the vector ω is known. Formula (1.3.10) gives the value at every point.

1.3.2 Screws

1.3.2.1 Basic definitions and properties

Enlarging the example of the velocity field of a rigid body described above, we propose the following definition.

Definition D1.6 *A vector field, H, on $\mathbf{R}^3$ is a screw if there exists a point O and a vector ω such that for all points M in $\mathbf{R}^3$,*

$$H(M) = H(O) + \omega \times OM. \qquad (1.3.11)$$

It is then easy to prove for every couple of points P, Q

$$H(P) = H(Q) + \omega \times PQ. \qquad (1.3.12)$$

1.3 SCREWS

Thus a screw is well defined by its value at a point and the vector ω. The vector ω is called the vector of the screw. In the following, a screw will be designated by a couple of vectors $(H(O), \omega)$ with the obvious notation.

Given two screws $H_1(H_1(O), \omega_1)$ and $H_2(H_2(O), \omega_2)$ we define the sum as the screw $(H_1(O) + H_2(O), \omega_1 + \omega_2)$. Similarly the product of a screw by a scalar λ is the screw $(\lambda H_1(O), \lambda \omega_1)$.

Proposition P1.11 *For every point P we have*

$$(H_1 + H_2)(P) = H_1(P) + H_2(P),$$
$$(\lambda H_1)(P) = \lambda H_1(P). \qquad (1.3.13)$$

The set S of screws is a vector space of dimension 6 on $\mathbf{R}$.

Proof. From the definition we get

$$(H_1 + H_2)(P) = H_1(O) + H_2(O) + (\omega_1 + \omega_2) \times OP$$
$$= (H_1(O) + \omega_1 \times OP) + (H_2(O) + \omega_2 \times OP). \qquad (1.3.14)$$

The sum of two screws defined previously is the natural one. The same holds for the product of a screw by a scalar.

If we go back now to formula (1.3.8) and express everything in terms of screws, v_{21} is the velocity of O_2 relatively to F_1 so that the velocity screw of F_2 relative to F_1 is

$$T_{21} = (T_{21}(O_2), \omega_{21}) = (v_{21}, \omega_{21}) \qquad (1.3.15)$$

and the velocity screw of F_1 relative to F_0 is

$$T_{10} = (T_{10}(O_1), \omega_1) = (v_1, \omega_1). \qquad (1.3.16)$$

Let now

$$T_{20} = (T_{20}(O_2), \omega_2) = (v_2, \omega_2) \qquad (1.3.17)$$

be the velocity screw of F_2 relative to F_0. Then eqns (1.3.8) and (1.3.9) translate into

$$\omega_2 = \omega_{21} + \omega_1 \qquad (1.3.18)$$
$$T_{20}(O_2) = T_{21}(O_2) + T_{10}(O_1) + \omega_1 \times O_1 O_2$$
$$= T_{21}(O_2) + T_{10}(O_2) \qquad (1.3.19)$$

giving the simple result

$$T_{20} = T_{21} + T_{10}. \qquad (1.3.20)$$

1.3.2.2 Product of screws and duality

Definition D1.7 *The product of two screws* H_1 $(H_1(O), \omega_1)$ *and* H_2 $(H_2(O), \omega_2)$ *is the scalar*

$$\langle \omega_1, H_2(O) \rangle + \langle \omega_2, H_1(O) \rangle. \tag{1.3.21}$$

The product of the two screws H_1 *and* H_2 *is denoted by* $H_1 \bullet H_2$.

Consider now a second point P. We get

$$H_i(P) = H_i(O) + \omega_i \times OP; \quad i = 1, 2, \tag{1.3.22}$$

from which we compute

$$\langle \omega_1, H_2(P) \rangle + \langle \omega_2, H_1(P) \rangle = \langle \omega_1, H_2(O) \rangle + \langle \omega_2, H_1(O) \rangle$$
$$+ \langle (\omega_1 \times OP), \omega_2 \rangle + \langle (\omega_2 \times OP), \omega_1 \rangle. \tag{1.3.23}$$

Since

$$\langle (\omega_1 \times OP), \omega_2 \rangle = - \langle (\omega_2 \times OP), \omega_1 \rangle, \tag{1.3.24}$$

we have proved that the product of two screws defined above is independent of the reference point O.

If a point O is chosen and if screws are assimilated to couples of vectors $(H(O), \omega)$ considered as vectors in $\mathbf{R}^6$, the product of two screws is the bilinear map associated with the matrix

$$\begin{bmatrix} 0 & Id_3 \\ Id_3 & 0 \end{bmatrix}.$$

This bilinear map is not degenerate, but *it is not a scalar product*. Nevertheless, since it is not degenerate, it induces an isomorphism between S and its dual S^*. If $(H(O), \omega)$ is a screw, the associated linear form is

$$\Phi_{(H(O), \omega)}(H_1(O), \omega_1) = (H(O), \omega) \bullet (H_1(O), \omega_1). \tag{1.3.25}$$

Thus the dual of S can be considered as a space of screws. It must be emphasized that the point O used in the definitions of the screws and of the elements of S^* must be the same.

As it is usual for bilinear maps, we can express the screw product in a matrix form:

$$(H_1, \omega_1) \bullet (H_2, \omega_2) = [H_1^T, \omega_1^T] \begin{bmatrix} 0 & Id_3 \\ Id_3 & 0 \end{bmatrix} \begin{bmatrix} H_2 \\ \omega_2 \end{bmatrix} \tag{1.3.26}$$

or equivalently

$$(H_1, \omega_1) \bullet (H_2, \omega_2) = [\omega_1^T, H_1^T] \begin{bmatrix} H_2 \\ \omega_2 \end{bmatrix}. \tag{1.3.27}$$

1.3 SCREWS

In matrix form, the screw product differs from the usual scalar product by the permutation of the two vectors H and ω before transposition.

Since the bilinear map is not degenerate, we can introduce the following orthogonality-like concept:

Definition D1.8 *Two screws are reciprocal if they have a null product.*

Two subspaces S_1 and S_2 are reciprocal if every element in S_1 is reciprocal to every element in S_2.

This notion will prove very useful in finding the remaining degrees of freedom when a rigid body is tied by linkages.

1.3.2.3 Particular screws and decompositions of the space of screws

Consider a finite set of couples (A_i, V_i) $i = 1, \ldots, n$, where A_i is a point in $\mathbf{R}^3$ and V_i a three-dimensional vector. We shall say that the vector V_i is linked to the point A_i. It is possible to define a vector field H associated with the set of couples (A_i, V_i) by

$$H(P) = \sum_{i=1}^{n} PA_i \times V_i. \tag{1.3.28}$$

Proposition P1.12 *The vector field defined above is a screw with vector*
$$V = \sum_{i=1}^{n} V_i.$$

The proof is straightforward.

An interesting case occurs when the set of linked vectors is $\{(A, V); (B, -V)\}$. The associated screw has 0 as vector and the value of the screw is constant over $\mathbf{R}^3$. The vector field is uniform.

Proposition P1.13 *The following properties are equivalent for a screw.*

1. *The vector of the screw is null.*

2. *The screw is a uniform vector field.*

3. *There exists a vector V and two points A, B such that the screw is associated with $\{(A, V); (B, -V)\}$.*

The implication (1) $\Rightarrow$ (2) is obvious.
Let H be a constant screw, $H = (H(O), \omega)$. Then for every couple of points (P, Q) we have

$$PQ \times \omega = 0 \tag{1.3.29}$$

and hence $\omega = 0$. Now let V be a non-null vector orthogonal to $H(O)$ (which is supposed to be non-null otherwise H is the null field and the case is trivial).

A necessary condition for the screw H to be associated with $\{(A, V)\}$ is

$$H(O) = OA \times V. \tag{1.3.30}$$

A solution to this equation is $OA = [V \times H(O)]/\|V\|^2$. The screw H is then associated with $\{(A, V); (O, -V)\}$.

Definition D1.9 *A screw with one of the equivalent properties (1)–(3) is called a couple.*

A couple is a screw reciprocal to itself. This example shows that the product of two screws is not a scalar product.

Proposition P1.14 *The set of couples C is a subvector space of the space of screws with dimension 3.*

This is an immediate consequence of property (1).

Another example of a screw is the screw associated with $\{(A, V)\}$. This screw has two nice properties: $H(A) = 0$, and, if B_1 and B_2 are points belonging to a straight line with direction $\mathbf{R}V$, we have

$$H(B_2) = H(B_1) + B_1 B_2 \times V = H(B_1). \tag{1.3.31}$$

So the value of this screw is constant along every straight line with direction $\mathbf{R}V$.

Proposition P1.15 *If the screw $(H(O), V)$ is zero at a point A, it is the screw associated with (A, V).*

Definition D1.10 *A screw associated with a couple (A, V) is called a slider through A.*

From Proposition P1.13 a slider is also a screw reciprocal to itself.

Proposition P1.16 *The set S_A of sliders associated with (A, V) with V a vector in $\mathbf{R}^3$ is a subvector space of the space of screws, isomorphic to $\mathbf{R}^3$.*

The proof of the last two properties is straightforward.

The intersection of the space of couples C and of the space, S_A of sliders through A, is reduced to the null screw, since a couple is a screw of constant value. Together with dimension considerations, this last fact reveals the following.

Proposition P1.17 *The space of screws is the direct sum of the subspace of couples C and any subspace of sliders S_A.*

1.3 SCREWS

Given a screw H and a point A, there exists a unique decomposition $H = G_A + C$ where G_A is a slider through A and C is a couple. This decomposition also provides a means of building a basis of the space of screws: for all points A let $\{v_1, v_2, v_3\}$ be a set of independent vectors and $\{c_1, c_2, c_3\}$ another set of independent vectors. A basis of S is $\{(A, v_i), c_i\}_{i=1, 2, 3}$ with the notation defined above.

1.3.2.4 A more abstract approach: the Lie group of displacements*

Let us return to the configuration space for frames. As we have seen previously, the Special Euclidean group SE_3 is a good configuration space for frames. When a reference frame is chosen we obtain a one-to-one mapping from the set of frames onto SE_3 and, moreover, we can associate a homogeneous matrix to every displacement in SE_3. This last mapping is a group homomorphism between SE_3 and the group of homogeneous matrices considered.

As differential relations are necessary in control analysis, we would like to examine more closely the differential structure on SE_3. It is common when dealing with displacements to consider the translation and the rotation separately. This is made possible by the natural bijection from $SO_3 \times \mathbf{R}^3$ onto SE_3 given by

$$\phi : (R, u) \longrightarrow \bar{R} = \begin{bmatrix} R & u \\ 0 & 1 \end{bmatrix}. \tag{1.3.32}$$

The Cartesian product $SO_3 \times \mathbf{R}^3$ is naturally endowed with a differential manifold structure. Given a chart (U_1, x_1) on SO_3, a chart (U, x) on $SO_3 \times \mathbf{R}^3$ is defined by

$$\left. \begin{array}{l} U = U_1 \times \mathbf{R}^3, \\ x(r, u) = (x_1(r), u). \end{array} \right\} \tag{1.3.33}$$

Thus SE_3 is a differential manifold with dimension 6 for the differential manifold structure carried on it by ϕ.

The componentwise natural group structure on $SO_3 \times \mathbf{R}^3$ does not turn the mapping ϕ into a group isomorphism. Nevertheless, the composition law

$$(r_1, u_1) T (r_2, u_2) = (r_1 r_2, r_1 u_2 + u_1) \tag{1.3.34}$$

does so. It is also straightforward to verify that this composition law is a differential mapping. The reader familiar with Lie groups will have recognized that SE_3 is a Lie group. For those less familiar with this theory, recall that a Lie group G is both a differential manifold and a group such that the group product is a differential mapping from $G \times G$ onto G. One important feature of Lie groups is that the tangent bundle is very simple. Everything is known about the tangent space at a point $\bar{r}$ if one knows the tangent space at the identity element for the group composition.

Let us consider again a homogeneous matrix $\bar{R} = \phi(r, u)$ and a differential path through $\bar{R}$: $c(t) = \bar{R}(t)$, $c(0) = \bar{R}$. This corresponds to a path in the set of frames provided a reference frame is chosen. Now, this path can be written

$$c(t) = \bar{R}(t) = \bar{R}_c(t)\bar{R} \tag{1.3.35}$$

where $R_c(t)$ is a path through the identity. Applying the composition of frame velocities we obtain

$$\left.\begin{aligned}\omega &= \omega_c \\ v &= v_c + \omega_c \times u\end{aligned}\right\}, \tag{1.3.36}$$

where (v_c, ω_c) is the screw representing $d\bar{R}_c/dt$ and (v, ω) represents $d\bar{R}(t)/dt$. Physically, any movement of a frame F can be considered as a time-varying displacement. The preceding formulae state that the velocity of the frame F can be deduced from the velocity of the reference frame if it were performing the same displacement. This remark allows us to consider the velocity of a solid as a vector field, as we have done. From a differential geometry point of view, (1.3.36) defines a mapping from the tangent space at the identity onto the tangent space at $\bar{R}$ and this mapping is nothing more than the differential of the right multiplication by $\bar{R}$.

Traditionally the tangent space at the identity of a Lie group is called the Lie algebra. The Lie algebra of SE_3 will be denoted by se_3 and can be assimilated to the set of screws. Considered physically, it is the set of all possible velocities of the reference frame for all possible displacements. If $\bar{r}(t)$ is a path on SE_3 we shall denote by $d\bar{r}/dt$ the screw representing the velocity.

1.3.2.5 Change of reference frame

In the preceding sections we have considered only one reference frame. Suppose now a reference frame F_0 is chosen and we want to change to a new reference frame F_1. This change of frame induces an isomorphism on SE_3. If $\bar{R}_{0i}$ represents the position of the frame F_i relative to F_0 then $\bar{R}_{1i} = \bar{R}_{10}\bar{R}_{0i}$ represents the position of the frame F_i relative to F_1. The isomorphism is simply left multiplication by $\bar{R}_{01}$.

For screws, things are a little more complicated. Given a screw (v_0, ω_0) belonging to se_3, it may be considered as the tangent vector to a path $t \to \bar{R}_0(t)$ such that $\bar{R}_0(0) = Id$. For a given t, $\bar{R}_0(t)$ represents a displacement.

Now if the reference frame is changed, the same displacement is represented by $\bar{R}_1(t)$ given by the standard change-of-basis formula

$$\bar{R}_1(t) = \bar{R}_{10}\bar{R}_0(t)\bar{R}_{01}. \tag{1.3.37}$$

1.3 SCREWS

Differentiating with respect to t and taking into account that

$$\left.\frac{d\bar{R}_0(t)}{dt}\right|_{t=0} = (v_0, \omega_0) \tag{1.3.38}$$

we obtain

$$\left.\frac{d\bar{R}_1(t)}{dt}\right|_{t=0} = (v_1, \omega_1) \tag{1.3.39}$$

with

$$\left.\begin{array}{l} \omega_1 = R_{10}\omega_0 \\ v_1 = R_{10}v_0 + R_{10}(\omega_0 \times [O_0 O_1]_0) \end{array}\right\}. \tag{1.3.40}$$

Hence the new expression of the frame velocity is the same screw computed at the origin of the new reference frame and expressed in the corresponding basis. From a differential point of view, the last two formulae express the differential of the conjugacy mapping:

$$\begin{array}{c} SE_3 \longrightarrow SE_3 \\ \bar{R} \longrightarrow \bar{R}_{10}\bar{R}(\bar{R}_{10})^{-1} \end{array} \tag{1.3.40}$$

often denoted $Ad_{R_{10}}$. The 6×6 matrix of the differential of $Ad_{R_{10}}$ can be deduced from (1.3.40)

$$dAd_{R_{10}} = \begin{bmatrix} R_{10} & -R_{10}AS([O_0, O_1]_0) \\ 0 & R_{10} \end{bmatrix} \tag{1.3.41}$$

again considering the screw (v, ω) as a vector $\begin{bmatrix} v \\ \omega \end{bmatrix}$.

1.3.2.6 Differentials of maps defined on SE_3

As we have already seen with rotations, it is necessary in robotics to consider functions depending on the location of a rigid body, for example, functions representing the output signal provided by an 'exteroceptive' sensor. If $f: SE_3 \to \mathbf{R}$ is a scalar differential function defined on SE_3, its differential is a linear mapping

$$df|_{\bar{r}}: se_3 \longrightarrow \mathbf{R};$$

thus it is an element of se_3^*, the dual of the screw space. $df|_{\bar{r}}$ can be represented by a screw (h_1, R_1) and

$$df|_{\bar{r}}(v, \omega) = (h_1(\bar{r}), R_1(\bar{r})) \bullet (v, \omega). \tag{1.3.42}$$

1 FUNDAMENTAL CONFIGURATION SPACES

Through the isomorphism from $SO_3 \times \mathbf{R}^3$, f may also be considered as a function of two variables r and u. A simple calculation yields

$$\left. \begin{aligned} \frac{\partial f}{\partial r}(r, u) &= h_1(r, u) \\ \frac{\partial f}{\partial u}(r, u) &= R_1(r, u) \end{aligned} \right\}. \quad (1.3.43)$$

Representing the screw (v, ω) as a vector $\begin{bmatrix} v \\ \omega \end{bmatrix}$ in $\mathbf{R}^6$, we obtain a matrix representation of df:

$$df|_{\bar{r}} = \left(\frac{\partial f}{\partial u}(\bar{r}) \quad \frac{\partial f}{\partial r}(\bar{r}) \right) \begin{bmatrix} v \\ \omega \end{bmatrix}. \quad (1.3.44)$$

Now if g is a function with values in se_3,

$$g(r, u) = \begin{bmatrix} g_1(r, u) \\ g_2(r, u) \end{bmatrix} \quad (1.3.45)$$

its differential is a linear map from se_3 into se_3 with matrix

$$dg(\bar{r}) = \begin{bmatrix} \dfrac{\partial g_1}{\partial u} & \dfrac{\partial g_1}{\partial r} \\[6pt] \dfrac{\partial g_2}{\partial u} & \dfrac{\partial g_2}{\partial r} \end{bmatrix}. \quad (1.3.46)$$

The modelling of elastic force sensors must also consider second-order derivatives of functions defined on SE_3 with values in $\mathbf{R}$. Since the differential $df(\bar{r}): SE_3 \to se_3^*$ maps a displacement $\bar{r}$ onto a screw in se_3^*, the second derivative maps a screw in se_3 onto a screw in se_3^*:

$$d^2 f|_{\bar{r}}(v, \omega) = (h_2(v, \omega), R_2(v, \omega)). \quad (1.3.47)$$

Using (1.3.26), we obtain the matrix form

$$d^2 f(\bar{r}) = \begin{bmatrix} \dfrac{\partial^2 f}{\partial u \partial r} & \dfrac{\partial^2 f}{\partial r^2} \\[6pt] \dfrac{\partial^2 f}{\partial u^2} & \dfrac{\partial^2 f}{\partial u \partial r} \end{bmatrix} \quad (1.3.48)$$

1.3 SCREWS

so that we obtain the bilinear form associated with $d^2f|_{\bar{r}}$.

$$\phi((v_1, \omega_1), (v_2, \omega_2)) = (d^2f_{\bar{r}}(v_1, \omega_1)) \bullet (v_2, \omega_2)$$

$$= \begin{bmatrix} \dfrac{\partial^2 f}{\partial u \partial r} & \dfrac{\partial^2 f}{\partial r^2} \\ \dfrac{\partial^2 f}{\partial u^2} & \dfrac{\partial^2 f}{\partial u \partial r} \end{bmatrix} \begin{bmatrix} v_1 \\ \omega_1 \end{bmatrix} \bullet \begin{bmatrix} v_2 \\ \omega_2 \end{bmatrix} \quad (1.3.49)$$

with an obvious abuse of notation.

A straightforward computation leads then to the more familiar matrix of the bilinear form $d^2f|_{\bar{r}}$:

$$\begin{bmatrix} \dfrac{\partial^2 f}{\partial r^2} & \dfrac{\partial^2 f}{\partial r \partial u} \\ \dfrac{\partial^2 f}{\partial u \partial r} & \dfrac{\partial^2 f}{\partial u^2} \end{bmatrix}. \quad (1.3.50)$$

Finally, when a change of reference frame is performed, the standard change of basis for bilinear forms applies with the change-of-basis matrix $dAd_{R_{10}}$.

1.3.3 Applications of screws

1.3.3.1 The virtual work approach in mechanics

Among the various formulations of the basic principles of mechanics, that of d'Alembert is very interesting for the modelling of static linkages. In d'Alembert's approach, forces are described in terms of their action on the mechanical system under consideration. In short, a configuration space is defined which describes all the 'positions' of the system. A virtual movement is then a vector field on the configuration space (this supposes that the configuration space is smooth enough). The tangent bundle of the configuration space is in a sense a space of virtual velocities. Then vector spaces of virtual movements can be considered at each time t. The forces (in a very general meaning) applied to the mechanical system are modelled by continuous linear forms on the vector space of virtual movements.

In the case of rigid bodies, the configuration space is the space of frames and the space of virtual movement is the space of continuous functions from **R** to the space of frame velocities described in section 1.3.1. As we have seen in section 1.3.2.2, every linear form on the space of screws can be considered as a screw itself, through the isomorphism associated with the bilinear form described in section 1.3.2.2. The following computation will show that the duality introduced above on the space of screws is natural.

Let $(H(O), \omega)$ be a screw representing a frame velocity and consider a linear form L on the screw space. From classical (Euclidean) duality relations, there exist vectors (F_O, m_O) such that

$$L((H(O), \omega)) = \langle F_O, H(O) \rangle + \langle m_O, \omega \rangle. \tag{1.3.51}$$

Let P be another point. We also have

$$L((H(P), \omega)) = \langle F_P, H(P) \rangle + \langle m_P, \omega \rangle. \tag{1.3.52}$$

Since

$$H(P) = H(O) + \omega \times OP, \tag{1.3.53}$$

we get

$$\langle F_O, H(O) \rangle + \langle m_O, \omega \rangle = \langle F_P, H(O) \rangle + \langle (m_P + OP \times F_P), \omega \rangle. \tag{1.3.54}$$

The equality is valid for every couple $(H(O), \omega)$ and we can deduce

$$F_P = F_O \tag{1.3.55}$$

$$m_P = m_O + F_O \times OP. \tag{1.3.56}$$

The conclusion is that every mechanical action on a rigid body is defined by a vector F (often called the general resultant) and a vector field m which is a screw $(m(O), F)$. The virtual power of this screw for a virtual movement $(H(O), \omega)$ is the product of the two screws:

$$L((H(O), \omega)) = (H, \omega) \bullet (m, F). \tag{1.3.57}$$

Definition D1.11 *The space of frame velocities is the space of twists or kinematic screws. The dual of the space of twists is the space of wrenches.*

1.3.3.2 Application of screws to the modelling of static or instantaneous constraints

A rigid body moving in space without constraint has six degrees of freedom. In other words, its configuration space (the space of frames) is a six-dimensional differentiable manifold. The velocities belong to the six-dimensional space of twists. The most natural example of a rigid body being subjected to constraints is when it is brought into contact with one or several rigid bodies during its movement. The movement is constrained as long as contact is effective. Moreover, the nature of the constraints may change with time as the relative positions of the bodies change. In general, a global description of the motion of a rigid body subject to constraints is difficult if not impossible. The set of possible movements has to be computed at each time and may change with time.

In robotics, the problem of constrained movements of rigid bodies also appears in situations other than contacts between rigid objects. In obstacle avoidance, surface following, or automatic grasping, different parts of the manipulator are subject to constraints on their movement. Constraints arise from the task itself and are independent of the manipulator. Chapter 7 is largely devoted to modelling this type of situation.

Since our basic intuition derives from constraints due to contacts between rigid bodies, we shall now concentrate on this kind of constraint. As we have seen in the brief discussion above, global analysis of the constraints during the motion of a body is impossible. Instead of searching for all possible movements compatible with the constraints, we shall look at the set of possible velocities at each time t. This implies that the analysis will be valid only for the instant under consideration. If things are smooth enough, results for velocities remain valid for small finite motions.

The essential modelling assumption put forward is that the set of velocities compatible with the constraints is a subvector field of the space of velocities: in our case, the space of kinematic screws or twists. This assumption can be discussed for the case of contact between bodies where half-spaces seem to be more appropriate. But if our purpose is not only to describe the effect of the contact but also to have movements maintaining the contact, subvector spaces are then appropriate. With the above hypothesis we can propose the following definition.

Definition D1.12 *A set of constraints on the motion of a rigid body is called a linkage. The class of the linkage is the dimension of the subspace of compatible velocities.*

As usual in linear algebra, the subspace of compatible velocities can be described by a set of linear equations or equivalently as the intersection of the kernels of linear forms $L_1, \ldots, L_p$. Again these linear forms can be represented by screws belonging to the space of wrenches. Thus it is equivalent to define a linkage by a subspace of wrenches. The compatible velocities are the set of twists reciprocal to every wrench associated with the linkage. If the class of the linkage is n, the subspace of wrenches associated with the linkage is of dimension $6 - n$.

This screw theory has been used in the analysis of articulated robot hands. In robot control it plays an essential role in the synthesis of controls using sensors.

1.3.3.3 From local to global: a possible approach*

For the modelling of constraints another point of view is to consider, at each time, all possible positions of a solid body in the configuration space. For each position, define a set (subvector space) of possible velocities compatible with the local constraints. In this way we obtain a local description of all the constraints. It is local in the sense that the constraints are described by differential conditions at each point of the configuration space. One would like to 'glue' these local conditions together to get some idea of the movements compatible with the constraints. Of course it is generally hopeless to search for a description on the whole configuration space. A geometric description valid only on an open subset of the configuration space would be satisfactory for modelling assembly tasks or surface-following tasks. Differential geometry provides the tools to move from a local description to a more global one.

Recall that a distribution is a function $\Delta: p \to \Delta_p$ where p is a point of a differentiable manifold M and Δ_p is a subvector space of the tangent space M_p of M at p. Δ is C^∞ if for any p in M there exist a neighbourhood U of p and C^∞ vector fields $X_1, \ldots, X_k$ such that $\{X_1(q), \ldots, X_k(q)\}$ is a basis of Δ_q for every q in U. The distribution Δ is of rank k at p if Δ_p is a k-dimensional subvector space. Given a vector field X, we say that X belongs to Δ if $X(q) \in \Delta_q$ for all q in M. A distribution Δ is involutive if for every pair X, Y of vector fields belonging to Δ, the Lie bracket $[X, Y]$ also belongs to Δ.

The first theorem providing some indication of the link between a global description of the constraints and a local one is the following.

Theorem 1.4 (Frobenius) *Let Δ be a C^∞ involutive distribution of constant rank k on M. For every $p \in M$ there is a coordinate system (x, U) with*

$$x(p) = 0, \quad x(U) \text{ is included in } (-\varepsilon, \varepsilon) \times \cdots \times (-\varepsilon, \varepsilon)$$

such that for each $a^{k+1}, \ldots, a^n$ with $|a^i| < \varepsilon$ the set $\{q \in U \mid x^{k+1}(q) = a^{k+1}, \ldots, x^n(q) = a^n\}$ is a submanifold of M with Δ_p as tangent space at p.

This theorem shows that under certain conditions, a local description of constraints may lead naturally to a (more) global one. In general, the distribution obtained by considering admissible velocities is not of constant rank. More recent theorems (see Section 1.4) deal with this more complex situation. We shall not deal further with this subject. Our purpose is only to provide some clues to bridge the gap between local descriptions of constraints and global descriptions. Such global descriptions can be found in the theory of C-surfaces used in the modelling of assembly tasks.

This informal commentary is given merely to show that a local or differential description of the constraints is more fruitful than it seems at first glance.

1.4 BIBLIOGRAPHIC NOTE

The first section of this chapter on frames in robotics is standard in geometry, computer graphics, and robotics. Further development may be found in the books by Paul (1982), Coiffet (1981), or Gorla and Renaud (1984). The modified Denavit–Hartenberg parameters are from Kleinfinger (1986).

Quaternions go back to Hamilton, and the subject seems to have found new favour with mechanical engineers. Some points in the study of the geometry of the space SO_3 are new, to our knowledge. Further details on quaternions may be found in Le Borgne (1987). The elements of differential geometry that it uses can be found in classical treatises such as Spivak (1970). The recent results mentioned after Theorem 1.4 are from Sussman (1973) and Nagano (1966).

Screws were used in robotics by Salisbury (1985) in the analysis of contacts occurring during gripping. The subject is also standard in mechanics. A classical reference in the theory of mechanisms is Siestrunck (1973). Theory of C-surfaces is presented for instance in Mason (1981, 1982).

APPENDIX A1.1 PROOF OF PROPOSITION P1.6

A1.1 The canonical map π does not admit any reasonable section.

In order to prove that there exists no section f of π which is a group homomorphism or continuous, consider the set Γ of rotations with axis $\mathbf{Ri}$. This set is a subgroup of SO_3 isomorphic to the group of the rotations of the plane and therefore to the group of complex numbers with unit norm. From the relations between the axis and angle of a rotation and a quaternion representing this rotation we can infer that

$$\pi^{-1}(\Gamma) = \Gamma' = S_1(\mathbf{R} + \mathbf{Ri})$$

is the unit circle of the plane $\mathbf{R} + \mathbf{Ri}$ included in $\mathbf{H}$. Moreover, it is easy to check that this plane is isomorphic to the field of complex numbers, since

$$(a + bi)(a' + b'i) = aa' - bb' + (ab' + ba')i.$$

So $\pi|_{\Gamma'}$ can be considered as a mapping from the complex numbers $\mathbf{C}$ into themselves, and again from section 1.2.3.4 we obtain that $\pi|_{\Gamma'}$ is the map $z \to z^2$.

Suppose now that there exists a section f of π and consider $f|_\Gamma$. If f is an isomorphism we must have $f(Id) = 1$ and $f(-1) = \pm i$ because $\pi(f(-1)) = f(-1)^2 = -1$ from the considerations above and $\pi \circ f = Id_{SO_3}$. But $f(-1) = i$ implies $(f(-1))^2 = f((-1)^2) = f(1) = 1 = -1$, a contradiction. The same holds if $f(-1) = -i$.

Suppose now that f is continuous. Then $\pi(f(z)) = (f(z))^2 = z$. But it is well known that there is no continuous map from $\mathbf{C}$ into itself with the above property.

2
BASIC MODELS OF ROBOTS

INTRODUCTION

Control design always relies on some modelling of the system to be controlled. Throughout this book, two types of modelling equations have been used: a differential equation in the state space describing the evolution of the state of the system, and a mapping from the state space to an output space modelling the observation of the system and/or the objective of the control. In the case of robot manipulators, a rough analysis shows that the modelling of a robot performing a task can be split into two parts which are very different from the user's point of view.

1. Given a manipulator with rigid links, a basic description of its behaviour uses the joint positions and velocities as state variables. The state evolution equation comes from the laws of mechanics, and the parameters, with the noticeable exception of the payload, are determined when the robot is manufactured. The user cannot, in general, modify this part of the robot model known as the dynamic model.

2. On the other hand, the output function is an essential feature for describing a robotic task. In general several options are available to the control designer. Chapter 3 is entirely devoted to output functions in general, while in Chapters 4 and 7 more specific functions are studied. In the beginning of this chapter we examine the model of the dynamics of the manipulator. Since many possible output functions and their differentials depend on the location (position and attitude) and the velocity, respectively, of the end-effector, we also examine the kinematic models of manipulators as first examples of output functions. In this study we have retained only the features relevant to control theory. For more details the reader is referred to the classical texts on the subject (see the bibliographic note at the end of the chapter).

2.1 THE DYNAMIC MODEL

Dynamic equations provide the essential part of the state evolution equations. They are also used in the design of manipulators to perform simu-

2.1 THE DYNAMIC MODEL

lations to test the performance of a robot or the relative merits of different control schemes.

Because of the importance of the problem and the combinatorial complexity of the equations, the subject has been an attractive area of study in recent years. A manipulator is a system of rigid bodies with natural generalized coordinates (chart): the *joint coordinates*. So the first attempts to establish dynamic equations used the Lagrange formulation with the traditional analytical approach. The inertia matrix was first computed in closed form and the Lagrange equations were then applied to derive the dynamic equations. The use of partial derivatives in complex expressions required a computer program to perform formal computations.

Later, another approach using the Newton–Euler formulation was developed. In contrast to the earlier method, the dynamic parameters (velocities, accelerations, torques, and forces) are computed numerically and recursively.

Many improvements have been made, especially in Lagrange methods. The equivalence of the Lagrange and the Newton–Euler formulations has been proved; the difficulty was to show that it is possible to derive the same algorithm by both approaches.

In the following, we shall consider the Newton–Euler formalism for *rigid* manipulators *with no closed kinematic chains*. Although the case of closed kinematic chains has been considered in several papers, we do not think the subject is mature enough to be included in this book.

A reference frame F_0 is chosen; let n be the number of joints of the manipulator. A frame F_i, $1 \leq i \leq n$, is associated with each link and the frame F_{n+1} is associated with the end-effector. The reference frame F_0 is considered in a first step as a Galilean frame. The ith joint variable is denoted by q_i and the vector of joint variables is denoted by q with components q_i, $i = 1, \ldots, n$. This set of variables is also a generalized coordinate system from a mechanical point of view. We shall refer to it as *joint coordinates*.

2.1.1 The general dynamic model

Before developing effective procedures to compute the dynamic model of a manipulator, let us derive the structure of this model. The form we shall obtain will be sufficient for control analysis. The Lagrangian formalism gives an elegant and fast method of establishing the state evolution equation.

A rigid link manipulator being a mechanical system invariant with time, its kinetic energy T is a function of joint velocity of the form

$$T(q, \dot{q}) = \tfrac{1}{2} \dot{q}^T M(q) \dot{q} \qquad (2.1.1)$$

where $M(q)$ is a symmetric $n \times n$ positive definite matrix for all q. The positivity of the matrix M will play a crucial role in the design of the controls of the mechanical system (cf. Chapter 6).

Let $Q = (Q_i)_{i=1,\ldots,n}$ be the vector of external torques or forces at each joint including gravity, friction, and forces due to the actuators. The Lagrange equations then give

$$Q_i = \frac{d}{dt}\left(\frac{\partial T}{\partial \dot{q}_i}\right) - \frac{\partial T}{\partial q_i}, \quad i = 1, \ldots, n. \tag{2.1.2}$$

Let $M = (m_{ij}); 1 \le i, j \le n$. Successive differentiations lead to

$$\frac{\partial T}{\partial \dot{q}_i} = M\dot{q} = \sum_{j=1}^{n} m_{ij}\dot{q}_j, \tag{2.1.3}$$

$$\frac{d}{dt}\left(\frac{\partial T}{\partial \dot{q}_i}\right) = \sum_{j=1}^{j=n}\left(m_{ij}\ddot{q}_j + \sum_{k=1}^{n}\frac{\partial m_{ij}}{\partial q_k}\dot{q}_j\dot{q}_k\right). \tag{2.1.4}$$

On the other hand the second term of the Lagrange equation is

$$\frac{\partial T}{\partial q_i} = \frac{1}{2}\sum_{j,k=1}^{n}\frac{\partial m_{jk}}{\partial q_i}\dot{q}_j\dot{q}_k. \tag{2.1.5}$$

Combining the expressions, we obtain

$$\frac{d}{dt}\left(\frac{\partial T}{\partial \dot{q}_i}\right) - \frac{\partial T}{\partial q_i} = \sum_{j=1}^{n} m_{ij}\ddot{q}_j + \sum_{j,k=1}^{n}\frac{1}{2}\left(\frac{\partial m_{ij}}{\partial q_k} + \frac{\partial m_{ik}}{\partial q_j} - \frac{\partial m_{jk}}{\partial q_i}\right)\dot{q}_j\dot{q}_k. \tag{2.1.6}$$

In a more compact form it is possible to write

$$Q = M(q)\ddot{q} + C(q,\dot{q}) \tag{2.1.7}$$

with

$$(C(q,\dot{q}))_i = \tfrac{1}{2}\dot{q}^T B_i \dot{q}. \tag{2.1.8}$$

The symmetric matrix B_i is defined by

$$(B_i)_{jk} = \frac{1}{2}\left(\frac{\partial m_{ij}}{\partial q_k} + \frac{\partial m_{ik}}{\partial q_j} - \frac{\partial m_{jk}}{\partial q_i}\right). \tag{2.1.4}$$

The vector $C(q,\dot{q})$ represents the effects of Coriolis and centrifugal forces at the joint level. Let G be the vector of gravitational effects on the joints, f the vector representing friction, and Γ the vector of torques or forces produced at the joint level and considered as the input of the dynamic system:

$$Q = \Gamma + G + f. \tag{2.1.10}$$

The equation giving the relation between the input Γ and the joint variables $q, \dot{q}, \ddot{q}$ is

$$\Gamma = M(q)\ddot{q} + N(q,\dot{q}) \tag{2.1.11}$$

2.1 THE DYNAMIC MODEL

with $N(q, \dot{q}) = C(q, \dot{q}) - G - f$. Equation (2.1.11) is known as the 'inverse dynamics' equation of the system.

2.1.2 Newton–Euler formalism and a recursive algorithm

Since the dynamic model represents at least a part of the state equation of a robot manipulator, it is necessary to have an effective procedure for computing the elements of this equation. For simulation purposes it is necessary to compute joint accelerations from joint positions and velocities and from the actuator torques or forces. Some control algorithms also involve the computation of the dynamic model. The common basic tool for this is the Newton–Euler algorithm. It takes its name from the formalism used when it was first obtained. As we pointed out above, a derivation using the Lagrange formalism is now available.

The use of Newton–Euler formalism implies two recursive computations: a forward recursion from the base to the end-effector of the manipulator which computes the velocity and acceleration (angular and linear) of each link, and a backward recursion from the effector to the base of the manipulator which computes forces and torques applied to each joint.

To simplify the formulae, we introduce some notation:

(1) $r_j = u_j$ is a unit vector of the rotation axis if the jth joint is a rotation;
(2) $r_j = 0$ if the jth joint is a translation;
(3) $t_j = u_j$ is a unit vector of the translation axis if the jth joint is a translation;
(4) $t_j = 0$ if the jth joint is a rotation.

Let ω_i be the angular velocity of link i (or frame F_i) relative to the reference frame F_0 and $\omega_{j,i}$ the angular velocity of link i with respect to link j.

We need also to abbreviate some vector notation:

$$a_i = O_0 O_i, \qquad b_i = O_{i-1} O_i = p_i + q_i t_i, \qquad (2.1.12)$$

where $p_i = O_{i-1} O_i$ when $q_i = 0$. G_i is the mass centre of link i and I_i the inertia matrix computed at the centre of mass with respect to the frame F_i associated with the ith link. Then let

$$g_i = O_0 G_i, \qquad h_i = O_i G_i. \qquad (2.1.13)$$

We shall distinguish between rotational joints and translational joints to improve the readability of the equations.

2.1.2.1 Kinematics (forward recursion)

1. *Joint i is a rotational joint.* Applying the composition of angular velocities for frames F_{i-1} and F_i we have

$$\omega_i = \omega_{i-1} + \omega_{i-1,i} = \omega_{i-1} + r_i \dot{q}_i, \qquad (2.1.14)$$

$r_i \dot{q}_i$ being the angular velocity of frame F_i relative to frame F_{i-1}. Differentiating with respect to time and taking into account that r_i is a vector linked to the moving frame F_i, we obtain

$$\dot{\omega}_i = \dot{\omega}_{i-1} + r_i \ddot{q}_i + \omega_i \times r_i \dot{q}_i. \qquad (2.1.15)$$

The acceleration of the centre of mass of link i is obtained from

$$g_i = a_i + h_i. \qquad (2.1.16)$$

Differentiating twice leads to

$$\ddot{g}_i = \ddot{a}_i + \dot{\omega}_i \times h_i + \omega_i \times (\omega_i \times h_i). \qquad (2.1.17)$$

Thus it is necessary to compute the acceleration of the origin, O_i, of the frame F_i. Differentiating twice the equality $a_i = a_{i-1} + b_i$, we get

$$\ddot{a}_i = \ddot{a}_{i-1} + \dot{\omega}_{i-1} \times b_i + \omega_{i-1} \times (\omega_{i-1} \times b_i). \qquad (2.1.18)$$

Neglecting the auxiliary equations, the recursive relations for a rotational joint i are

$$\left.\begin{aligned}
\omega_i &= \omega_{i-1} + r_i \dot{q}_i, \\
\dot{\omega}_i &= \dot{\omega}_{i-1} + r_i \ddot{q}_i + \omega_i \times r_i \dot{q}_i, \\
\ddot{a}_i &= \ddot{a}_{i-1} + \dot{\omega}_{i-1} \times b_i + \omega_{i-1} \times (\omega_{i-1} \times b_i), \\
\ddot{g}_i &= \ddot{a}_i + \dot{\omega}_i \times h_i + \omega_i \times (\omega_i \times h_i).
\end{aligned}\right\} \qquad (2.1.19)$$

2. *Joint i is a translational joint.* In this case the angular velocity and acceleration of the ith joint are the same as those of the joint $i - 1$. The computation of the acceleration of the centre of mass of link i is just the same as in the case of a rotational joint. The equation giving the acceleration of the origin of the frame is slightly different. From

$$a_i = a_{i-1} + p_i + q_i t_i \qquad (2.1.20)$$

and differentiating twice again, we get

$$\ddot{a}_i = \ddot{a}_{i-1} + t_i \ddot{q}_i + \dot{\omega}_i \times b_i + 2\omega_i \times \dot{q}_i + \omega_i \times (\omega_i \times b_i). \qquad (2.1.21)$$

Let us summarize the equations in the case of a translational joint:

$$\left.\begin{aligned}
\omega_i &= \omega_{i-1}, \\
\dot{\omega}_i &= \dot{\omega}_{i-1}, \\
\ddot{a}_i &= \ddot{a}_{i-1} + t_i \ddot{q}_i + \dot{\omega}_i \times b_i + 2\omega_i \times \dot{q}_i + \omega_i \times (\omega_i \times b_i), \\
\ddot{g}_i &= \ddot{a}_i + \dot{\omega}_i \times h_i + \omega_i \times (\omega_i \times h_i).
\end{aligned}\right\} \qquad (2.1.22)$$

2.1 THE DYNAMIC MODEL

The initial conditions must be indicated to complete the set of equations. The frame F_0 is the reference frame so its velocity is zero, giving

$$\omega_0 = \dot{\omega}_0 = 0. \tag{2.1.23}$$

If the manipulator is on the Earth, it is possible to simulate the effect of gravity by considering that the reference frame has an acceleration

$$\ddot{a}_0 = (0, 0, 9.81)^T. \tag{2.1.24}$$

In this case F_0 is no longer Galilean but the forces and torques computed in the backward recursion will include the gravity effects.

2.1.2.2 Dynamics (backward recursion)

We now have to write the equations which relate the moments and forces exerted on the different links during a movement.

Let f_i be the force exerted on link i by link $i - 1$. This force includes among others the reaction forces from the axis of the joint i and the force resulting from the action of the actuator of the ith joint. With the hypothesis of no frictional forces, or if we include friction forces in the actuator action, the projection of f_i onto t_i (when the ith joint is a translational joint) is the force τ_i resulting from the action of the ith actuator:

$$\tau_i = \langle f_i, t_i \rangle. \tag{2.1.25}$$

According to the above definition and Newton's first law, $-f_{i+1}$ is the force exerted on link i by link $i + 1$.

Let F_i be the total force exerted on link i. Then the application of Newton's second law gives

$$F_i = m_i \ddot{g}_i \tag{2.1.26}$$

where m_i is the mass of the ith link. Since the only forces applied to the ith link are f_i and $-f_{i+1}$ we also get

$$F_i = f_i - f_{i+1}. \tag{2.1.27}$$

To derive the equation governing the rotational dynamics, let N_i be the total moment (computed at G_i) exerted on link i and I_i the ith link inertia tensor computed at the centre of mass. The Euler theorem of rotational dynamics states

$$N_i = \frac{d}{dt}(I_i \omega_i) = I_i \dot{\omega}_i + \omega_i \times I_i \omega_i. \tag{2.1.28}$$

Let n_i be the moment, computed at O_i, exerted on link i by link $i - 1$. If joint i is a rotation and the friction forces are neglected or included in the model of the actuator, the torque exerted by the actuator at joint i is

$$\tau_i = \langle n_i, r_i \rangle. \tag{2.1.29}$$

On the other hand N_i is equal to the sum of the moments exerted by links $i-1$ and $i+1$ on link i. The moment exerted by link $i-1$, computed at G_i, is

$$n_i - O_i G_i \times f_i = n_i - h_i \times f_i. \qquad (2.1.30)$$

Similarly the moment exerted on link i by link $i+1$, computed at G_i, is

$$-n_{i+1} + O_{i+1} G_i \times f_{i+1} = -n_{i+1} - b_{i+1} \times f_{i+1} + h_i \times f_{i+1}. \qquad (2.1.31)$$

Adding these two moments we get

$$n_i = N_i + h_i \times F_i + n_{i+1} + b_{i+1} \times f_{i+1}. \qquad (2.1.32)$$

Combining the equations of the backward recursion gives:

$$\left.\begin{aligned}
F_i &= m_i \ddot{g}_i, \\
N_i &= \frac{d}{dt}(I_i \omega_i) = I_i \dot{\omega}_i + \omega_i \times I_i \omega_i, \\
f_i &= F_i + f_{i+1}, \\
n_i &= N_i + h_i \times F_i + n_{i+1} + b_{i+1} \times f_{i+1}, \\
\tau_i &= \langle f_i, t_i \rangle + \langle n_i, r_i \rangle.
\end{aligned}\right\} \qquad (2.1.33)$$

The initial conditions are:

$$f_n = 0, \quad n_n = 0. \qquad (2.1.34)$$

Notice that these initial conditions can be used to take into account forces acting on the end-effector. For that purpose, f_n and n_n have to be set equal to the values of these forces.

2.2 DIRECT AND INVERSE KINEMATICS

2.2.1 Direct kinematics

2.2.1.1 Introduction

The geometric model of a manipulator is usually defined as the relation between the joint coordinates and the location of the end-effector. (Of course the relation between joint coordinates and location of any link of the manipulator can also be considered.) Insofar as direct measurements of the location of the end-effector are not available, this model is the only way of obtaining it. Since the range of the application is the space of frames, the practical computation of the geometric model depends on the representation of rotations adopted. We shall distinguish between the matrix representation and the quaternion representation. A chart representation can be derived from a matrix representation if such a representation is needed.

2.2 DIRECT AND INVERSE KINEMATICS

2.2.1.2 Geometric model with homogeneous matrices

The homogeneous matrix $\bar{R}_{0,n+1}$ defines the position of the frame F_{n+1} relative to the frame F_0. Let $\bar{R}_{i,i+1}$ be the homogeneous matrix defining the position of the frame F_{i+1} relatively to the frame F_i. Then using recursively the change-of-frame formula from (1.1.6) we get

$$\bar{R}_{0,n+1} = \bar{R}_{0,1} \bar{R}_{1,2} \ldots \bar{R}_{n,n+1}. \tag{2.2.1}$$

It remains to compute the matrices $\bar{R}_{i,i+1}$. If the position of the frame F_{i+1} relative to the frame F_i is given by the Denavit–Hartenberg parameters, $\bar{R}_{i,i+1}$ is computed as in (1.1.7). If the joint $i+1$ is a rotation defined by its axis and the angle q_{i+1}, the rotation matrix is given by formula (1.2.44):

$$\bar{R}_{i,i+1} = \cos(q_{i+1}) Id + \sin(q_{i+1}) AS(u_{i+1}) + (1 - \cos(q_{i+1})) u_{i+1} u_{i+1}^T \tag{2.2.2}$$

where u_{i+1} is a unitary vector on the rotation axis such that q_{i+1} is the measure of the angle of the rotation with the orientation induced by u_{i+1}. If the joint $i+1$ is a translation, the rotation matrix of $\bar{R}_{i,i+1}$ is the identity. In both cases the translation vector is $[O_i O_{i+1}]_i$.

2.2.1.3 Geometric model with quaternions

Using notation analogous to that used for the homogeneous matrices above, let $\lambda_{i,j}$ represent the attitude of frame F_j relative to frame F_i. Then $\lambda_{i,i+1}$ is a unitary quaternion representing the rotation which defines the attitude of the frame F_{i+1} relative to the frame F_i. From the correspondence between the rotation product and the quaternion product we can deduce

$$\lambda_{0,n+1} = \lambda_{01} \lambda_{12} \ldots \lambda_{n,n+1}. \tag{2.2.3}$$

The computation of $\lambda_{i,i+1}$ is easy if the joint $i+1$ is a rotation given by u_{i+1}, q_{i+1}. Then by (1.2.43),

$$\lambda_{i,i+1} = (\cos\tfrac{1}{2}(q_{i+1}), \sin\tfrac{1}{2}(q_{i+1})[u_{i+1}]_i). \tag{2.2.4}$$

If the joint $i+1$ is a translation we obviously have $\lambda_{i,i+1} = (1, 0, 0, 0)$.

The computation of the translation vector $[O_0 O_{n+1}]_0$ is based on the relation

$$[O_0 O_{i+1}]_0 = [O_0 O_i]_0 + R_{0i}[O_i O_{i+1}]_i, \tag{2.2.5}$$

with

$$O_0 O_0 = 0; \quad R_{00} = Id. \tag{2.2.6}$$

Translated into quaternion formulation this gives

$$[O_0 O_{i+1}]_0 = [O_0 O_i]_0 + \lambda_{0i}[O_i O_{i+1}]_i \bar{\lambda}_{0i}, \tag{2.2.7}$$

where λ_{0i} is computed recursively by

$$\lambda_{0i} = \lambda_{0,i-1} \lambda_{i-1,i}; \quad \lambda_{00} = (1, 0, 0, 0). \tag{2.2.8}$$

2.2.2 The inverse kinematics problem

The geometric or direct kinematic model of a manipulator is a complex non-linear function from the joint space to the space of displacements. Even if the rotational joints are limited to moving within an interval of length 2π, the function is not one-to-one, or onto. It is a challenging problem in robot design to determine the subset of SE_3 reachable by the end-effector. In general this subset, being the image of a connected subset of $\mathbf{R}^n$ by a continuous map, is a connected set. But, because of bounds on the joint movements, this subset is in general not simply connected. Intuitively this means that the image of this output function is a subset of SE_3 with 'holes' inside. The property of single connectivity is important in connection with Theorem 0.5 and will find applications in control.

The inverse kinematics problem is: given a frame F or a homogeneous matrix in the reachable subset, find the corresponding values of the joint variables. In contrast to the direct kinematics problem, the inverse kinematics problem does not have a general analytic solution. If the geometry of the manipulator is too complex, it may be impossible to solve the problem explicitly. In this case it is possible to use a Newton–Raphson or gradient-type iterative method to solve

$$f(q) = \bar{R} \qquad (2.2.9)$$

where f accounts for the geometric model of the manipulator and $\bar{R}$ is the desired position of the end-effector. An essential drawback from a control point of view is that this kind of iterative method does not provide any insight into the structure of the geometric model. Another flaw of these methods is the necessity for a starting point.

Fortunately there exists a large class of geometric configurations for which the inverse kinematics problem admits an explicit solution. Most manipulators currently encountered belong to this class, for example, the 6-jointed manipulators, the last three rotation axes of which are concurrent. Although there is no general solution to the problem, various methods have been developed, each one valid for a restricted class of manipulator geometries.

We shall not survey these methods, since they have no influence on the basic control problem and are not the subject of this book. Moreover, as we shall see later, it is better to consider the control of the end-effector in SE_3. With this approach it is not necessary to have an inverse kinematic model. Nevertheless the direct geometric model is in this case an output function. Since many output functions in robotics depend on the location of some link, the direct geometric model can be considered as a prototype for output functions. The study of the inverse kinematics problem gives some insight into the type of difficulty encountered with output functions in robot control.

Let us consider a simple example. The manipulator considered has three degrees of freedom with three rotational joints (Fig. 2.1). The modified Denavit–Hartenberg parameters of this manipulator are given in Table 2.1.

2.2 DIRECT AND INVERSE KINEMATICS

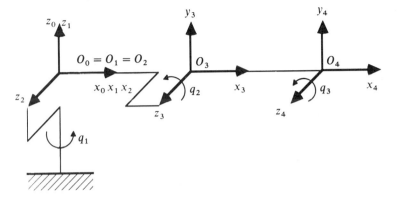

Fig. 2.1 A manipulator with three rotational joints

TABLE 2.1 Denavit–Hartenberg parameters

i	α	d	θ	r
1	0	0	θ_1	0
2	$\frac{1}{2}\pi$	0	θ_2	0
3	0	d	θ_3	0
4	0	d	0	0

Links 2 and 3 are taken to be of equal length to ease the computations and the subsequent analysis. Now let L_{04}^* be a homogeneous transformation matrix representing a desired location of the end-effector:

$$L_{04}^* = \begin{bmatrix} t_{11} & t_{12} & t_{13} & t_{14} \\ t_{21} & t_{22} & t_{23} & t_{24} \\ t_{31} & t_{32} & t_{33} & t_{34} \\ 0 & 0 & 0 & 1 \end{bmatrix}. \qquad (2.2.10)$$

Since our manipulator has only three degrees of freedom, we shall consider only the position of the origin of the end-effector and search for joint positions such that

$$\bar{R}_{04} = \begin{bmatrix} * & * & * & t_{14} \\ * & * & * & t_{24} \\ * & * & * & t_{34} \\ 0 & 0 & 0 & 1 \end{bmatrix}. \qquad (2.2.11)$$

From the Denavit–Hartenberg parameters we obtain four transformation matrices:

$$\bar{R}_{01} = \begin{bmatrix} c_1 & -s_1 & 0 & 0 \\ s_1 & c_1 & 0 & 0 \\ 0 & 0 & 1 & 0 \\ 0 & 0 & 0 & 1 \end{bmatrix}, \quad \bar{R}_{12} = \begin{bmatrix} c_2 & -s_2 & 0 & 0 \\ 0 & 0 & -1 & 0 \\ s_2 & c_2 & 0 & 0 \\ 0 & 0 & 0 & 1 \end{bmatrix},$$

$$\bar{R}_{23} = \begin{bmatrix} c_3 & -s_3 & 0 & d \\ s_3 & c_3 & 0 & 0 \\ 0 & 0 & 1 & 0 \\ 0 & 0 & 0 & 1 \end{bmatrix}, \quad \bar{R}_{34} = \begin{bmatrix} 1 & 0 & 0 & d \\ 0 & 1 & 0 & 0 \\ 0 & 0 & 1 & 0 \\ 0 & 0 & 0 & 1 \end{bmatrix},$$

(2.2.12)

where s_i stands for $\sin(q_i)$ and c_i stands for $\cos(q_i)$. Then we obtain

$$\bar{R}_{24} = \begin{bmatrix} c_3 & -s_3 & 0 & c_3 d + d \\ s_3 & c_3 & 0 & s_3 d \\ 0 & 0 & 1 & 0 \\ 0 & 0 & 0 & 1 \end{bmatrix}, \quad (2.2.13)$$

$$\bar{R}_{14} = \begin{bmatrix} c_2 c_3 - s_2 s_3 & -s_3 c_2 - s_2 c_3 & 0 & d(c_2(c_3+1) - s_2(s_3+1)) \\ 0 & 0 & -1 & 0 \\ s_2 c_3 + c_2 s_3 & -s_2 s_3 + c_2 c_3 & 0 & d(s_2(c_3+1) + c_2(c_3+1)) \\ 0 & 0 & 0 & 1 \end{bmatrix}.$$

(2.2.14)

We shall use the following lemma intensively in the computation of the desired joints.

Lemma 2.1 *The equation* $\alpha \cos(\Theta) + \beta \sin(\Theta) = \gamma$ *has a solution if and only if* $\alpha^2 + \beta^2 \geq \gamma^2$. *In this case the solution is*

$$\left. \begin{aligned} \sin(\Theta) &= \frac{\beta \gamma + \varepsilon \alpha \sqrt{\alpha^2 + \beta^2 - \gamma^2}}{\alpha^2 + \beta^2} \\ \cos(\Theta) &= \frac{\alpha \gamma - \varepsilon \beta \sqrt{\alpha^2 + \beta^2 - \gamma^2}}{\alpha^2 + \beta^2} \end{aligned} \right\} \quad (2.2.15)$$

where $\varepsilon = \pm 1$.

The proof is left to the reader.

2.2 DIRECT AND INVERSE KINEMATICS

The algorithm used to solve the inverse kinematics problem is based on the resolution of the successive equations

$$L_{i4}^* = \bar{R}_{4i-1} L_{i-14}^* = \bar{R}_{i4}. \qquad (2.2.16)$$

This algorithm is quite general and can be used in more complex situations. For the first step we get

$$L_{14}^* = \begin{bmatrix} * & * & * & c_1 t_{14} + s_1 t_{24} \\ * & * & * & -s_1 t_{14} + c_1 t_{24} \\ * & * & * & t_{34} \\ 0 & 0 & 0 & 1 \end{bmatrix} = \bar{R}_{14}, \qquad (2.2.17)$$

where * replaces entries that are not necessary for the computations. We can extract the equation

$$t_{24} c_1 - t_{14} s_1 = 0. \qquad (2.2.18)$$

Using the above lemma we get:

$$\left.\begin{aligned} \sin(q_1) = s_1 &= \frac{\varepsilon_1 t_{24} \sqrt{t_{14}^2 + t_{24}^2}}{t_{14}^2 + t_{24}^2} = \frac{\varepsilon_1 t_{24}}{\sigma}, \\ \cos(q_1) = c_1 &= \frac{\varepsilon_1 t_{14}}{\sigma}, \\ \sigma &= \sqrt{t_{14}^2 + t_{24}^2}, \end{aligned}\right\} \qquad (2.2.19)$$

which gives two solutions for q_1:

$$q_1 = \arctan\left(\frac{t_{24}}{t_{14}}\right), \qquad q_1 = \arctan\left(\frac{t_{24}}{t_{14}}\right) + \pi. \qquad (2.2.20)$$

q_1 cannot be computed and is indeterminate if $\sigma^2 = t_{14}^2 + t_{24}^2 = 0$. The corresponding position of O_4 is on the z_0 axis.

Replacing s_1 and c_1 by the values found above gives

$$L_{14}^* = \begin{bmatrix} * & * & * & \varepsilon_1 \sigma \\ * & * & * & 0 \\ * & * & * & t_{34} \\ 0 & 0 & 0 & 1 \end{bmatrix}. \qquad (2.2.21)$$

In order to compute q_2 we set up the equation

$$L_{24}^* = \begin{bmatrix} * & * & * & c_2 \varepsilon_1 \sigma + s_2 t_{34} \\ * & * & * & -s_2 \varepsilon_1 \sigma + c_2 t_{34} \\ * & * & * & 0 \\ 0 & 0 & 0 & 1 \end{bmatrix} = \bar{R}_{24}. \qquad (2.2.22)$$

Again we extract the following two scalar equations:

$$\varepsilon_1 \sigma c_2 + t_{34} s_2 - d = c_3 d, \qquad (2.2.23)$$

$$t_{34} c_2 - \varepsilon_1 \sigma s_2 = s_3 d. \qquad (2.2.24)$$

Eliminating c_3 and s_3 we get

$$\varepsilon_1 \sigma c_2 + t_{34} s_2 = \frac{\|t\|^2}{2d}, \qquad (2.2.25)$$

where $\|t\|^2 = t_{14}^2 + t_{24}^2 + t_{34}^2$.

From Lemma 2.1, solutions of this equation exist if $\|t\| \leq 2d$. Geometrically, a point with coordinates (t_{14}, t_{24}, t_{34}) is not in the reachable space if it is outside the sphere with centre O_0 and radius $2d$. If this is not the case we get

$$\left. \begin{array}{l} \sin(q_2) = s_2 = \dfrac{t_{34} \|t\| + \varepsilon_1 \varepsilon_2 \sigma \sqrt{4d^2 - \|t\|^4}}{2d \|t\|} \\[2mm] \cos(q_2) = c_2 = \dfrac{\varepsilon_1 \sigma \|t\| - \varepsilon_2 t_{34} \sqrt{4d^2 - \|t\|^4}}{2d \|t\|} \end{array} \right\} \qquad (2.2.26)$$

Again, for each choice of q_1 there are two solutions for q_2. The coordinate of the second joint cannot be computed if $\|t\| = 0$, corresponding to $O_4 = O_0$. Although this configuration is generally physically impossible for this kind of manipulator, it should be noted that it implies $q_3 = \pm \pi$. We should also like to emphasize that the two solutions for q_2 are equal if $\|t\| = 2d$, i.e. if the point is on the border of the reachable space. In this case, q_3 is still entirely determined and must be equal to 0.

The computation of q_3 follows the same path:

$$L_{34}^* = \begin{bmatrix} * & * & * & c_3 \tau_1 + s_3 \tau_2 - c_3 d \\ * & * & * & -s_3 \tau_1 + c_3 \tau_2 + s_3 d \\ * & * & * & 0 \\ 0 & 0 & 0 & 1 \end{bmatrix} = \bar{R}_{34}, \qquad (2.2.27)$$

with

$$\tau_1 = \frac{\|t\|^2}{2d}, \qquad \tau_2 = -\varepsilon_2 \|t\| \sqrt{4d^4 - \|t\|^4}. \qquad (2.2.28)$$

The second entry in the fourth column gives the equation

$$s_3(d - \tau_1) + c_3 \tau_2 = 0 \qquad (2.2.29)$$

2.2 DIRECT AND INVERSE KINEMATICS

with solutions

$$\left.\begin{aligned} s_3 &= \frac{\varepsilon_3 \tau_2 \sqrt{(d - \tau_1^2) + \tau_2^2}}{(d - \tau_1)^2 + \tau_2^2}, \\ c_3 &= \frac{\varepsilon_3 (\tau_1 - d) \sqrt{(d - \tau_1^2) + \tau_2^2}}{(d - \tau_1)^2 + \tau_2^2}. \end{aligned}\right\} \qquad (2.2.30)$$

The first entry in the fourth column gives a second equation which completely determines ε_3. Computation of the inverse kinematics of this simple robot is then completed.

The first thing illustrated by this simple example is that the geometric model is not one-to-one or onto, as previously stated. Given a point in the reachable space of the manipulator, there are in general four sets of joint coordinates corresponding to that point. In particular cases, the number of solutions may be less than four or infinite.

Now consider the singular case for q_1. In this case, q_1 cannot be determined and every value of q_1 is valid (Fig. 2.2 (a)). The necessary and sufficient condition for O_4 to be on the z_0 axis is $2q_2 + q_3 = \pm \pi$.

Now given one value for q_1, the set of points reachable in the plane defined by q_1 is split into two regions corresponding to $-\pi < 2q_2 + q_3 < \pi$ and the complement (Fig. 2.2(b)). If a reachable point is in one of these regions, there are in general two couples of values for q_2 and q_3 corresponding to this point. The exceptions are when $q_3 = \pm \pi$ or $q_3 = 0$. Since the first case is unrealistic we shall suppose that $|q_3| < \pi$.

From the above analysis, it is possible to decompose the joint coordinates space into four subspaces with the planes $2q_2 + q_3 = \pm \pi$ and $q_3 = 0$ as

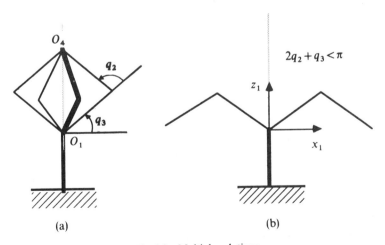

Fig. 2.2 Multiple solutions

boundaries (Fig. 2.3). These four regions correspond to the four classes of solutions found in the general case. In Fig. 2.3 some regions are in two parts because we have represented the joint space as a Cartesian space, although it is better represented as the product of three unit circles.

The geometric model, when restricted to each region, is a one-to-one application onto its image. For a real manipulator, the joint variables are bounded for mechanical reasons. A consequence is that the regions described above are larger than what is allowed by mechanical bounds. Let us now consider the restricted regions obtained as the intersection of the previous ones with the true joint space (taking bounds into account). Let us also call the image of each region by the geometric model map an 'aspect'. There is a one-to-one correspondence between each aspect and the corresponding region of the joint coordinates. But, owing to the bounds, the different aspects may not be equal subsets of the reachable space. It may happen that two points in the reachable space do not belong to the same aspect. If we wish to move the end-effector from one point to the other, it is necessary to cross the boundary between two regions in the joint space. We have then at least two control problems:

(1) detecting that two points are not in the same aspect;
(2) crossing the boundary between two regions. As the geometric model behaves poorly at these boundaries, control problems may arise.

This short analysis of the inverse kinematics problem shows the kinds of difficulties which may be encountered in non-linear control, in particular with a non-linear output function. In Chapter 3, we shall generalize the description of robot tasks to other than purely positioning tasks. Nevertheless, the

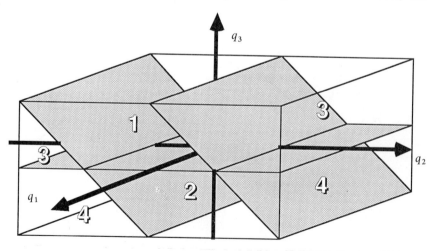

Fig. 2.3 Aspects in the joint space

2.3 THE DIFFERENTIAL MOTION MODEL

geometric model of a manipulator will remain the prototype of the output functions found in robot control.

2.3 THE DIFFERENTIAL MOTION MODEL

2.3.1 Introduction

As the geometric model of a manipulator maps the joint coordinates onto the location of a link, the differential motion model maps the joint velocities onto the velocity of a link (often the end-effector). Basically, the differential motion model is obtained by differentiating the geometric model, but the latter is very complex and the differential motion model is never derived from the geometric model expression, but only by direct computation of the velocities concerned.

Another problem is the need, in control algorithms, to have a parametrization of the rotation space. Differentiating some parametrizations was done in Section 1.2.5.4, and relations have been established between the angular velocity of a frame and the corresponding variations of the attitude parameters. Hence the differential motion model is also a basic tool when dealing with differentials of output functions depending on attitudes and more generally on the location of a link. As we shall see later, the differential of the output function plays a crucial role in control analysis.

Since the position of the end-effector (or of another link) is characterized by the relative positioning of a frame F_{n+1} linked to the end-effector, with respect to a reference frame F_0, the velocity of the end-effector is a vector in the tangent space of the frame space. We have seen in Section 1.3.1.1 that this vector is an element (v, ω) of se_3, with v the linear velocity of the origin of the frame F_{n+1} and ω the angular velocity of F_{n+1}.

Let us now provide a more precise terminology:

Definition D2.1 *The translational Jacobian of a manipulator is the mapping $J_L(q)$ such that $v = J_L(q)\dot{q}$. The rotational Jacobian is the mapping $J_R(q)$ such that $\omega = J_R(q)\dot{q}$, where (v, ω) is the velocity of the end-effector.*

Of course, the expression of these applications as matrices depends on the frame chosen for writing the velocities ω and v. If the frame F_i is chosen, we shall denote by $J_{L,i}$ and $J_{R,i}$ the two matrices and we obtain the relations

$$[v]_i = J_{L,i}(q)\dot{q}, \qquad [\omega]_i = J_{R,i}(q)\dot{q}. \tag{2.3.1}$$

We shall also denote

$$J_i = \begin{bmatrix} J_{L,i} \\ J_{R,i} \end{bmatrix}, \tag{2.3.2}$$

omitting the subscript when it is not necessary. By default, the Jacobian matrix will be assumed to be expressed in terms of the reference frame, F_0. Note also that since (v, ω) is a linear combination of the columns of the Jacobian, each column may be considered as a screw. Therefore, when changing the frame (origin and basis) related to v and ω, the expression (1, 3.4 1) can also be used for transforming the Jacobian matrix.

In connection with the differential of the output function, singular points play an important role in control problems. This justifies the following definition:

Definition D2.2 *The manipulator has a kinematic singularity at q if $J(q)$ is not full rank.*

2.3.2 Computation of the Jacobian

2.3.2.1 General setting

We know that the velocity of the frame F_{n+1} linked to the end-effector of the manipulator is given by the linear velocity of a reference point O_{ref} of the frame F_{n+1} and the angular velocity of the frame. In general the reference point is the origin of the frame but it is possible to choose another point of the frame. Moreover, this point may vary with time and may for example coincide at each time with the origin O_i of the ith frame associated with the ith link. Of course if M is another point of the frame F_{n+1}, the velocity of M is

$$V(M) = V(O_{\text{ref}}) + \omega \times O_{\text{ref}}M, \qquad (2.3.3)$$

where $\omega = J_R \dot{q}$ is the angular velocity of the frame F_{n+1}.

The computation of the angular velocity of a joint has already been done, when we established the kinematic relations for the Newton–Euler algorithm. Recalling these results we have, since $\omega_{n,n+1} = 0$,

$$\omega_{n+1} = \sum_{j=0}^{n-1} \omega_{j, j+1} \qquad (2.3.4)$$

and

$$\omega_{n+1} = \sum_{j=1}^{n} r_j \dot{q}_j. \qquad (2.3.5)$$

To compute the linear velocity of O_{n+1} we could use the formulae already established for the dynamic model. Instead, we have chosen to derive them independently. Recall that

$$O_k O_{k+1} = p_{k+1} + q_{k+1} t_{k+1}. \qquad (2.3.6)$$

With the above notation we get

$$O_0 O_{n+1} = \sum_{k=0}^{n-1} (p_{k+1} + q_{k+1} t_k) + p_{n+1}. \qquad (2.3.7)$$

2.3 THE DIFFERENTIAL MOTION MODEL

Differentiating with respect to time gives the linear velocity of O_{n+1} with respect to O_0:

$$\frac{dO_0 O_{n+1}}{dt} = \sum_{k=0}^{n-1} \left(\frac{dp_{k+1}}{dt} + \frac{d(q_{k+1} t_{k+1})}{dt} \right) + \frac{dp_{n+1}}{dt}$$

$$= \sum_{k=0}^{n-1} (\omega_k \times p_{k+1} + \dot{q}_{k+1} t_{k+1} + \omega_k \times q_{k+1} t_{k+1}) + \omega_n \times p_{n+1}. \quad (2.3.8)$$

Replacing ω_k by the value found above and letting $t_{n+1} = 0$ leads to

$$\frac{dO_0 O_{n+1}}{dt} = \sum_{k=1}^{n} \left[\sum_{i=1}^{k} \dot{q}_i r_i \times (p_{k+1} + q_{k+1} t_{k+1}) \right] + \sum_{k=1}^{n} \dot{q}_k t_k$$

$$= \sum_{i=1}^{n} \left[\dot{q}_i r_i \times \left(\sum_{k=1}^{n} p_{k+1} + q_{k+1} t_{k+1} \right) \right] + \sum_{k=1}^{n} \dot{q}_k t_k$$

$$= \sum_{i=1}^{n} \dot{q}_i r_i \times O_i O_{n+1} + \sum_{k=1}^{n} \dot{q}_n t_k. \quad (2.3.9)$$

Now consider a reference point O_{ref}. As seen previously, we have

$$V(O_{\text{ref}}) = V(O_{n+1}) + \omega \times O_{n+1} O_{\text{ref}}$$

$$= \sum_{i=1}^{n} \dot{q}_i r_i \times O_i O_{i+1} + \sum_{k=1}^{n} \dot{q}_k t_k + \sum_{i=1}^{n} \dot{q}_i r_i \times O_{n+1} O_{\text{ref}}. \quad (2.3.10)$$

Finally,

$$V(O_{\text{ref}}) = \sum_{i=1}^{n} \dot{q}_i (r_i \times O_i O_{\text{ref}} + t_i). \quad (2.3.11)$$

2.3.2.2 Algorithms

The various algorithms described in this section compute the matrices J_R and J_L recursively. The basic idea is at step i to compute the ith column of each matrix by evaluating the velocity of the frame F_{n+1} with $\dot{q}_i = 1$ and $\dot{q}_k = 0$ for $k \neq i; k = 1, \ldots, n$. The necessary matrices and vectors are computed recursively. This phase of the computation gives more or less explicitly the direct kinematic model of the manipulator. In the following, $J_{R,j}^i$ and $J_{L,j}^i$ are the ith columns of the rotational and translational components of J_j.

1. *Waldron algorithm*. The reference point is O_0 and the computed Jacobian is J_0. It gives the velocity coordinates in frame F_0. The vectorial equations are then

$$\omega = \sum_{i=1}^{n} \dot{q}_i r_i, \quad (2.3.12)$$

$$V = \sum_{i=1}^{n} \dot{q}_i r_i \times O_0 O_i + \sum_{k=1}^{n} \dot{q}_k t_k. \quad (2.3.13)$$

Putting these equations in recursive matrix form, let:

$$R_{0,0} = Id,$$
$$R_{0,i} = R_{0,i-1} R_{i-1,i}, \quad i = 1, \ldots, n,$$
$$J_{R,0}^i = R_{0,i} [r_i]_i, \quad i = 1, \ldots, n,$$
$$[O_0 O_0]_0 = 0,$$
$$[t_0]_0 = 0,$$
$$[t_i]_0 = R_{0,i} [t_i]_i, \quad i = 1, \ldots, n,$$
$$[O_0 O_i]_0 = [O_0 O_{i-1}]_0 + R_{0,i-1} [p_i]_{i-1} + q_i [t_i]_0, \quad i = 1, \ldots, n,$$
$$J_{L,0}^i = -J_{R,0}^i \times [O_0 O_i]_0 + [t_i]_0, \quad i = 1, \ldots, n. \qquad (2.3.14)$$

Remark. This algorithm explicitly computes the geometric model of the manipulator in the frame F_0.

2. *Renaud algorithm.* The reference point here is O_k, the origin of the frame of the middle link of the manipulator. The Jacobian is computed in frame F_k. The vector equations are

$$\omega = \sum_{i=1}^n \dot{q}_i r_i \qquad (2.3.15)$$

$$V = \sum_{i=1}^n \dot{q}_i r_i \times O_i O_k + \sum_{i=1}^n \dot{q}_i t_i. \qquad (2.3.16)$$

In recursive form we have:

$$R_{k,k} = Id,$$
$$R_{k,i} = R_{k,i-1} R_{i-1,i}, \quad i = k+1, n,$$
$$R_{k,i-1} = R_{k,i} R_{i,i-1}, \quad i = k, 2,$$
$$J_{R,k}^i = R_{k,i} [r_i]_i, \quad i = 1, \ldots, n,$$
$$[t_i]_k = R_{k,i} [t_i]_i, \quad i = 1, \ldots, n,$$
$$[O_i O_k]_k = [O_{i-1} O_k]_k - R_{k,i-1} [p_i]_{i-1} - q_i [t_i]_k, \quad i = k+1, \ldots, n,$$
$$[O_{i-1} O_k]_k = [O_i O_k]_k + R_{k,i-1} [p_i]_{i-1} + q_i [t_i]_k, \quad i = k, \ldots, 2,$$
$$J_{L,k}^i = J_{R,k}^i \times [O_i O_k]_k + [t_i]_k.$$

$$(2.3.17)$$

Remark. This algorithm uses two recursive schemes which may be processed in parallel. In general the interesting Jacobians are J_0 or J_{n+1}. The matrices necessary to compute J_0 or J_{n+1} from J_k are also produced by the algorithm. This algorithm is advantageous in analytic computations. It gives simple expressions and simplifies a study of the points where the rank of the Jacobian changes. As we shall see later, these points play an important role in control

3. *Orin and Shrader algorithm.* The reference point is O_{n+1} and the Jacobian is computed in frame F_n. The computation is similar to those in the Waldron algorithm, except that it proceeds backwards from the effector to the reference frame F_0. The vector equations are the same as in the Waldron algorithm. The recursive equations are:

$$\left.\begin{aligned}
& R_{n+1,n+1} = Id, \\
& R_{n+1,i-1} = R_{n+1,i} R_{i,i-1}, \quad i = n+1,\ldots,1, \\
& J^i_{R,n+1} = R_{n+1,i}[r_i]_i, \quad i = 1,\ldots,n, \\
& [t_i]_{n+1} = R_{n+1,i}[t_i]_i, \quad i = 1,\ldots,n, \\
& [O_{n+1}O_{n+1}]_{n+1} = 0, \\
& [O_{n+1}O_{i-1}]_{n+1} = [O_{n+1}O_i]_{n+1} - R_{n+1,i-1}[p_i]_{i-1} - q_i[t_i]_{n+1}, \\
& J^i_{L,n+1} = -J^i_{R,n+1} \times [O_{n+1}O_i]_{n+1} + [t_i]_{n+1}.
\end{aligned}\right\}$$

(2.3.18)

Remark. The geometric model of the manipulator is implicitly computed.

We have presented different algorithms for computing the dynamic, the geometric and the differential motion models of a manipulator. Of course the choice of an algorithm depends on the robotic task to be performed and on the control algorithm chosen. Although we have presented the algorithms separately, we have already observed that they share some of the recursive equations. Advantage can be taken of these redundancies to realize an efficient implementation when two or more of the models are utilized.

2.4 BIBLIOGRAPHIC NOTE

The models presented in this chapter are among the earliest established in robotics. The direct and inverse kinematics problems had to be solved by pioneers in the field. They are now considered as standard material and appear in the initial chapters of robotics books: Coiffet (1981), Paul (1982), Vukobratovic and Potkonjak (1982a), Vukobratovic and Stokic (1982b),

Gorla and Renaud (1984), and Dombre and Khalil (1988). See also Brockett (1984) or Desa and Roth (1985) for other modelling approaches.

The computation of the Jacobian of a manipulator has also given rise to many publications (e.g. Renaud (1981)). Our exposition closely follows a survey paper by Orin and Schrader (1983), which includes a presentation of the Waldron method, and is also detailed in Le Borgne (1985).

Increasing the performance of the robot implies taking into account the dynamic effects. As stated in this chapter, the first computations of the dynamic model of a manipulator used Lagrange formalism and required symbolic computations. The models obtained this way were in 'closed form' and very heavy (Khalil 1978). Later, the Newton–Euler formalism was used to obtain a model in a recursive form (Luh et al. 1980a), while the model derived from Lagrange's equations was also put into a recursive form by Hollerbach (1980). Finally it was proved by Silver (1982) that the two formalisms can lead to the same recursive equations.

In Dombre and Khalil (1988) and Kleinfinger (1986), for example, the modelling problems of tree-structure robots and of robots with closed kinematic chains is addressed. The problem of finding a model with minimal computations has been widely studied throughout the 1980s (see, for example, Renaud (1983) or Dombre and Khalil (1988)).

The geometric model has not been studied very much from a control point of view. Work has been done to find all the singular points for the most common geometries of manipulators. The notion of 'aspect' was introduced by Borrel (see Dombre and Khalil (1988) and Aldon (1986)) who defined it in a more general setting.

Because the dynamic and differential motion models appear as essential parts of some control algorithms, much attention has been paid to their efficient implementations. These studies take various forms: efficient simulations performed on various computers (from microcomputers to mainframes), dedicated VLSI architectures, and implementations on multiprocessor architectures (Luh and Lin 1982; Lilly and Orin 1986; Orin et al. 1986; Lathrop 1984; Borrelly and Le Borgne 1988).

Enlarging the domain of robotic applications also makes it necessary to use manipulators with more than six joints (see Chapter 4). The inverse kinematic problem has also been considered for this kind of robot (Wampler 1987).

3

THE TASK FUNCTION CONCEPT

OUTLINE

This chapter is divided into two parts: in section 3.1 modelling assumptions and model structures are discussed informally, and in sections 3.2–3.4 formal definitions are given to reduce the class of all possible output functions to the class of so-called ρ-admissible functions.

3.1 AN INFORMAL APPROACH

3.1.1 Applications and tasks

The subject of this book is the treatment of low-level control problems. Low-level control is to be understood within the meaning of automatic control theory. This suggests the existence of a higher level of control taking into account aspects of robot control which are not necessarily relevant to automatic control theory. Let us consider for example 'pick-and-place' applications: the high-level control provides a sequence of points in the joint space and a low-level controller carries out the positioning of the robot at these points. A more elaborate application, such as assembly of two parts, may use a positioning task as before, followed by a navigation task using some kind of sensor for obstacle avoidance and/or fine positioning, and then a third task involving force feedback.

This rough analysis of a possible assembly application shows a decomposition of the application into three tasks. The control objectives differ from one task to another. In the first one we are interested in following a desired trajectory in the joint space as closely as possible. This trajectory is provided by the higher level of control. In the second task, measurements from some kind of optical reflectance sensor or magnetic sensor are taken into account and the control objective is to maintain (for example) a function of the sensor signals as small as possible. In the last motion of the application, a force sensor is used and a possible control objective would be to regulate the sensor output around a given value while moving the effector along a given direction.

If we call the robot plus the sensors a *robotic system*, in the application described above we are interested in three different outputs of the robotic system. Automatic control techniques may be used for monitoring any of these outputs. But the decision for switching from one task to another must be taken by a higher level controller.

Clearly the state equations of the robotic system are independent of the task which is pursued, since the robotic system remains the same. What differs when switching from one task to another is the output function of the system and the control objectives. Thus the output function appears as the essential feature in the description of a robotic task.

Let us discuss the term 'robotic task' further. For many authors a robotic task is what we have called an application. Throughout this book a robotic task will be defined by an *output function* and a *control objective*.

The modelling of a robotic task by an output function gives rise to two problems.

1. How do we design control laws to match the control objective? This low-level control problem is essentially the subject of this book.

2. At a higher level, how do we choose an output function to achieve the desired objective. This may prove to be a very difficult question. For example, what are the relevant parameters for inserting a pin in a hole? How do we make a correct arc-welded seam? How are distance sensors used to avoid obstacles in a given environment?

The answers to these questions require modelling of the corresponding jobs, and such models are loosely dependent on the robotic system. To formalize this in a general sense is very difficult. The situations involved are generally very complex, and simulations are often the only means of validating an answer to the second question. Although the book is not centred on this issue, some clues can be found throughout the text concerning the choice of adequate output functions.

3.1.2 State-space model and task functions

In order to apply automatic control techniques one has to start from a mathematical description of the controlled system. Such a description necessarily involves a state-space representation of the robotic system. The natural and minimal state for a robot is $\begin{bmatrix} q \\ \dot{q} \end{bmatrix}$, the vector of joint positions and velocities. The hypothesis of rigid links is essential to justify this choice of the state space. The evolution of the state is then described by a differential equation

$$\begin{bmatrix} \dot{q} \\ \ddot{q} \end{bmatrix} = f(q, \dot{q}, u), \qquad (3.1.1)$$

3.1 AN INFORMAL APPROACH

u being a control vector. The basic model for the function f was given in Chapter 2. Components may be added to the state vector if we wish to take into account the dynamics of actuators or flexible transmissions, for example.

As the dynamics of most sensors are much faster than the dynamics of the mechanical part of a robotic system, they are not taken into account in the state-space model. For this reason, it is possible to consider that all the tasks share the same state-space model.

The output function is an essential part of the description of a robot task. We shall not consider general output functions but rather restrict our attention to a class of output functions designated as *task functions*. More precisely we shall consider only functions $e(q, t)$ of the position vector q and the time variable t taken as an independent parameter. The values of the task functions are taken in $\mathbf{R}^n$.

The dependence of the output on time is necessary to model tasks such as tracking a target with the end-effector. More generally, the time parameter is necessary to take into account a smooth variation in the objective of the control. This kind of variation is in general linked to variations in the environment.

As is usual in control theory, we shall also reduce the control problem to an output regulation problem. That is, the task considered is performed perfectly during $[0, T]$ if $e(q(t), t) = 0$ for all t in $[0, T]$. The reduction to a regulation problem is fairly natural, and unifies the control objectives. For instance suppose that a clever analysis of a task leads to a mathematical description using a state-space formulation with an output function $e'(q, t)$. The mathematical description is not complete if the desired behaviour of the output is not specified. So one must also define a desired (ideal) trajectory $e'_d(t)$ for the output. Taking $e(q, t) = e'(q, t) - e'_d(t)$ as the new output, the control problem is now reduced to the regulation of this output function $e(q, t)$.

It is still possible to object that this way of stating the control problem restricts the class of control objectives. For instance, we shall not consider any kind of optimal control techniques. One reason is that we only wish to consider control techniques which are adapted to real-time implementations. However optimal control techniques may still be applied off-line, for example, to determine an optimal ideal trajectory of the robot. In this way the problems of optimality and robustness of the control can be separated.

The task functions we shall consider depend only on the joint positions and not on the whole state, as is usual in general control theory. One reason is that most robotic tasks can be analysed as positioning problems. Another reason is that the regulation of a more general output function $e(q, \dot{q}, t)$ of both position and velocity underlies an implicit differential equation $e(q, \dot{q}, t) = 0$ for which two cases have to be considered.

1. $\partial e / \partial \dot{q}$ is nonsingular; then the Implicit Function Theorem applies and the implicit differential equation is equivalent (at least locally) to an explicit

one. This equation defines a desired trajectory $q_d(t)$ and we are brought back to the case of an output function depending only on q and t.

2. $\partial e/\partial q$ is not nonsingular; this situation is much more complex. For instance, singular integrals may appear and as a result the theorem of uniqueness of the solution of a differential equation does not apply any more. Starting from one point (q_0, t_0) there may be two or more trajectories in the state space corresponding to the achievement of the control objective $e(q, \dot{q}, t) = 0$. We shall consider such a control problem to be ill-posed, and the task must be reformulated.

3.1.3 The initial condition

Let us suppose that we have managed to pose the control problem as an output regulation problem with a task function $e(q, t)$ depending only on the joint positions and the independent parameter t. If t represents a time evolution of the output, a solution $q_r(t)$ of equation $e(q, t) = 0$ can be interpreted, when it is twice differentiable, as an ideal trajectory in the joint space. In general $e(q, t)$ is a complex non-linear function. In Chapter 2 we studied an example of such a function: the geometric model of a manipulator. Consider again the simple manipulator described in the section devoted to the inverse kinematics problem (section 2.2.2). Suppose one wants to move the point O_4 along a desired trajectory $e_d(t)$. If $f(q)$ represents the position of O_4, a task function is

$$e(q, t) = f(q) - e_d(t). \tag{3.1.2}$$

Now for a fixed t, the equation $e(q, t) = 0$, equivalent to $f(q) = e_d(t)$, has in general four solutions belonging to the four regions described in section 2.2.2. If t varies within $[0, T]$, it is in general possible to find four trajectories $q_r^i(t)$, $i = 1, \ldots, 4$, each included in one of the four regions and such that $e(q_r^i(t), t) = 0$ for all t in $[0, T]$. This is not possible if for some t, $e_d(t)$ is a point such that $f(q) = e_d(t)$ has either an infinite number of solutions or none. When the four trajectories exist they are completely disjoint.

The situation that we have just described is fairly general with non-linear output functions. One consequence is that a control problem is not completely specified until an initial condition q_0 such that $e(q_0, t_0) = 0$ is given. The role of q_0 is to ensure (along with some other conditions) the uniqueness of the ideal trajectory corresponding to the solution of the regulation problem. This is done by imposing the equality $q_r(0) = q_0$. q_0 is not necessarily the initial position of the robot.

Remark. As pointed out above, a well-defined task is a couple: $\{e(q, t), q_0\}$ such that

$$e(q_0, 0) = 0 \tag{3.1.3}$$

3.1 AN INFORMAL APPROACH

q_0 being the position in which the robot should ideally stay at $t = 0$. Moreover, if e is time-independent, q_0 is also the objective position, q^*, to be reached by the robot.

3.1.4 The feasibility of a task: existence of solutions

A necessary condition for a task defined by a task function and an initial condition to be feasible is the existence of a solution to the equation $e(q, t) = 0$ for each t within $[0, T]$. It may happen that no solution exists except the initial condition: it is impossible to track perfectly a target outside the attainable space, for example. If a solution $q_r(t)$ exists, it may not be unique. With a sufficiently smooth function $e(q, t)$ three situations may occur in a neighbourhood of a solution $q_r(t)$:

1. $(\partial e/\partial q)(q_r(t), t)$ is invertible: again the Implicit Function Theorem (Theorem 0.3) implies the existence of a local function $t \to q_r(t)$ ensuring the uniqueness of the solution locally.

2. $(\partial e/\partial q)(q_r(t), t)$ is not invertible but onto. Then Proposition P0.8 applies and the set of solutions $\{q | e(q, t) = 0\}$ for a fixed t is locally a submanifold of the joint space with dimension the co-rank of $(\partial e/\partial q)(q_r(t), t)$. The control problem is ill-posed because several trajectories (an infinite number in general) in the joint space correspond to a zero output. But in a sense the problem is also underdetermined: the output function does not put enough constraints on the state. This is typically the redundant case (see Chapter 4).

3. $(\partial e/\partial q)(q_r(t), t)$ is not onto. This case is the worst one; we shall refer to it as a singular case. Various situations are possible and we shall study some of them later in this chapter.

3.1.5 A simple example, and more on state trajectories

Let us take an example: suppose that an application is defined as the minimization of the distance between the origin of a frame linked to the end-effector and a pin-point target, the position of which is known at each time. Consider this application for a very simple plane robot with two rotational joints (Fig. 3.1).

If $x_r(t) = \begin{bmatrix} x_{r1}(t) \\ x_{r2}(t) \end{bmatrix}$ is the coordinate vector of the target and $x(q) = \begin{bmatrix} x_1(q) \\ x_2(q) \end{bmatrix}$ the coordinate vector of the origin O_3 of the frame linked to

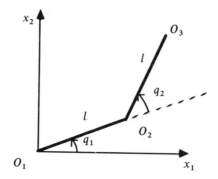

Fig. 3.1 A plane robot with two rotational joints

the end-effector, a straightforward computation gives

$$x_1(q) = lc_1 + lc_{12} \atop x_2(q) = ls_1 + ls_{12} \qquad (3.1.4)$$

where c_i stands for $\cos q_i$, s_i stands for $\sin q_i$, c_{12} for $\cos(q_1 + q_2)$ and s_{12} for $\sin(q_1 + q_2)$. This type of shorthand notation will be used throughout the book.

A necessary condition for minimizing the distance between the end-effector and the target is that the gradient, relative to q, of the square of the distance $h(q, t) = \frac{1}{2}(x_1 - x_{r1})^2 + (x_2 - x_{r2})^2$ is equal to zero. Since

$$e(q, t) = \frac{\partial h}{\partial q} = \begin{bmatrix} x_1(x_2 - x_{r2}) - x_2(x_1 - x_{r1}) \\ lc_{12}(x_2 - x_{r2}) - ls_{12}(x_1 - x_{r1}) \end{bmatrix} \qquad (3.1.5)$$

it is easy to check that this gradient is zero in two situations:

(1) $x = x_r$: tracking is perfect;

(2) O_1, O_2, O_3 and the target point are on the same straight line.

If the target moves in the working space of the robot, perfect tracking is realizable, whereas the other solutions are not satisfactory and correspond to points where $h(q, t)$ is maximal or stationary. Although many robotic tasks are naturally specified as minimizing (or maximizing) a function of the joint positions, translation into an output regulation control problem may lead to ambiguous descriptions. More generally, among the solutions of the equation $e(q, t) = 0$, some may not correspond to the achievement of a desired task. Choosing an initial condition may be a way to force a correct choice.

Let us now consider the problem of minimizing the distance from the end-effector to the target when the target is outside the attainable space of the robot. Perfect tracking is then impossible and only one solution among the

3.1 AN INFORMAL APPROACH

second set of solutions of grad $(h(q, t)) = 0$ is satisfactory. The corresponding task may be understood as a pointing task.

Consider again our simple robot and the same application with a target moving uniformly along a straight line orthogonal to the boundary of the attainable space (the x_1 positive axis for example). It seems natural to split this application into two tasks: perfect tracking as long as the target is in the attainable space, and pointing when the target goes outside the attainable space.

Possible task functions are, respectively,

$$e_1(q, t) = \begin{bmatrix} x_{r1}(r) - l(c_1 + c_{12}) \\ l(s_1 + s_{12}) \end{bmatrix}, \qquad (3.1.6)$$

$$e_2(q, t) = \begin{bmatrix} q_1 \\ q_2 \end{bmatrix}, \qquad (3.1.7)$$

where $x_{r1}(r)$ is the position of the target on the x_1 axis. The different trajectories of the joint variables are shown in Fig. 3.2. Two trajectories q_{i1} and q_{i2} are possible depending on the initial condition q_0. Let us compute the joint velocities, assuming perfect realization of each task. Knowing that $e_1(q, t) = 0$ is equivalent to

$$2q_1 + q_2 = 0, \qquad q_1 = \pm \arccos\left(\frac{x_{r1}(t)}{2l}\right) \qquad (3.1.8)$$

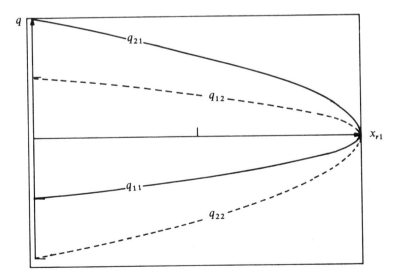

Fig. 3.2 Ideal trajectories

the required joint velocities are:

(1) inside the attainable space

$$\dot{q}_1 = \frac{V}{-2ls_1}, \qquad \dot{q}_2 = -2\dot{q}_1 \qquad (3.1.9)$$

where V is the constant velocity of the target along the x axis;

(2) outside the attainable space:

$$\dot{q}_1 = \dot{q}_2 = 0. \qquad (3.1.10)$$

When the target reaches the boundary of the attainable space from inside, we get

$$\lim \dot{q}_1 = \lim_{q_1 \to 0} \frac{V}{-2ls_1} = \pm \infty. \qquad (3.1.11)$$

This shows that it is physically impossible to achieve the objective exactly, since it would require infinite joint velocities immediately followed by zero joint velocities. A way of overcoming this difficulty would be to modify the application by abandoning perfect tracking of the target near the boundary. With this less restrictive specification of the application it may be possible to produce a mathematical description involving only one task. This example shows that the existence of a solution $q_r(t)$ to the equation $e(q, t) = 0$ for each t is far from sufficient to give a well-posed regulation problem in robotics. It is also necessary to have a smooth enough corresponding trajectory (at least C^2 for mechanical reasons).

All the regularity conditions considered in this section and the previous ones will be gathered into the concept of ρ-admissible function tasks. Intuitively, a robotic control problem is well-posed if the corresponding output function results in a feasible task and a smooth state trajectory. This does not mean that it is impossible to find a good control law when the task function is not ρ-admissible, but the problem is still open in the general case.

Remark. Dimension of the Output Space. So far $e(q, t)$ is a vector in a vector space $\mathbf{R}^n$. However, it may happen that a p-dimensional output vector function aimed to describe the task spans a differential manifold with dimension smaller than p. For example, consider the task of control of the attitude of the end-effector of a manipulator. This can be done by taking a unitary quaternion $\lambda(q)$ representing the attitude of the effector and a reference path $\lambda_r(t)$ in the set of unitary quaternions. Choose $f(q,t) = \lambda_r(t)\bar{\lambda}(q) - (1,0,0,0)$ so that $f(q,t) = 0$ if and only if $\lambda(q(t)) = \lambda_r(t)$. We know from section 1.2.3 that $\lambda_r(t)\bar{\lambda}(q)$ is a unitary quaternion. So the values of $f(q,t)$ belong to the unit sphere S^3. Even though $f(q,t)$ is written as an $\mathbf{R}^4$ vector, its range is a manifold of dimension lower than 4. Therefore,

3.2 p-ADMISSIBLE FUNCTIONS AND TASKS

$f(q, t)$ is not a suitable task function, and it is necessary to change it, which is generally possible. For example, when the differential manifold containing the range of the task function is independent of t as in the above case, taking a chart ϕ of the manifold allows us to obtain a new output $\phi(e(q, t))$ with the desired property.

In the general situation, a theorem like the Rank Theorem 0.4 may apply to transform the output function into a task function with the desired properties.

3.2 p-ADMISSIBLE FUNCTIONS AND TASKS

Now, let us keep in mind that we would like to reduce the definition of a task to the problem of maintaining a task function $e(q, t)$ as close to zero as possible. We are now going to define sufficiency properties of the task function for the control problem to be well-posed.

3.2.1 p-Admissible functions

3.2.1.1 A basic definition

Let $e(q, t)$ be a vector application of class C^k, $k \geq 1$, from an open subset Ω of $\mathbf{R}^n \times \mathbf{R}$ to $\mathbf{R}^n$.

Definition D3.1 $e(q, t)$ *is a p-admissible function on the set $C_{p,T}$ during the time interval $[0, T]$ if and only if the function $F(q, t) = (e(q, t), t)$ is a C^k class-diffeomorphism (i.e. a C^k class-bijection, the reciprocal of which is also C^k) from $C_{p,T}$ onto the closed sphere $B_p \times [0, T]$ centred at 0, of radius p.*

This property will be referred to as the p-admissibility property.

3.2.1.2 Some comments

The interest of this diffeomorphism lies in the coordinate transformation $(q, t) \leftrightarrow (e, t)$, where (q, t) is a point of $C_{p,T}$, and e a point of B_p. If $(e, t) \to (\phi_1(e, t), \phi_2(e, t))$ is the inverse function, since

$$\phi_1(e(q, t), t) = q; \quad \phi_2(e(q, t), t) = t \quad (3.2.1)$$

and F is surjective, for all (e, t) we necessarily have $\phi_2(e, t) = t$. Hence $e \to \phi_1(e, t)$ is, for each t, the C^k inverse of $q \to e(q, t)$ and the section of $C_{p,T}$, $C_{p,T} \cap \Omega \times \{t\}$ is diffeomorphic to B_p. We have also shown that the canonical projection of Ω onto the time coordinate contains $[0, T]$.

The admissibility property justifies the following assertion: given an initial point $(q_0, 0)$ of $C_{p,T}$, and a C^k-path $x(t)$, $t \in [0, T]$, belonging to B_p, and such that $x(0) = e(q_0, 0)$, there exists a *unique* C^k-path $q(t)$ such that, $q(0) = q_0$ and, for every $t \in [0, T]$, $e(q(t), t) = x(t)$.

Moreover, if $e_1(t)$ and $e_2(t)$ are any two C^k-paths in B_ρ, then the two reciprocal paths $q_1(t)$ and $q_2(t)$ in the projection of $C_{\rho, T}$ on $\mathbf{R}^n$ are such that

$$\|q_1(t) - q_2(t)\| \leq \beta \|e_1(t) - e_2(t)\|, \quad \beta > 0, \tag{3.2.2}$$

i.e. F^{-1} is Lipschitz on $B_\rho \times [0, T]$. The constant β depends only on e and T. This means that if the paths $e_1(t)$ and $e_2(t)$ are close (in the meaning of the distance defined by the sup norm on every compact set for example) then the paths $q_1(t)$ and $q_2(t)$ are also close.

A trivial consequence of the admissibility property is that, if $e(q, t)$ is ρ-admissible on $C_{\rho, T}$ during $[0, T]$ it is also ρ-admissible on $C_{\rho', T'}$ during $[0, T']$, with

$$\rho' \leq \rho, \quad T' \leq T, \quad C_{\rho', T'} \subset C_{\rho, T}.$$

It should be noted that the set $C_{\rho, T}$ is one of the connected components of the set $F^{-1}(B_\rho \times [0, T]) = e^{-1}(B_\rho) \cap \mathbf{R}^n \times [0, T]$.

Let $C'_{\rho, T}$ be another connected component of $F^{-1}(B_\rho \times [0, T])$; the fact that $e(q, t)$ is ρ-admissible on $C_{\rho, T}$ during $[0, T]$ *does not mean* that $e(q, t)$ is ρ-admissible on $C'_{\rho, T}$ during $[0, T]$. A counter-example is $e(q, t) = (2 - t)q^2 \sin(1/q)$ with $T = 1$. Figure 3.3 shows the intersections of two connected components $C_{\rho, T}$ and $C'_{\rho, T}$ with the planes $\{t = 0\}$

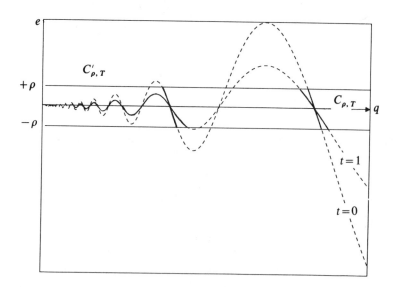

Fig. 3.3 Admissible and non-admissible sets

3.2 p-ADMISSIBLE FUNCTIONS AND TASKS

and $\{t = 1\}$. Although the fibre of $C'_{\rho, T}$ over $t = 0$ is homeomorphic to $B_\rho = [-\rho, +\rho]$, the fibre over $t = 1$ is no more homeomorphic to B_ρ since the restriction of e to it is not one-to-one.

However, if, during $[0, T]$ $e(q, t)$ is:

(1) ρ-admissible on $C_{\rho, T}$;
(2) ρ'-admissible on $C'_{\rho', T}$,

with $\rho' > \rho$ and $C_{\rho, T} \cap C'_{\rho', T} = \emptyset$, we shall say that $e(q, t)$ is *more admissible* on the connected set $C'_{\rho', T}$ than on $C_{\rho, T}$ during $[0, T]$.

Finally, T may be taken to $+\infty$ in the previous definition.

3.2.1.3 A second formulation of the p-admissibility property

Theorem 3.1 *If* $T < +\infty$ *the ρ-admissibility property is equivalent to the following assertions*:

1. $C_{\rho, T}$ *is a non-empty arcwise connected component of* $F^{-1}(B_\rho \times [0, T])$.

2. $B_y(C_{\rho, T}) \cap B_y(\Omega) = \emptyset$ (*i.e.* $C_{\rho, T}$ *is a closed part of* $\mathbf{R}^n \times \mathbf{R}$), *where* $B_y(.)$ *means boundary*.

3. *For every* (q, t) *of* $C_{\rho, T}$:

   (a) $\det\left(\dfrac{\partial e}{\partial q}(q, t)\right) \neq 0$;

   (b) $\left\|\left(\dfrac{\partial e}{\partial q}(q, t)\right)^{-1}\right\| < m_{\rho, T} < \infty$;

   (c) $\left\|\dfrac{\partial e}{\partial t}(q, t)\right\| < m'_{\rho, T} < \infty$;

where $\|.\|$ denotes the Euclidean norm or the associated matrix norm (spectral norm).

If $T = +\infty$ *the above conditions are only sufficient*.

Proof. The necessity of the conditions is obvious if $T < +\infty$ since $C_{\rho, T}$ is compact and e is C^1 at least.

Sufficiency: The property (3a) implies that the Jacobian matrix of $F(q, t)$ is invertible in any point X_0 of $C_{\rho, T}$, since

$$\frac{\partial F}{\partial X}(X_0) = \begin{pmatrix} \dfrac{\partial e}{\partial q}(X_0) & \dfrac{\partial e}{\partial t}(X_0) \\ 0 & 1 \end{pmatrix}, \quad (X = (q, t)). \qquad (3.2.3)$$

Thus

$$\det\left(\frac{\partial F}{\partial X}(X_0)\right) = \det\left(\frac{\partial e}{\partial q}(X_0)\right). \quad (3.2.4)$$

Moreover, it is easy to verify that the inverse of $(\partial F/\partial X)(X_0)$ is

$$\left[\frac{\partial F}{\partial X}(X_0)\right]^{-1} = \begin{bmatrix} \left[\dfrac{\partial e}{\partial q}(X_0)\right]^{-1} & -\left[\dfrac{\partial e}{\partial q}(X_0)\right]^{-1}\dfrac{\partial e}{\partial t}(X_0) \\ 0 & 1 \end{bmatrix}. \quad (3.2.5)$$

Hence the assertions (3b) and (3c) imply that the inverse of the Jacobian matrix of $F(q, t)$ is bounded over $C_{\rho, T}$. Since B_ρ is arcwise and simply connected and $C_{\rho, T}$ has the desired properties, we can apply Theorem 0.5 and thus complete the proof.

The following example shows that the conditions in the above theorem are not necessary if $T = +\infty$.

Example 3.1 Take $e(q, t) = (q - t)/(t + 1)$; $q_0 = 0$ with $\Omega = \mathbf{R} \times \mathbf{R}^+$. Then $F(q, t) = ((q - t)/(t + 1), t)$ has $\mathbf{R} \times \mathbf{R}^+$ as range. Now if (e, t) is an element of $\mathbf{R} \times \mathbf{R}^+$, a direct computation shows $F^{-1}(e, t) = (t + (t + 1)e, t)$. So F is a bijection from Ω onto $\mathbf{R} \times \mathbf{R}^+$. It is easy to check that F and F^{-1} are C^∞ and then are diffeomorphisms on their domains. So $e(q, t)$ is ∞-admissible ($\rho = \infty$) during $[0, \infty[$. Nevertheless, $(\partial e/\partial q)^{-1} = t + 1$ is not bounded on $C_{\infty, \infty} = \mathbf{R} \times \mathbf{R}^+$.

3.2.2 Admissible tasks

Let us suppose that a preliminary step has led to the definition of the task as a regulation problem of $e(q, t)$ around zero; $e(q, t)$ can be considered as an n-dimensional error vector. It is assumed that its value is known at each time, either by direct measurement or by estimation.

In the most general case, $e(q, t)$ depends both on the n internal joint positions, q, of the robot and on external variables, represented by the independent time variable t. The aim of the controller then is to keep the vector $e(q(t), t)$ as close as possible to zero, during the task execution time $[0, T]$.

We are now able to define precisely the considerations developed in the first part of this chapter.

Definition D3.2 (ideal trajectory) $\{q_r(t)\}$ *is an ideal trajectory for the task* $\{e(q, t), q_0\}$ *if* $e(q_r(t), t) = 0$ *for every* $t \in [0, T]$ *and* $q_r(0) = q_0$.

3.2 p-ADMISSIBLE FUNCTIONS AND TASKS

For physical reasons, owing to the second-order nature of mechanical equations, we must add a restriction to the concept of trajectory.

Definition D3.3 (realizable task) *A robotic task* $\{e(q, t), q_0\}$ *is realizable if there exists a twice-differentiable ideal trajectory* $q_r(t)$ *on* $[0, T]$.

Then there exists an ideal control torque function $\Gamma_r(t)$ which allows the robot to track the realizable trajectory exactly; using the model presented in Section 2.1 of Chapter 2, $\Gamma_r(t)$ is, for example, the ideal open-loop control scheme:

$$\Gamma_r(t) = M(q_r(t))\ddot{q}_r(t) + N(q_r(t), \dot{q}_r(t)), \quad t \in [0, T]. \quad (3.2.6)$$

Definition D3.4 (weakly admissible task) *A task* $\{e(q, t), q_0\}$ *is said to be weakly admissible on* $[0, T]$ *if it is realizable and the associated trajectory* $q_r(t)$ *is unique.*

The assumption of uniqueness is required in order to ensure repeatability of the motion in practice and the uniqueness of the ideal control.

3.2.3 p-Admissible tasks

As seen previously, a first requirement for obtaining a meaningful task function, i.e. such that there exists a unique ideal control yielding its realization, is the weak admissibility property.

This condition is unfortunately not sufficient to ensure well-conditioned properties for the regulation problem: indeed, $\Gamma_r(t)$ is not practically accessible for many reasons: modelling errors, disturbances, technological limitations, etc., even if the ideal trajectory is available. Moreover, in most cases $q_r(t)$ cannot in practice be computed despite its proven existence. This means that in practice the best expected behaviour would be such that $e(q, t)$ is kept *small* instead of zero. Furthermore, smallness of $e(q, t)$ is not yet a sufficient condition on its own; additionally we should like to ensure that:

(1) the actual trajectory of the robot stays close to the ideal;

(2) convergence of $e(q, t)$ to zero implies convergence of q to the ideal trajectory $q_r(t)$;

(3) a slow variation in $e(q, t)$ does not involve a very fast variation in $q(t)$.

Suppose that $e(q, t)$ is of class C^2 everywhere that it is defined. Then

$$\dot{e}(q, t) = \frac{\partial e}{\partial q}(q(t), t) \dot{q}(t) + \frac{\partial e}{\partial t}(q(t), t). \quad (3.2.7)$$

and thus

$$\|\dot{q}(t)\| \leq \left\|\frac{\partial e}{\partial q}(q(t), t)^{-1}\right\| \left(\|\dot{e}(t)\| + \left\|\frac{\partial e}{\partial t}(q(t), t)\right\|\right). \quad (3.2.8)$$

This shows that, in order for a slow variation in $e(t)$ ($\|\dot{e}(t)\|$ bounded) to correspond to a slow variation in $q(t)$ ($\|\dot{q}(t)\|$ bounded), it is sufficient that $(\partial e/\partial q)(q(t), t)$ is non-singular with bounded inverse, and that $(\partial e/\partial t)(q(t), t)$ is bounded. If one of these conditions is violated, it is always possible to find an output trajectory $e(q, t)$ such that $\|\dot{q}(t)\|$ becomes unbounded.

In this sense, these conditions are necessary to obtain a well-conditioned control problem. Recalling that the same conditions characterize the ρ-admissible functions of section 3.2.1, we may finally connect the idea of a well-conditioned task with the property of ρ-admissibility of $e(q, t)$ through the following definition.

Definition D3.5 *A task $\{e(q, t); q_0)\}$ is said to be ρ-admissible during $[0, T]$ if there exists a set $C_{\rho, T}$, containing $(q_0, 0)$ such that $e(q, t)$ is of class C^2 and is a ρ-admissible function during $[0, T]$ on the set $C_{\rho, T}$.*

The set of ρ such that the task is ρ-admissible, if not empty, is clearly a positive interval containing 0. This justifies the following definition.

Definition D3.6 *The upper bound ρ_M of the values of ρ for which the task is ρ-admissible on $[0, T]$ is the admissibility radius of the task during $[0, T]$.*

As long as $e(t)$ stays inside a sphere $B(0, \rho)$, with $\rho < \rho_M$ the convergence of $e(t)$ to zero at any time involves the convergence of $q(t)$ to $q_r(t)$. If, at a given time, $\|e(t)\| > \rho_M$, a risk does occur of being unable to bring $q(t)$ back to $q_r(t)$. The admissibility radius thus determines in a way the degree of *robustness* of the task from the controls point of view. The robustness problem of *control* itself comes later, and has no meaning when the task is not admissible.

Lemma 3.1 *A 0-admissible task is weakly admissible.*

The $C_{0, T}$ diffeomorphic to $\{0\} \times [0, T]$ is clearly the graph of the unique C^2 ideal trajectory of the task. We shall use these two concepts interchangeably.

3.2.3.1 Some Simple Examples

Example 3.2 Let $\Omega = \mathbf{R} \times \mathbf{R}^+$ and $e(q, t) = q(t-1)$, $q_0 = 0$. $C_{0, T} = F^{-1}(\{0\} \times \mathbf{R}^+)$ is as shown in Fig. 3.4. The set $\{q = 0\}$ is the unique C^2 ideal trajectory of the task but $C_{0, T}$ is not diffeomorphic to $\{0\} \times \mathbf{R}^+$. Since

3.2 p-ADMISSIBLE FUNCTIONS AND TASKS

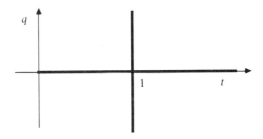

Fig. 3.4 An isolated singularity

$(\partial e/\partial q) = t - 1$, property (3) of Theorem 3.1 is not verified because $(\partial e/\partial q)(q, 1)$ is not invertible. This example shows that a task may be weakly admissible without being admissible.

Example 3.3 Let us again take the example of the two-rotational robot given in section 3.1.5, with the following supplementary elements:

$$\left.\begin{array}{l}\Omega = \mathbf{R}^2 \times \mathbf{R}; \quad q^T = (q_1 \quad q_2)^T \\ e(q, t) = x(q) - x_r(t)\end{array}\right\} \quad (3.2.9)$$

where

$$x(q) = \begin{pmatrix} l(s_1 + s_{12}) \\ l(c_1 + c_{12}) \end{pmatrix} \quad (3.2.10)$$

and $x_r(t)$ is a two-dimensional C^2-function with bounded derivatives. Then

$$\frac{\partial e}{\partial q}(q, t) = \frac{\partial x}{\partial q}(q) = \begin{pmatrix} l(c_1 + c_{12}) & lc_{12} \\ -l(s_1 + s_{12}) & -ls_{12} \end{pmatrix} \quad (3.2.11)$$

which yields

$$\det\left(\frac{\partial e}{\partial q}(q, t)\right) = -l^2 s_2 \quad (3.2.12)$$

which vanishes if and only if $q_2 = k\pi$, $k \in Z$.

In order to have 0-admissibility of $e(q, t)$ during $[0, T]$, it is necessary that the end-effector, represented by the O_2 point, avoids during $[0, T]$ the points x_r^* such that the solutions q^* of

$$x(q^*) - x_r^* = 0 \quad (3.2.13)$$

are of the form $q^{*T} = (q_1^* \quad k\pi)^T$.

This imposes on O_2 avoidance of the origin O_0 and the circle $C(O_0, 2l)$. In the same way, one may easily verify that ρ-admissibility of $e(q, t)$ during $[0, T]$ is obtained if the trajectory $O_2(t)$ lies inside the ring defined by the circles $C_1(O_0, \rho)$ and $C_2(O_0, 2l - \rho)$.

A necessary and sufficient condition for ρ-admissibility is then

$$\rho < \|x_r(t)\| < 2l - \rho \quad \text{for every } t \in [0, T]. \tag{3.2.14}$$

This relation shows that in such a case, the admissibility radius (i.e. the size of B_ρ) depends on $x_r(t)$, and that this radius is bounded from above by l.

To conclude this section, we should like to emphasize the intuitive idea that task admissibility and ease of completing the task are connected: the more admissible a task is (size of $B(0, \rho_M)$), the less constrained is the control problem. Of course, this is a general feeling, and we shall have to go into details on this assertion in subsequent chapters. However, we may here make the following remarks, which derive from a little commonsense:

1. If the same task (i.e. leading to the same ideal trajectory) may be described by two functions $e_1(q, t)$ and $e_2(q, t)$, the most admissible one should be preferred. For example, among two parametrizations of the attitude of the end-effector, one should use the one for which the desired trajectory is the furthest from the singular points of the parametrization.

2. For robustness reasons, it is sometimes preferable to replace a task that exactly represents the user's wishes by a slightly different one if it is much more admissible.

3. Obviously, in order to avoid any subsequent control problem, it is recommended that the admissibility of a task is checked in advance. However, this is not always possible. A typical case arises when the ideal trajectory to be tracked is generated on-line, for example from sensory information. In this case, it is necessary to monitor the admissibility of the task continuously and to switch from the current control to execute 'emergency' tasks in case of insufficient admissibility.

3.3 NON-ADMISSIBILITY AND TASK SINGULARITIES

3.3.1 Introduction

If we look at admissibility during a bounded time interval $[0, T]$, it is easy to see that most of the work of verifying the admissibility property is done once it is proved that the connected component of $F^{-1}(B_\rho \times [0, T])$ containing $(q_0, 0)$ is compact in $\mathbf{R}^n \times \mathbf{R}$. In this case conditions (3b) and (3c) of Theorem 3.1 are automatically satisfied and the only property which may be missing is the non-singularity of $(\partial e/\partial q)(q, t)$.

When $C_{\rho, T}$ is not bounded, it is useless to go further, since the compactness of $C_{\rho, T}$ is a necessary condition for ρ-admissibility. The following example illustrates this case.

3.3 NON-ADMISSIBILITY AND TASK SINGULARITIES 93

Example 3.4 $e(q, t) = q/(q - t - 1)$; $\rho = 1$. It is easy to show that $C_{1, T}$ is as in Fig. 3.5. $C_{1, T}$ is not bounded, so the function e is not admissible. A straightforward computation leads to

$$\left(\frac{\partial e}{\partial q}\right)^{-1} = \frac{(q - t - 1)^2}{-(t + 1)} \qquad (3.3.1)$$

so condition (3b) of Theorem 3.1 also fails.

The other way $C_{\rho, T}$ fails to be a compact subset of $\mathbf{R}^n$ is when it is not a closed subset of $\mathbf{R}^n$. It is always a closed subset of the domain Ω of $e(q, t)$ and fails to be a closed subset of $\mathbf{R}^n$ essentially when the boundary of $C_{\rho, T}$ as a subset of $\mathbf{R}^n$ and the boundary of Ω have some point in common. This may happen when $e(q, t)$ is a function which is not defined for all q in the joint space.

A typical case is when one wants to define a tracking task for the attitude of the end-effector using Euler angles. If $(\theta(q), \phi(q), \psi(q))$ are the three Euler angles and $(\theta_r(t), \phi_r(t), \psi_r(t))$ is the desired trajectory, a reasonable task function would be

$$e(q, t) = \begin{bmatrix} e_1(q, t) \\ e_2(q, t) \end{bmatrix} \begin{matrix} \downarrow 3 \\ \downarrow n - 3 \end{matrix}, \qquad (3.3.2)$$

with

$$e_1(q, t) = (\theta(q) - \theta_r(t), \phi(q) - \phi_r(t), \psi(q) - \psi_r(t))^T, \qquad (3.3.3)$$

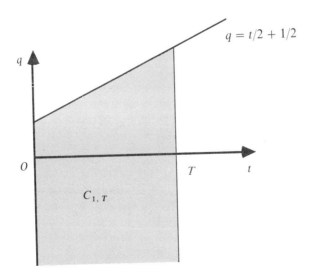

Fig. 3.5 A non 1-admissible set

and where $e_2(q, t)$ represents additional constraints which we do not need to specify here.

But it is well known that Euler angles are not defined if the z axes of the two frames are colinear. So $e(q, t)$ is not defined for all q and t and the values of (q, t) for which $e(q, t)$ is not defined must be excluded from Ω.

Let us examine this case more closely. Suppose that the desired trajectory $t \to (\theta_r(t), \phi_r(t), \psi_r(t))$ is such that for some t_1, $|\theta_r(t_1)|$ is small. With $\rho \geq |\theta_r(t_1)|$, B_ρ contains points such that $\theta = 0$. These points are not reached by the mapping $q \to e(q, t_1)$ so the task is not ρ-admissible. The points of the joint space such that $\theta(q) = 0$ are on the boundary of Ω and also on the boundary of $C_{\rho, T}$. This shows that the task may be "little"-admissible, even if $\theta_r(t) \neq 0$ for all t in $[0, T]$, and the ideal trajectory exists. The admissibility radius depends on how close $\theta_r(t)$ approaches to zero.

3.3.2 Formal definitions

A necessary and sufficient condition for a task $\{e(q, t); q_0\}$ to be admissible for some $\rho > 0$ is that it is 0-admissible. Moreover, if $C_{0, T}$ is the connected component of $F^{-1}(\{0\} \times [0, T])$ containing $(q_0, 0)$, it contains all the possible ideal trajectories $(q_r(t), t)$ of the given task. We have seen (Example 3.2) that a task may be weakly admissible without being admissible. So $C_{0, T}$ may contain a unique C^2 trajectory despite the fact that the task is not 0-admissible. Nonetheless if a task is 0-admissible, $C_{0, T}$ coincides with the corresponding C^2 ideal trajectory and is a compact set in $\mathbf{R}^n$.

For the time being we shall consider that the output function is defined on an open set Ω of $\mathbf{R}^n \times \mathbf{R}$.

Definition D3.7 *A point (q_s, t_s) in Ω is a Jacobian singularity if the matrix $(\partial e/\partial q)(q_s, t_s)$ is singular.*

The Jacobian singularities are the main source of control problems and the major obstacles to admissibility. The following theorem makes this remark more explicit.

Theorem 3.2 *Let $\{e(q, t); q_0)\}$ be a task admitting an ideal trajectory $q_r(t)$ defined on $[0, T]$. If $e(q, t)$ is C^2 and the ideal trajectory has no Jacobian singularity then the task is 0-admissible, and if $T < +\infty$ there exists $\rho > 0$ such that the task is ρ-admissible on $[0, T]$.*

Proof The existence of an ideal trajectory implies that the equation $e(q, t) = 0$ has a solution for all t. Let $t_1 \in [0, T]$. Since $(\partial e/\partial q)(q_r(t_1), t_1)$ is invertible, by the Implicit Function Theorem there exists a unique function

3.3 NON-ADMISSIBILITY AND TASK SINGULARITIES

ϕ defined around t_1 such that

$$\left.\begin{array}{l}\phi(t_1) = q_r(t_1) \\ e(\phi(t), t) = 0\end{array}\right\} \quad (3.3.4)$$

So $\phi(t) = q_r(t)$. This implies:

1. The ideal trajectory is unique.

2. Since $e(q, t)$ is C^2, the function ϕ is also C^2.

3. The two sets $\{(q_r(t), t) | t \in [0, T]\}$ and $\{0\} \times [0, T]$ are diffeomorphic by F. Hence the task is 0-admissible.

Now suppose $T < +\infty$. $C_{0,T} = \{(q_r(t), t) | t \in [0, T]\}$ is a compact set. By the local inverse theorem, for each t there exist ρ_t, ε_t and a neighbourhood V_t of $(q_r(t), t)$ such that F is a C^2 diffeomorphism from V_t to $B_{\rho_t} \times [t - \varepsilon_t, t + \varepsilon_t]$. Since $C_{0,T}$ is compact, it is possible to choose a finite number of points t_i; $i = 1, \ldots, l$ such that $[0, T]$ is included in $\cup [t_i - \varepsilon_i, t_i + \varepsilon_i]$ and $C_{0,T}$ is included in $\bigcup_{i=1}^{l} V_{t_i}$. Let $\rho = \inf \rho_{t_i}; i = 1, \ldots, l$. It is then easy to verify that F is a diffeomorphism from $F^{-1}\left((B_\rho \times [0, T]) \cap \left(\bigcup_{i=1}^{l} V_{t_i}\right)\right)$ onto $B_\rho \times [0, T]$.

Remark. Theorem 3.1 is a very powerful theorem. It states in particular that the equation $e(q, t) = 0$ has a solution for all $t \in [0, T]$ if $e(q, t)$ has the required properties (1)–(3) and if there exists a solution q_0 for $t = 0$. It is a global implicit function theorem in the sense that it proves the existence of the function $q_r(t)$ on a domain given in advance. In practice, it is often possible to directly verify the existence of an ideal trajectory. We have then to look for Jacobian singularities. Nevertheless, Theorem 3.1 is useful for proving ρ-admissibility ($\rho > 0$) for a given ρ and/or for $T = \infty$. It is also a means of checking if the control problem, posed as an output regulation problem, is correctly modelled.

3.3.3 Examples of singularities in robotics

The earliest singularities to be described in robotics were kinematic singularities. The output function, which is implicitly considered in this case, involves the location (position/attitude) of the end-effector. We have already encountered this type of singularity in the inverse kinematics problem. They occur at points where the inverse kinematics problem has an infinite number of solutions, showing the existence of Jacobian singularities at these points. Let us now look again at these singularities from a control point of view. In order to keep the computational complexity to a manageable level, let us examine separately the singularities occurring in the positioning of the end-effector and those occurring in its attitude. The practical importance of the Jacobian of a manipulator comes, at first, from the fact that for all positioning and trajectory following problems a natural task function is $e(q, t) = x(q) - x_d(t)$ where $x(q)$ represents the position/attitude of the end-effector and $x_d(t)$ the desired trajectory. Then

$$\frac{\partial e}{\partial q}(q, t) = \frac{\partial x}{\partial q}(q)$$

and, when a good parametrization of the attitude is chosen, the points where $\partial e/\partial q$ is singular are precisely the points where the Jacobian of the manipulator is singular. For that reason we shall focus on the kinematic singularities of the manipulator in the following examples.

3.3.3.1 Examples of positioning singularities

Consider a manipulator with the first three joints as in Fig. 2.1 and a three-degrees-of-freedom wrist with concurrent axes (Fig. 3.6). If we suppose that the centre of the wrist is the extremity O_4 of the third link, the position of the end-effector depends only on the first three degrees of freedom. We already know that the points of the z_0 axis are singular positions for O_4. In order to determine all the Jacobian singularities, let us compute the Jacobian matrix of the manipulator. The position function is

$$x(q) = c_1(c_2 + c_{23})$$
$$y(q) = s_1(c_2 + c_{23}) \quad (3.3.5)$$
$$z(q) = s_2 + s_{23}$$

with $q = (q_1, q_2, q_3)^T$. The Jacobian matrix of the position function is then

$$J(q) = \begin{bmatrix} -s_1(c_2 + c_{23}) & -c_1(s_2 + s_{23}) & -c_1 s_{23} \\ c_1(c_2 + c_{23}) & -s_1(s_2 + s_{23}) & -s_1 s_{23} \\ 0 & c_2 + c_{23} & c_{23} \end{bmatrix} \quad (3.3.6)$$

with determinant

$$\det(J(q)) = -(c_2 + c_{23})s_3. \quad (3.3.7)$$

Hence the singular points are:

$$2q_2 + q_3 = \pi \quad \text{or } q_3 = 0 \quad \text{or } q_3 = \pm \pi. \quad (3.3.8)$$

The first set of points corresponds to the z_0 axis and the second to positions where links 2 and 3 are aligned and the position of O_4 is on the boundary of

3.3 NON-ADMISSIBILITY AND TASK SINGULARITIES

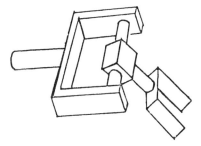

Fig. 3.6 A wrist with three concurrent axes, or Cardan's shaft

the attainable space. The value $q_3 = \pm \pi$ is less realistic, so it will not be considered further.

Other manipulator geometries present similar singularities except for manipulators where the position of the origin of the end-effector is obtained by means of three orthogonal translations. For manipulators with at least one rotational joint, singularities appear whenever the origin of the end-effector is positioned on the axis of a rotational joint.

If we consider now an attainable point on the z_0 axis, the Jacobian matrix of the position function is then

$$J(q) = \begin{bmatrix} 0 & -2c_1 s_2 & -c_1 s_2 \\ 0 & -2s_1 s_2 & -s_1 s_2 \\ 0 & 0 & c_2 \end{bmatrix} \qquad (3.3.9)$$

showing that $(\dot{q}_1, 0, 0)^T$ is in the null-space of the Jacobian: as already noted, the position is independent of q_1. The admissible velocities of O_4 are in the range of $J(q)$ which is of dimension 2. It is easy to verify that these velocities are in the plane defined by the links C_2 and C_3.

The second kind of singularity corresponding to $q_3 = 0$ is different in nature from the one examined above. The first thing to be noticed is that, for every point on the boundary of the attainable space, there exist only two points $(q_1, q_2, 0)$ and $(q_1 + \pi, \pi - q_2, 0)$ in the joint coordinates corresponding to that position of O_4. The second difference is that the set of attainable points around a boundary point is not an open set as it was for points of the z_0 axis inside the attainable domain.

It would be very useful to distinguish between these two kinds of singularity for general output functions $e(q, t)$. Second-order derivatives give some information on the kind of singularity encountered. Unfortunately, mixed situations occur, as in our example for the point O_4 corresponding to $q_2 = q_3 = 0$. The information given by the second derivatives is then more difficult to interpret.

3.3.3.2 Singularities of the attitude function

A very common wrist model is that involving the three classical Euler angles. This mechanical configuration goes back to Cardan's shaft (Fig. 3.6). As it is a realization of the Euler chart, it necessarily has singular points. Recall that, for topological reasons, it is impossible to mechanically realize a wrist with three degrees of freedom without a singular point. Of course, mechanical bounds on the joints are not taken into account. In order to examine the singularities of this wrist, let us compute the angular velocity of the frame (O, x, y, z) in the basis associated with that frame,

$$\omega(q) = \dot{q}_1 z_0 + \dot{q}_2 u + \dot{q}_3 z \quad (3.3.10)$$

from which we get the Jacobian matrix

$$J(q) = \begin{bmatrix} s_2 s_3 & c_3 & 0 \\ s_2 c_3 & -s_3 & 0 \\ c_2 & 0 & 1 \end{bmatrix} \quad (3.3.11)$$

and the Jacobian determinant

$$\det(J(q)) = -s_2. \quad (3.3.12)$$

The singular points are the points for which $q_2 = 0$ or $q_2 = \pi$. They correspond to the situation where the Euler angles are not uniquely defined (cf. section 1.2.2). Again we get singular points for which the output function is not one-to-one. At a singular point the Jacobian matrix is

$$J(q) = \begin{bmatrix} 0 & c_3 & 0 \\ 0 & -s_3 & 0 \\ \pm 1 & 0 & 1 \end{bmatrix}. \quad (3.3.13)$$

No angular velocity collinear with $\begin{bmatrix} s_3 \\ c_3 \\ 0 \end{bmatrix}$ is attainable.

3.3.3.3 Control problems with singularities

In the above sections on admissibility we have reduced every control problem in robotics to a regulation problem. The admissibility condition is a sufficient condition under which the regulation problem is well-posed. Let us now consider some problems where singularities occur. Take again our first example of a manipulator with three degrees of freedom and consider the task of tracking an ideal trajectory or a moving target in the Cartesian space. An

3.3 NON-ADMISSIBILITY AND TASK SINGULARITIES

output function modelling this task is

$$e(q, t) = x(q) - x_d(t), \qquad (3.3.14)$$

where $x(q)$ is the vector of coordinates of O_4 relative to the frame F_0 and $x_d(t)$ is the desired trajectory. We shall suppose from here on that the desired trajectory is realizable, so that an ideal C^2 trajectory $q_r(t)$ exists.

Several cases of task singularities may occur, depending on the desired trajectory $x_d(t)$. The simplest is when an isolated point $q_r(t)$ is a singular point of the manipulator. This occurs when the desired trajectory crosses a set of singular points 'transversally', i.e. as shown in Fig. 3.7. An output feedback, involving in general the inverse of $\partial e/\partial q$, would not work in the vicinity of the singular point. An intuitive solution to the singularity problem is then to replace the task $e(q, t)$ locally by another giving an ideal trajectory close to the initial one. A possible choice is

$$e'(q, t) = q - q'_r(t) \qquad (3.3.15)$$

where $q'_r(t)$ is a function defined on an interval $]t_s - \delta, t_s + \delta[$ in order to extrapolate $q(t)$ over the singularity occurring at t_s. A simple choice would be:

$$q'_r(t) = q(t_s - \delta) + (t - t_s + \delta)\hat{q}_v(t_s - \delta) \qquad (3.3.16)$$

with $\hat{q}_v(t_s - \delta)$ an estimation of the joint velocities at time $t_s - \delta$.

Although $e'(q, t)$ is ∞-admissible, its main drawback lies in the weak reliability of such a prediction. When the initial task becomes admissible again at time $t > t_s + \delta$, there is no guarantee that $q'_r(t)$ belongs to the admissibility domain of $e(q, t)$. Or the error $q'_r(t_s + \delta) - q_r(t_s + \delta)$ may

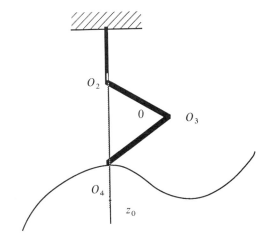

Fig. 3.7 A trajectory transversal to a set of singular points

be of unacceptable size in practice. Moreover, switching from the task $e(q, t)$ to the new task $e'(q, t)$ must be done, for numerical reasons, before time t_s. This supposes, especially when the trajectory is not known in advance, a criterion for detecting the appearance of a singular point. But the main problem is that in general, without further information, the trajectory is completely unpredictable as we shall now see.

Consider the case where the desired trajectory follows the z_0 axis between instants t_1 and t_2. Moreover, suppose it joins the z_0 axis while moving in the plane P_1 and leaves it and then moves in the plane P_2 (Fig. 3.8). Because q_1 is undetermined as long as O_4 is on the z_0 axis, the value of q_1 corresponding to the motion in plane P_2 is unpredictable. If q_1 does not have the correct value when $x_d(t)$ leaves the z_0 axis, it has to change instantaneously, which is physically impossible. The task is not weakly admissible. The situation is similar to that encountered at the beginning of this chapter where we examined the tracking of a moving target leaving the attainable space. In the presence of such singularities, in practice we have to accept a tracking error in the neighbourhood of the singular point. This is a sort of 'wait and see' policy, making possible the necessary adjustments when moving from singular points to non-singular ones. It is not always easy to make it work. If the relative position of the moving target $x_d(t)$ is measured by means of a sensor of limited range, we have also to ensure that the tracking error remains within the range of the sensor.

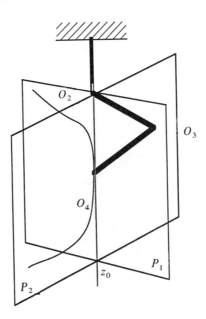

Fig. 3.8 Unpredictable trajectory on leaving a singular domain

3.3 NON-ADMISSIBILITY AND TASK SINGULARITIES

We have chosen the preceding example because it is easy to explain and draw. The same singular control problem may happen with a wrist such as the one described in section 3.3.3.2. Consider a desired trajectory described by the differential equation

$$\frac{dR}{dt} = \tilde{\omega}(t)R(t), \qquad (3.3.17)$$

where $\omega(t)$ is a differentiable vector function such that $\omega(t) = \lambda(t)z_0$, $(\lambda(t) \in \mathbf{R})$ from instant t_1 to instant t_2. The situation is entirely similar to that illustrated in Fig. 3.8. Exactly as in the previous case, q_1 and q_2 are not uniquely determined between t_1 and t_2. They may also have the wrong value when the desired trajectory leaves the singular set (for example, if the orthogonal projection of the angular velocity onto $(s_3, c_3, 0)^T$ is not zero).

3.3.4 A regularization technique

When singularities occur one has to redefine the task function either by replacing it with an equivalent one, or by modifying it enough but not too much, to make the singularities disappear to ensure behaviour of the robot similar to that implied by the unmodified task function. Roughly speaking, two task functions are equivalent if

$$e_1(q, t) = 0 \Leftrightarrow e_2(q, t) = 0. \qquad (3.3.18)$$

A more detailed study of this concept is carried out in the next chapter. A simple example of equivalent task functions is given when it is possible to compute the ideal trajectory $q_r(t)$ such that $e_1(q_r(t), t) = 0$. An equivalent task is then $e_2(q, t) = q - q_r(r)$. Such a task is ∞-admissible (cf. section 3.4.1) if $q_r(t)$ is C^2.

Another way to regularize a task function is more technical. The idea is to add to the first output function $e_p(q, t)$ a second task function $e_s(q, t)$ such that

$$e_p(q_0, 0) = e_s(q_0, 0) = 0, \qquad (3.3.19)$$

the new task being

$$e(q, t) = e_p(q, t) + e_s(q, t) \qquad (3.3.20)$$

with the associated Jacobian matrix

$$\frac{\partial e}{\partial q} = \frac{\partial e_p}{\partial q} + \frac{\partial e_s}{\partial q}. \qquad (3.3.21)$$

This matrix may be non-singular without its two components being necessarily non-singular. The main difference in replacing the task function with an equivalent one is that the perfect execution of the initial task is no longer granted since the ideal trajectory must now cancel the sum of the two task functions. The function $e_s(q, t)$ should be chosen such that it fulfils the above mentioned regularization without causing substantial alteration to the initial task objectives.

As many examples of singularities appear with physically impossible tasks involving unbounded joint velocities, a possible choice, among others, would be to minimize

$$h_s^1(q, \dot{q}, t) = \tfrac{1}{2} \gamma(t) \|\dot{q}\|^2. \tag{3.3.22}$$

But task functions depend only on joint positions, so it is necessary to replace h_s^1 by another function with a similar effect:

$$h_s(q, t) = \tfrac{1}{2} \gamma(t) \|q - y(t)\|^2, \qquad \gamma(t) \geq 0, \tag{3.3.23}$$

where the vector $y(t)$ is obtained by filtering the robot position $q(t)$ through a low-pass filter:

$$\left. \begin{array}{l} \dot{y}(t) + \alpha(t) y(t) = \alpha(t) q(t) \\ y(0) = q(0) \end{array} \right\} \tag{3.3.24}$$

The parameter $\alpha(t)$ may be chosen so as to ensure the boundedness of the filtered velocity $\dot{y}(t)$. For example (see also Chapter 4, section 4.3.4.2),

$$\alpha(t) = \frac{\lambda_1}{1 + \lambda_2(\lambda_3 + \|q(t) - y(t)\|)^{1/2}}. \tag{3.3.25}$$

The added task function is then

$$e_s(q, t) = \gamma(t)(q - y(t)) \tag{3.3.26}$$

and the new Jacobian matrix is

$$\frac{\partial e}{\partial q}(q, t) = \frac{\partial e_p}{\partial q}(q, t) + \gamma(t) I_n. \tag{3.3.27}$$

A sufficient condition for having $\partial e / \partial q$ non-singular along the ideal trajectory $q_r(t)$ is

$$\left\| \frac{\partial e_p}{\partial q}(q_r(t), t) \right\| < \gamma(t). \tag{3.3.28}$$

Since this condition is only sufficient, a smaller value of γ may also give good results in practice.

The method is best justified when the initial task is itself a minimization task, because we can then be sure that $e_p(q, t)$ is the gradient of a function $h_p(q, t)$ and the Jacobian of the compounded task reads

$$\frac{\partial e}{\partial q} = \frac{\partial^2 h_p}{\partial q^2}(q, t) + \gamma(t) I_n. \tag{3.3.29}$$

It is then possible to choose $\gamma(t)$ such that the task Jacobian is not only invertible but also positive along the ideal trajectory. A sufficient condition is:

$$\gamma(t) + \lambda_{\min}(q_r(t), t) > 0 \qquad (3.3.30)$$

where $\lambda_{\min}(q_r(t), t)$ denotes the smallest eigenvalue of $(\partial^2 h_p / \partial q^2)(.)$.

3.4 ADMISSIBILITY OF USUAL ROBOTIC TASKS

In this final section of Chapter 3, we give some examples of classical robotic tasks, analysed from the point of view of their admissibility. However, two classes of task will be passed over here and will be examined in Chapters 4 and 7 respectively: *redundant* tasks, and *sensor-based* tasks. This choice arises from both their practical importance and their specificity, which requires special developments: the former are characterized by their non-uniqueness, while in the latter, data provided by relative exteroceptive sensors are used in real time.

For the present, let us consider the two basic classes of robotics tasks, which are defined according to the *spaces* in which they are performed.

3.4.1 Trajectory tracking in joint space

Let $\{q_r(t); t \in [0, T]\}$ be any desired C^2-trajectory, with bounded time derivative:

$$\|\dot{q}_r(t)\| < k < \infty, \qquad t \in [0, T]. \qquad (3.4.1)$$

A natural, but non-unique, choice of task is then

$$e(q, t) = q - q_r(t) \qquad (3.4.2)$$

with $\Omega = \mathbf{R}^n \times [0, T]$ and $q_0 = q_r(0)$.

This yields

$$\frac{\partial e}{\partial q}(q, t) = I \qquad (3.4.3)$$

and

$$\frac{\partial e}{\partial t}(q, t) = \dot{q}_r(t). \qquad (3.4.4)$$

It is easy to verify that $\{e(q, t), q_0\}$ is ∞-admissible on $[0, T]$, which obviously confirms the known fact that this task is not singular, for any kind of robot: its robustness with regard to the control problem is excellent. Then, $q_r(t)$ is the ideal trajectory itself.

3.4.2 Trajectory tracking in SE_3

Let

$$x = \begin{pmatrix} x_1 \\ x_2 \end{pmatrix} \qquad (3.4.5)$$

(x_1) be the position and (x_2) a chart-type (cf. section 1.2.2) parametrization of the attitude (cf. section 1.2.2) of a fixed frame F with regard to F_0 (or to any fixed reference frame). This question of parametrization is also studied in Chapter 5, section 5.3.1.2.2 and in Chapter 6, section 6.6.1.3.

Suppose that $x(q)$ is allowed to move in the whole frames configuration space with the restriction that the attitude must remain within the domain of the chart x_2. Hence $\dim(x(q)) = 6$. Suppose also that the number n of joints of the manipulator is 6. The function $x(q)$ is assumed to be C^2 and defined on an open set Ω_x of $\mathbf{R}^6$ (generally $x(q)$ will be C^∞).

The complement of Ω_x in $\mathbf{R}^n$ is the set of the points where the parametrization is not defined. As already noted, such points always exist if x_2 is a chart on SO_3.

Let $x_r(t)$, $t \in [0, T]$, be a C^2 desired trajectory, with bounded time derivative. The tracking problem is to find a control which ensures that the error

$$\varepsilon(q, t) = x(q) - x_d(t) \qquad (3.4.6)$$

is small during $[0, T]$. This type of task has been considered in the previous examples. Let us refine the analysis a little more.

In the subsequent discussion, we distinguish the situation where $x(q)$ is measured from the one where it is only estimated.

3.4.2.1 The ideal case: $x(q)$ is known

We assume in this subsection that the values of $x(q)$ are directly accessible at each time (or may be accurately estimated) from sensory data (this sets the problem of endpoint sensing). Then the tracking error is measurable, and we may choose

$$e(q, t) = \varepsilon(q, t); \qquad \Omega = \Omega_x \times [0, T] \qquad (3.4.7)$$

which leads to

$$\frac{\partial e}{\partial q}(q, t) = \frac{\partial x}{\partial q}(q, t) \qquad (3.4.8)$$

$$\frac{\partial e}{\partial t}(q, t) = -\dot{x}_d(t). \qquad (3.4.9)$$

Given q_0 such that $x(q_0) = x_d(0)$, we know from previous sections that $\{e(q, t), q_0\}$ is an admissible task on $[0, T]$ if the ideal trajectory $\{x_d(t)\}$

3.4 ADMISSIBILITY OF USUAL ROBOTIC TASKS

is such that the corresponding trajectory $\{q_r(t)\}$ (i.e. $x_d(t) = x(q_r(t))$, $q_r(0) = q_0$) never crosses a singularity during $[0, T]$.

The translational and angular velocity vectors $V_6(t)$ and $\omega_6(t)$ are linked to $\dot{x}(t)$ by

$$\dot{x}_1(t) = V_6(t) \tag{3.4.10}$$

$$\dot{x}_2(t) = J_p(q)\omega_6(t) \tag{3.4.11}$$

where the 3×3 matrix $J_p(q)$ depends only on the set of parameters which has been chosen to represent the attitude of F_6 with regard to F_0.

Recalling that

$$\begin{pmatrix} V_6(t) \\ \omega_6(t) \end{pmatrix} = J_6(q(t))\dot{q}(t) \tag{3.4.12}$$

and using the relations above, we obtain

$$\dot{x}(t) = J_W(q)J_6(q)\dot{q}(t) \tag{3.4.13}$$

where

$$J_W(q) = \begin{pmatrix} I & 0 \\ 0 & J_p(q) \end{pmatrix}. \tag{3.4.14}$$

$J_p(q)$ represents the differential of the attitude parametrization. Various expressions for $J_p(q)$ corresponding to the usual parametrizations have been computed in section 1.2.5.4. Knowing that

$$\dot{x}(t) = \frac{\partial e}{\partial q}(q, t)\dot{q}(t), \tag{3.4.15}$$

this finally gives

$$\frac{\partial e}{\partial q}(q, t) = J_W(q)J_6(q). \tag{3.4.16}$$

Task singularities thus may arise only from two sources:

1. $J_6(q)$. As defined in Chapter 2, points q^s such as $\det(J_6(q^s)) = 0$ are known as *kinematic singularities*. These singularities depend only on the geometric structure of the robot itself.

2. $J_W(q)$. As we have already seen, points q_s where $J_p(q)$ is not invertible are points where the corresponding parametrization is not defined. The existence of such points in $C_{\rho, T}$ prevents this subset from being closed in $\mathbf{R}^n$.

The *a priori* knowledge of singularities allows admissibility of the task to be checked in the following way: if $x(q)$ is known and $x_d(t)$ predefined, the test may be performed off-line by computing $q_r(t)$ and verifying that $q_r(t)$ does not come near q^s or q_s. If the distance to singularities becomes too small, it is preferable to modify the task and thus the task function.

3.4.2.2 The usual case: $x(q)$ is unknown

In most cases, $x(q)$ is not measured but only approximated by a function $\hat{x}(q)$, which is a model of the kinematics of the robot. Unless the model is perfect, the error $\varepsilon(q, t)$ may not be accurately controlled. Some practical solutions to this problem are as follows:

1. *Learning.* In some cases, accuracy is not needed all along the trajectory, but only in the final position and possibly at a small set of intermediate points.

A *learning* stage may then consist of finding joint values $\{q_j^*; j = 1, \ldots, p\}$ which give the correct positions when transformed by the model $\hat{x}$. Then a C^2 desired trajectory $x_d(t)$ passing by these positions $\hat{x}(q_j^*)$ may be computed, by various methods. A possible task function is then

$$e(q, t) = \hat{x}(q) - x_d(t) \qquad (3.4.17)$$

2. *Control in $\hat{x}$-space.* Since measurement of $\varepsilon(q, t)$ is impossible, let us fall back upon the problem of controlling

$$\hat{\varepsilon}(q, t) = \hat{x}(q) - x_d(t). \qquad (3.4.18)$$

A first possible task function is as in (3.4.17)

$$e(q, t) = \hat{\varepsilon}(q, t), \qquad (3.4.19)$$

which gives, in a manner analogous to eqn (3.4.16):

$$\frac{\partial e}{\partial q}(q, t) = \frac{\partial x}{\partial q}(q, t) = J_W(q)\hat{J}_6(q) \qquad (3.4.20)$$

in which J_W is the same as in (3.4.13), for a parametrization is only the user's choice (i.e. does not depend on the kinematics), and $\hat{J}_6$ is the basic Jacobian matrix associated with model $\hat{x}(q)$.

The singularities coming from the parametrization are not changed, but the Jacobian ones do not remain the same, except in the case

$$\det J_6 = 0 \Leftrightarrow \det \hat{J}_6 = 0. \qquad (3.4.21)$$

However, in the general case, singularities q^s and $\hat{q}^s$ will be different. In practice, if a good identification of $x(q)$ has been performed, it may be hoped that x and $\hat{x}$ will be close, and also their singularities.

3. *Use of inverse kinematics.* A method frequently used consists of transforming the problem of controlling $\varepsilon(q, t)$ into a problem of trajectory tracking in joint space by again setting

$$e(q, t) = q - q_r(t), \qquad (3.4.22)$$

3.4 ADMISSIBILITY OF USUAL ROBOTIC TASKS

where $q_r(t)$ is the result of an inverse kinematics computation, such that

$$\hat{x}(q_r(t)) = x_d(t). \tag{3.4.23}$$

If $q_r(t)$ is C^2 and $\dot{q}_r(t)$ is bounded this new task is ∞-admissible. This may exceptionally be the case even if the task function $\varepsilon(q, t) = \hat{x}(q) - x_r(t)$ is not admissible, as is the case in Fig. 3.8 when the trajectory is 'flat' enough and transverse to the z_0 axis when crossing the singularity.

Generally, however, the computation of the desired trajectory in the joint space using the inverse kinematics does not lead to a C^2 trajectory with bounded time derivative when kinematic singularities have to be crossed. Moreover, we have seen in section 2.2.2 that the inverse kinematics are not well defined at a singular point.

This illustrates the fact that, in general, singularities cannot be circumvented by simply replacing the non-admissible task function by an equivalent one. Therefore, it is usually necessary to alter the task itself so as to obtain a new, nonequivalent, admissible task function.

The admissibility conditions also appear directly in the inverse kinematics when an iterative scheme is used to solve the equation

$$\hat{x}(q) = x_d(t). \tag{3.4.24}$$

A common way of implementing this solution is the following.

Let $\{t_k; k = 1, 2, \ldots\}$ be an increasing sequence of times, generally corresponding to a constant sampling period. At each time t_k, we wish to find a point $q_r(t_k)$ of the reference trajectory such that the cost function

$$f_k = \tfrac{1}{2}\|x_d(t_k) - \hat{x}(q_r(t_k))\|^2 \tag{3.4.25}$$

is minimum.

Knowing that

$$\frac{\partial f_k}{\partial q}(q) = \frac{\partial \hat{x}^T}{\partial q}(q)(\hat{x}(q)) - x_r(t_k)), \tag{3.4.26}$$

the most generally used recursion is

$$q_{k-1,j+1} = q_{k-1,j} + P_{k-1,j} \frac{\partial f_k}{\partial q}(q_{k-1,j}) \tag{3.4.27}$$

with

$$q_{k-1,0}(0) = q_r(t_{k-1}). \tag{3.4.28}$$

The recursion is stopped at step N when $f_k(q_{k-1,N})$ is smaller than a given threshold and the definitive result is

$$q_r(t_k) = q_{k-1,N}. \tag{3.4.29}$$

$P_{k-1,j}$ is a general positive gain. The choice of a scalar gain leads to a gradient method, and using $\partial^2 f_k^{-1}/\partial q^2$ yields a Newton method.

Remark. We have

$$\frac{\partial^2 f_k}{\partial q^2}(q) = \frac{\partial \hat{x}^T}{\partial q}\frac{\partial \hat{x}}{\partial q}(q) + \sum_{i=1}^{n}(\hat{x}_i(q) - x_{r,i}(t_k))\frac{\partial^2 \hat{x}_i}{\partial q^2}(q). \quad (3.4.30)$$

In the vicinity of the minimum of f_k, the second term of this equation may be assumed to be small with respect to the first, if q is far from Jacobian singularities of $\partial \hat{x}/\partial q$. Then, an algorithm frequently encountered uses

$$P_{k-1,j}^{-1} = \left(\frac{\partial \hat{x}^T}{\partial q}\frac{\partial \hat{x}}{\partial q}(q_{k-1,j})\right)^{-1} \approx \frac{\partial^2 f}{\partial q^2}(q_{k-1,j}) \quad (3.5.31)$$

which leads to the recursion

$$q_{k-1,j+1} = q_{k-1,j} + \left(\frac{\partial \hat{x}}{\partial q}\right)^{-1}(q_{k-1,j})(\hat{x}(q_{k-1,j}) - x_r(t_k)). \quad (3.5.32)$$

When the sampling period is small, this algorithm is often simplified to

$$q_{k-1,j+1} = q_{k-1,j} + \left(\frac{\partial \hat{x}}{\partial q}\right)^{-1}(q_r(t_{k-1}))(\hat{x}(q_{k-1,j}) - x_r(t_k)). \quad (3.5.33)$$

Of course, all these algorithms require the boundedness of $(\partial \hat{x}/\partial q)^{-1}$ at each point, which refers again to the problem of admissibility of $\varepsilon(q,t) = \hat{x}(q) - x_d(t)$.

3.5 BIBLIOGRAPHIC NOTE

As pointed out at the beginning of this chapter, the first problem to be solved in robot control was a modelling problem. This is more or less implicit in many papers dealing with various applications of manipulators. Much remains to be done in this domain and every new application of robotics will bring its field of research in modelling.

Although singularities have been encountered in non-linear control theory, they have not yet been analysed very deeply. The material in this chapter appears to be new and has been developed from previous work by one of the authors (Samson 1987a).

4

TASK REDUNDANCY

4.1 LOCALLY REDUNDANT SYSTEMS

4.1.1 Introduction: an intuitive approach to redundancy

The word 'redundancy' in robotics may be used in many different frameworks and applied to various problems. We shall concentrate here on the concept of redundancy in relation to the *control* problem.

From a general point of view, any robotic system in which the way of achieving a given task is not unique may be called 'redundant'. The concept of redundancy is thus related to the definition of the task and is not an intrinsic feature of the robot's structure. This fact is not always well understood in practice, and it is common to say that a robot is 'redundant' when it has more than six joints. This point of view implicitly supposes that the primary task consists of controlling the three-dimensional position and the attitude of the end-effector. This common definition is not precise enough for the present analysis, and we shall illustrate this fact by simple examples.

In the first place, there may be redundancy even when the robot has less than six joints. An example is provided by considering a two-jointed robot, with a primary task specified in terms of only the x direction (Fig. 4.1):

$$e(q, t) = x(q) - x_r(t). \qquad (4.1.1)$$

For a given value $x_r(t)$, $|x_r(t)| < l_1 + l_2$, there exists an infinite number of solutions $\{q_1, q_2\}$. However, even in this very simple case, the situation may be complicated by the existence of problems related to the definition of the task (cf. the discussion in section 3.1.1). For instance:

(1) when $|x_r(t)| = l_1 + l_2$, there is only one solution, $\{q_1 = 0, q_2 = 0\}$, and the robot no longer seems to be redundant;

(2) when $|x_r(t)| > l_1 + l_2$, perfect tracking is impossible, since e cannot be equal to zero; the specific task cannot be completed and the concept of redundancy obviously has no meaning in this case.

It is also important not to confuse redundancy with the existence of multiple solutions for the completion of a task. For example, Fig. 4.2 shows two possible robot configurations for the same position of the robot's tip. This results from the existence of distinct solutions to the inverse kinematics problem (cf. Chapter 2). However, if the task consists of maintaining the tip of

110   4 TASK REDUNDANCY

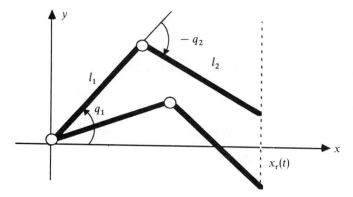

Fig. 4.1 Two-joint redundant robot

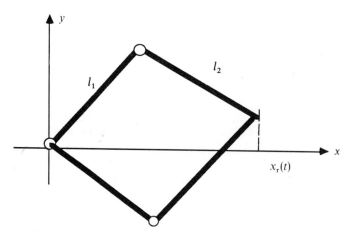

Fig. 4.2 Two solutions for the two-joint robot

the robot at the same position then, given an initial configuration of the robot, there is no redundancy in the way of achieving this task because the robot cannot instantaneously jump from one solution to the other. It can only stay in the initial configuration since there is no continuous path in the joint space which connects one solution to the other and along which the robot's tip is motionless. Generalizing this example, we may intuitively see that a robotic system is truly redundant for a given task only when there exists an infinite number of solutions forming a dense set in the joint space, i.e. such that the passage from any solution to another can be done by following a continuous path without leaving this set.

4.1 LOCALLY REDUNDANT SYSTEMS

Thus 'redundancy' is not an intrinsic feature of a robot. It depends on the task and may change with time. More exactly, the *redundancy rate* may vary with time. Before a more precise definition is formulated, the redundancy rate may intuitively be understood as being the number of 'degrees of freedom' that are superfluous for completion of the task; this is often (but not always) the difference between the dimension of the vector describing the task and the number of joints. Let us consider for example a multibar planar mechanism with its endpoint fixed on the x axis (Fig. 4.3). The redundancy rate is the maximum number of independent constraints that we may add to the system before it becomes motionless (see for example the clamped elements in Fig. 4.3). The system is then no longer redundant, and this illustrates the fact that the 'initial' task (endpoint on the x axis) together with the supplementary constraints constitutes an example of a non-redundant new task. We shall develop this point further.

It is to be emphasized that the effect of added constraints may be variable. For example, in the four-link mechanism of Fig. 4.4, the initial redundancy rate for the endpoint fixed in D is 2. Let us clamp another point, S, of the robot. It is obvious that if $S \in [OA]$ or $[CD]$, the redundancy rate becomes 1, while there is no redundancy if $S \in]AB]$ or $[BC[$.

A final point concerns the nature of redundancy. Consider the system shown in Fig. 4.5; intuitively it is 'more redundant' in the plane orthogonal to the rotational axis of joints $q_2 - q_6$ than elsewhere. This means that the redundancy rate is not sufficient to describe accurately the 'redundancy configuration' characterized by certain preferential directions.

To summarize this introduction, let us recall some facts that we have emphasized: redundancy has to be defined with respect to a *task*; it is a concept which may *evolve* during the completion of the task. Redundancy is not entirely characterized by the *redundancy rate*, because there are preferential directions of redundancy. In what follows, we shall try to associate these intuitive concepts with a mathematical formulation.

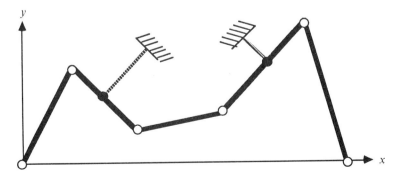

Fig. 4.3 Clamping in a redundant plane multibar mechanism

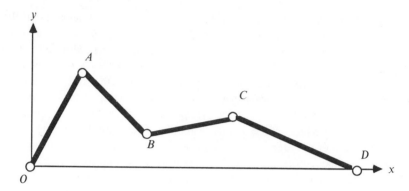

Fig. 4.4 Four-bar mechanism

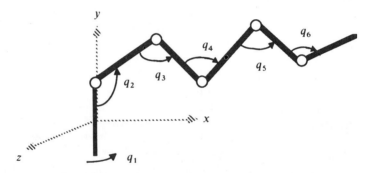

Fig. 4.5 Preferential redundant directions

In order to distinguish the tasks for which a *given* robotic system is redundant from those for which the same system is not redundant, it is convenient (although semantically incorrect) to talk of *redundant tasks* as opposed to *non-redundant tasks* and to gather all related problems under the banner of *task redundancy*. While doing so, it is however important to keep in mind that a certain task may be redundant with respect to a certain system and non-redundant with respect to another one.

Finally, let us mention that, from the user's point of view, a way of coping with redundancy is to modify the task, for example by adding constraints, until a new non-redundant problem is obtained. In the following, this idea will allow us to link the problem of redundancy to the task analysis of Chapter 3. It will also be shown that adding explicit constraints is not the only way of modifying a redundant task.

4.1.2 Basic definitions

Let us consider, as usual, a task function $e(q, t)$ from $\mathbf{R}^n \times \mathbf{R}$ to $\mathbf{R}^n$. As in Chapter 3, we are interested in the properties of the connected set $C_{0,T} \subseteq e^{-1}\{0\}$ of the *zeros* of $e(q, t)$, which contains the initial condition $(q_0, 0)$.

Studying the zeros of a function is known to be an extensive and difficult problem, and the structure of $e^{-1}\{0\}$ may take various forms. Let us illustrate this with a few simple examples.

Example 4.1

1. $$e(q_1, q_2) = (q_1 - 2)(q_1^2 + q_2^2 - 1). \tag{4.1.2}$$

   $e^{-1}\{0\}$ is made up of two distinct connected sets.

2. $$e(q_1, q_2) = (q_1 - 1)(q_1^2 + q_2^2 - 1). \tag{4.1.3}$$

   $e^{-1}\{0\}$ consists of a single connected set, but is not a submanifold.

3. $$e(q_1, q_2, q_3) = (q_1^2 + q_2^2 - 1)(q_3^2 + (q_2 - 2)^2). \tag{4.1.4}$$

   $e^{-1}\{0\}$ is made up of two distinct connected sets which are submanifolds with different dimensions.

4. Also, when the time variable appears, more complicated phenomena may occur. For instance

   $$e(q_1, q_2, t) = tq_1^2 + (1 - t)q_2^2 - 1. \tag{4.1.5}$$

   Here, the number and the structure of the connected components of $e^{-1}\{0\}$ change with time.

In the case where $C_{0,T}$ has the properties given in section 3.2, the function is admissible. In this chapter, we are interested in the case where the function is not admissible, but still presents some regularity properties. An intuitive idea for characterizing purely *geometric* aspects of redundancy consists of fixing the time index t and studying the properties of the set $C_0^* = \{q | (q, t^*) \in C_{0,T}\}$. This leads to the following preliminary definition:

Definition D4.1 *A task $\{e(q, t); q_0\}$ is said to be geometrically redundant at time t^* around the position q^* if, in an open neighbourhood U^* of q^*, C_0^* includes a p-dimensional submanifold ($p \geq 1$) containing q^*. The geometric redundancy rate at time t^* around q^* is the supremum of the dimensions of the submanifolds with the above property.*

However, this definition is of little practical interest because it does not take the temporal aspects into consideration. Indeed a robot is not only a geometric system but also a dynamic one. Some time interval has therefore to be considered.

Since we are looking for smooth paths in $C_{0,T}$, it is natural to try to characterize redundancy by the existence of a local submanifold contained in $C_{0,T}$ with dimension larger than one. However this is not sufficiently restrictive. On the other hand imposing a local submanifold structure on $C_{0,T}$ itself is too strong because many redundant tasks would then be declared not redundant. One may also be tempted to extend the definition of geometric redundancy to a similar one including the time parameter, but this is a complex problem because of the special nature of time. It is physical nonsense to go back and forth on the time axis. As a result, a purely geometric definition of local redundancy would involve a lot of technicalities borrowed from the theory of fibre bundles. Instead we propose:

Definition D4.2 *A task function $e(q,t)$ is locally redundant around $(q^*, t^*) \in C_{0,T}$ if there exists $\varepsilon > 0$, an open set U_1 containing q^* and a diffeomorphism Ψ from $U = U_1 \times]t^* - \varepsilon, t^* + \varepsilon[$ to $\mathbf{R}^n \times \mathbf{R}$ such that:*

(1) $\Psi(q, t) = (\phi(q, t), t); \Psi(q^*, t^*) = (0, t^*)$;

(2) *there exists a vector subspace V in $\mathbf{R}^n$ such that*

$$\Psi^{-1}(V \times]t^* - \varepsilon, t^* + \varepsilon[) \subseteq C_{0,T}.$$

The supremum p of the dimensions of the vector subspaces V with the above property is the local redundancy rate of the task around (q^*, t^*).

This definition is still of a geometric nature. It may be interpreted as meaning that a task is locally redundant if the set of zeros of $e(q, t)$ (restricted to the connected component containing $(q_0, 0)$ contains a smoothly time-varying submanifold. This definition also suggests that the task can be locally characterized by $n - p$ independent equations. Some definitions and lemmas will make this remark more precise.

Definition D4.3 *A task function $e'(q, t)$ is a restriction of $e(q, t)$ on Ω if, for any $(q, t) \in \Omega$*

$$e'(q, t) = 0 \Rightarrow e(q, t) = 0.$$

Then e is called an extension of e'.

According to this definition, replacing a given task by one of its restrictions has the effect of restricting the ways the task can be achieved. However, this replacement may be advantageous when the restriction possesses 'nicer' properties than the initial task function.

4.1 LOCALLY REDUNDANT SYSTEMS

Definition D4.4 *If e' is a restriction of e and e a restriction of e' on Ω then the task functions e and e' are said to be equivalent on Ω.*

In other words, two task functions are equivalent on Ω when they share the same set of zeros in Ω. The perfect completion of one task is equivalent to the perfect completion of the second task.

We are now ready to formulate a second definition of local redundancy which points out the existence of a particular task function whose set of zeros presents a strong regularity property.

Definition D4.5 *A task $\{e(q, t); q_0\}$ is locally redundant around $(q^*, t^*) \in C_{0,T}$ if there exists a task function e' of the form*

$$e' = \begin{pmatrix} e_1 \\ 0 \end{pmatrix},$$

with $\partial e_1 / \partial q$ of full rank in an open neighbourhood U of (q^*, t^*), which is a restriction of e in U.

This definition is equivalent to Definition D4.2 as we now show.

Proof of the equivalence Starting with Definition D4.2, consider a set of independent linear equations $f_1, \ldots, f_k$ defining the subspace V. Then the function

$$e_1(q, t) = \begin{bmatrix} f_1(\phi(q, t)) \\ \ldots \\ f_k(\phi(q, t)) \end{bmatrix} \quad (4.1.6)$$

is full rank on U and $e_1(q, t) = 0$ if and only if (q, t) belongs to $\Psi^{-1}(V \times]t^* - \varepsilon, t^* + \varepsilon[)$. Thus $\begin{pmatrix} e_1 \\ 0 \end{pmatrix}$ is a restriction of e with the desired properties.

Conversely, given $e_1(q, t) = \begin{bmatrix} f_1(q, t) \\ \ldots \\ f_k(q, t) \end{bmatrix}$, we define a mapping $\Psi(q, t) = (\Psi_i(q, t))_{i=1, \ldots, n+1}$ by:

$$\Psi_i(q, t) = f_i(q, t) \quad i = 1, \ldots, k$$
$$\Psi_i(q, t) = q_i \quad i = k+1, \ldots, n$$
$$\Psi_{n+1}(q, t) = t.$$

The Jacobian matrix of Ψ is then

$$\begin{bmatrix} \dfrac{\partial e_1}{\partial q} & | & \dfrac{\partial e_1}{\partial t} \\ -- & | & -- & | & -- \\ 0 & | & \text{Id} & | & 0 \\ -- & | & -- & | & -- \\ 0 & | & 0 & | & 1 \end{bmatrix}$$

and is non-singular in a neighbourhood of (q^*, t^*). Narrowing U if necessary, the Inverse Function Theorem shows that Ψ is a diffeomorphism. Moreover, $e_1(q, t) = 0$ is equivalent to $\Psi(q, t) \in \mathbf{R}^{n-k} \times \mathbf{R}$.

Starting with a general task function $e(q, t)$, there is no technical means of checking the local redundancy in general. However, in some cases, simple techniques may be applied to find a restriction of $e(q, t)$.

Example 4.2 Consider the robot of Fig. 2.1 in Chapter 2. The coordinates of the endpoint are

$$x = c_1(c_2 + c_{23}) \tag{4.1.7}$$

$$y = s_1(c_2 + c_{23}) \tag{4.1.8}$$

$$z = s_2 + s_{23} \tag{4.1.9}$$

and let us assume that the desired trajectory is such that $x_r(t) = y_r(t) = 0$ with $0 < z_r(t) < 2$. The first two components of the task function

$$e(q, t) = \begin{pmatrix} x - x_r(t) \\ y - y_r(t) \\ z - z_r(t) \end{pmatrix} \tag{4.1.10}$$

are not independent in the sense that their differentials are not independent on $C_{0,T}$. Dropping the second line gives a new task function

$$\begin{bmatrix} e_1(q, t) \\ 0 \end{bmatrix} = \begin{pmatrix} x - x_r(t) \\ z - z_r(t) \\ 0 \end{pmatrix} \tag{4.1.11}$$

equivalent to the previous one and with $\partial e_1/\partial q$ of full rank.

Another case is when the 'constant rank' theorem (cf. Theorem 0.4) applies:

Proposition P4.1 *If $\partial e/\partial q$ is of constant rank $m < n$ in a neighbourhood of (q^*, t^*), then $e(q, t)$ is locally redundant around (q^*, t^*) with a local redundancy rate equal to $n - m$.*

This is not the case in the previous example: the rank of the Jacobian matrix of the task function is 3 everywhere except on $C_{0,T}$ where it drops to 2.

Notice that it is usually not sufficient to have a constant rank m less than n along a *single* ideal trajectory passing through q^* in order to be able to conclude the discussion about the local redundancy property of the task. The reason is that the rank of $e(q, t)$ may jump up as soon as we leave the ideal trajectory. In the next section we shall focus on a subset of task functions for which this rank difficulty does not exist.

4.2 REDUNDANT TASKS

4.2.1 Definitions

The concept of local redundancy is not highly exploitable for control design purposes and for this reason a logical next step is to look for regularity properties valid along the whole of $[0, T]$ and leading to a more global concept of redundancy. We shall proceed in this direction by proposing the following definition:

Definition D4.6 *A canonical redundant task $\{e(q, t); q_0\}$ is a realizable task such that:*

1. *$e(q, t)$ has the form*

$$e(q, t) = \begin{pmatrix} e_1(q, t) \\ 0 \end{pmatrix}, \quad t \in [0, T]; \quad (4.2.1)$$

2. *$J_1(q, t) = (\partial e_1/\partial q)(q, t)$ is of full rank m for any (q, t) belonging to the connected set $C_{0,T}$ which contains $(q_0, 0)$.*

Remarks

1. The *global redundancy rate* is $n - m$; it is also the dimension of the *null space* of the mapping with matrix J_1. The structure of this null space characterizes at each (q, t) the preferential directions of redundancy.

2. Definition D4.6 sets strong properties for a canonical redundant task. These properties are in fact required only around the ideal trajectory that

we are interested in. The problem, at this stage, is that this trajectory is not defined until further constraints are added to the initial redundant task.

3. Complementary to D4.6, the function e' defined in D4.5 may be called 'locally canonical redundant'.

Note that $e(q, t)$ defined in this way is not yet admissible in the sense of Chapter 3 since the task Jacobian matrix is singular. Therefore the task function still has to be modified in order to become admissible.

Let us now apply the previous definitions to two simple examples.

Example 4.3 Let us return to the robot of Fig. 2.1 with the task function (4.1.10). When an ideal trajectory exists, avoiding the singular points given in Chapter 3, this task function is admissible. If we now choose $x_r(t) = y_r(t) = 0$, and $0 < z_r(t) < 2$ for any $t \in [0, T]$, the task function is geometrically and locally redundant at every t, according to Definitions D4.1 and D4.5. As seen in Example 4.2 there exists a task function e' which is a restriction of e. It is easy to verify that this function e' is a canonical redundant task function *equivalent* to e. This function is not unique. Another example is

$$e' = \begin{pmatrix} 2q_2 + q_3 - \pi \\ z(q) - z_r(t) \\ 0 \end{pmatrix}. \qquad (4.2.2)$$

It is easy to verify that around $e' = 0$ the rank of $\partial e'/\partial q$ remains equal to 2.

Example 4.4 Let us choose the *minimization* of the following scalar cost function at each instant as a goal:

$$h(q, t) = \tfrac{1}{2} \| x(q) - x_r(t) \|^2 \qquad (4.2.3)$$

where x is an m-dimensional vector, $m < n$, representing for example the position and the parametrized orientation of the end-effector. In the best case, it is possible to make h everywhere equal to zero if there exists $q_r(t)$ such that

$$x(q_r(t)) = x_r(t). \qquad (4.2.4)$$

Let us then define the task function

$$e(q, t) = \frac{\partial h}{\partial q}(q, t) = \frac{\partial x^T}{\partial q}(q)(x(q) - x_r(t)). \qquad (4.2.5)$$

Thus

$$J_T(q, t) = \frac{\partial x^T}{\partial q}(q)\frac{\partial x}{\partial q}(q) + \sum_{k=1}^{m} \frac{\partial^2 x_k}{\partial q^2}(q)(x_k(q) - x_{r_k}(t)) \qquad (4.2.6)$$

where the subscript k denotes the kth component of a vector.

4.2 REDUNDANT TASKS

When $q_r(t)$ satisfies (4.2.4):

$$J_T(q_r(t), t) = \frac{\partial x^T}{\partial q}(q_r(t)) \frac{\partial x}{\partial q}(q_r(t)) \qquad (4.2.7)$$

and since $\partial x/\partial q$ is a $(m \times n)$ matrix,

$$\text{rank } J_T(q_r(t), t) \leq m < n.$$

From this last relationship we may intuitively see that $e' = \begin{pmatrix} x(q) - x_r(t) \\ 0 \end{pmatrix}$ is not necessarily an equivalent canonical task function, because $C_{0,T}$ may contain points where the rank of $\partial x/\partial q$ is less than m.

For example, let us consider the case of Fig. 4.6, where $m = 2$ and $n = 3$. Everywhere inside the open disk (O, 3), it is possible to find a q^* such that e' given above is geometrically redundant at q^*, and, moreover, locally redundant at (q^*, t^*). However, due to the global character of the concept of canonical redundant task function, e' is canonical only if $x_r(t)$ lies inside the open disk (O, 3) and never crosses the circle (O, 1) where $\partial x/\partial q$ is of rank 1 at $q_1 = (2k + 1)\pi$. If we want the tip of the robot to stay on the circle, for example with $x_r(t) = \begin{pmatrix} x_r^1(t) \\ x_r^2(t) \end{pmatrix} = \begin{pmatrix} \cos t \\ \sin t \end{pmatrix}$, $t \in [0, 2\pi]$, then a possible canonical function is

$$e = \begin{pmatrix} q_1 + q_2 - t \\ q_2 + q_3 + \pi \\ 0 \end{pmatrix}. \qquad (4.2.8)$$

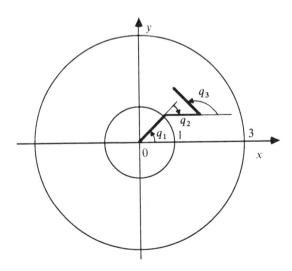

Fig. 4.6 Three-bar planar mechanism

Remarks

1. An interesting feature of the cost function (4.2.3) is that it can still be used when perfect tracking is not possible, for example when $x_r(t)$ is not in the range of $x(q)$. In that case, (4.2.4) is not satisfied and the minimum of h, $q_r(t)$, is such that

$$e(q_r(t)) = \frac{\partial x^T}{\partial q}(q_r(t))(x(q_r(t)) - x_r(t)) = 0 \qquad (4.2.9)$$

with

$$x(q_r(t)) - x_r(t) \neq 0.$$

This already shows that $(\partial x/\partial q)(q_r(t))$ is not full rank. In our previous example, this case corresponds to $x_r(t)$ outside the disk (O, 3) and $\partial x/\partial q$ of rank 1. However, in this case, the system is not redundant because the solution $q_r(t)$ is locally unique: the arm must be fully extended to point toward the target at coordinates $x_r(t)$. In fact, it can be shown from (4.2.6) that despite the rank deficiency of $\partial x/\partial q$, the task Jacobian matrix $J_T(q_r(t), t)$ is non-singular and the task function (4.2.5) is admissible. An equivalent task function, which may also be used to achieve the same objective, is

$$e'' = \begin{pmatrix} q_2 \\ q_3 \\ s_1 x_r^1(t) - c_1 x_r^2(t) \end{pmatrix}. \qquad (4.2.10)$$

2. This example demonstrates also the fact that tasks described by the same function $e(q, t)$ may be redundant or not according to the initial condition q_0, i.e. to the related connected component of $C_{0,T}$. Indeed, when perfect tracking is not possible, *or* when it is desired to maximize $h(q, t)$ instead of minimizing it, we have

$$x(q_0) \neq x_r(0) \qquad (4.2.11)$$

and

$$e(q_0, 0) = 0. \qquad (4.2.12)$$

Again, it may result from (4.2.6) that this task is admissible. On the other hand, in the case where (4.2.4) is satisfied, we must initially have:

$$x(q_0) = x_r(0) \qquad (4.2.13)$$

and, as previously seen, the task is redundant. This simply illustrates the fact that changing the initial condition q_0 may be enough to completely change the nature as well as the admissibility properties of the task.

4.2.2 Exploitation of redundancy

Since a non-admissible task is ill-conditioned with respect to the control problem, and since, as previously seen, a redundant task is not admissible, we need to modify it or at least to complement it in order to make it admissible.

When the initial task function is the gradient of a cost function $h(q, t)$ which we wish to minimize, a first possibility already evoked in Chapter 3 consists of adding to e a vector valued function which tends to make the resulting task admissible. For example:

$$e'(q, t) = e(q, t) + \lambda(q - y(t)).$$

This is in fact a regularization procedure whose only drawback is that it modifies the primary objective. But modifying the task is sometimes the only solution, when the complementation method, proposed next, is of no help. The complementation method is more specifically adapted to the treatment of redundant tasks, and consists of trying to take advantage of the unused degrees of freedom in order to achieve a secondary objective.

4.2.2.1 Constrained minimization of a secondary cost function

Let us assume that we are able to find a canonical redundant task function $e = \begin{pmatrix} e_1 \\ 0 \end{pmatrix}$ representing the initial (or primary) objective. A tempting idea is to consider a new task vector of the form

$$e(q, t) = \begin{pmatrix} e_1(q, t) \\ e_2(q, t) \end{pmatrix}. \qquad (4.2.14)$$

In (4.2.14), $e_2(q, t)$ is a new vector with dimension $n - m$, such that the resulting task function $e(q, t)$ must satisfy the admissibility conditions given in Chapter 3. Obviously, a basic requirement is that the equations $e_1(q, t) = 0$ and $e_2(q, t) = 0$ must be independent and compatible. Compatibility means that the intersection of the sets of zeros of e_1 and e_2 must be non-empty and also must have a regular structure. However, in practice, task e_2 is not often given directly in the form of a set of $n - m$ supplementary independent constraints, but is rather expressed as a *secondary minimization goal*. The global objective thus becomes a *constrained* minimization problem, which can be stated as follows:

for all $t \in [0, T]$ minimize a secondary C^2 cost function $h_s(q, t)$ with respect to q under the constraint

$$e_1(q, t) = 0. \qquad (4.2.15)$$

The need to have h_s of class C^2 will appear later.

The reader accustomed to constrained minimization problems or to the Implicit Functions Theorem will have guessed that solving this problem requires a projection operation somewhere. More precisely, it is well known that necessary conditions related to problem (4.2.15) are

$$e_1(q, t) = 0 \qquad (4.2.16)$$

$$\frac{\partial H}{\partial q}(q, t, \lambda) = g_s(q, t) + J_1^T(q, t)\lambda = 0 \qquad (4.2.17)$$

where $H = h_s + \lambda^T e_1$ is the associated Lagrangian, λ a vector of m Lagrange multipliers, and

$$g_s(q, t) = \frac{\partial h_s}{\partial q}(q, t) \qquad (4.2.18)$$

$$J_1(q, t) = \frac{\partial e_1}{\partial q}(q, t). \qquad (4.2.19)$$

We may now recall the following well-known result, which is an equivalent expression of (4.2.16) and (4.2.17).

Lemma 4.1 *If J_1 is of full rank, a solution q^* of the constrained minimization problem* (4.2.15) *satisfies the conditions:*

(1)
$$e_1(q^*, t) = 0 \qquad (4.2.20)$$

(2)
$$g_s(q^*, t) \text{ is orthogonal to } N(J_1(q^*, t)) \qquad (4.2.21)$$

where $N(J_1)$ is the null space of J_1.

Reciprocally, if q^* satisfies the above two conditions, it also satisfies the necessary conditions (4.2.16) and (4.2.17) associated with the problem. The proof is given in Appendix A4.1.1.

Intuitively, the second condition means that to keep e_1 equal to zero, we have to move along the null space of J_1. As long as g_s is not orthogonal to $N(J_1)$, h_s can be further reduced by moving along $N(J_1)$. At the optimum, g_s should be orthogonal to $N(J_1)$. This is equivalent to saying that g_s should belong to the range space $R(J_1^T)$ of J_1^T.

We now have to express these conditions within the form of a task function. A first idea would be to use condition (2) to form a $(n - m)$-dimensional vector, e_2, as the result of the orthogonal projection of $g_s(q, t)$ onto the null space of $J_1(q, t)$ and consider the following vector function

$$e(q, t) = \begin{pmatrix} e_1(q, t) \\ e_2(q, t) \end{pmatrix}. \qquad (4.2.22)$$

4.2 REDUNDANT TASKS

To satisfy condition (2), e_2 may be taken as

$$e_2 = R_1^T g_s \tag{4.2.23}$$

where R_1 is a matrix, the columns of which form a basis of $N(J_1)$. For example, if the partition

$$J_1 = (J_{11} \quad J_{12}) \tag{4.2.24}$$

is such that J_{11} is an $(m \times m)$ regular matrix, a basis of $N(J_1)$ is

$$R_1 = \begin{pmatrix} -J_{11}^{-1} J_{12} \\ I_{n-m} \end{pmatrix} \tag{4.2.25}$$

A more general choice for e_2 is:

$$e_2 = G R_1^T g_s = 0, \tag{4.2.26}$$

where G is an $(n-m) \times (n-m)$ regular matrix. This gives a particular solution to the problem of finding an n-dimensional output function.

The essential drawback to this solution is that it depends on the choice of the basis of $N(J_1)$ which may vary with time. This means for example that J_{11} may become singular while the rank of J_1 remains m. A way of avoiding this kind of difficulty is to compute the projection of g_s onto $N(J_1)$ in the surrounding space $\mathbf{R}^n$ without specifying a basis of $N(J_1)$. It then remains to map e_1 into $\mathbf{R}^n$ in a suitable way and add the two vectors to obtain the desired n-dimensional output. Let us just do this.

The orthogonal projection operator P_1 onto $N(J_1)$ is known to be

$$P_1 = I - W^T(WW^T)^{-1}W \tag{4.2.27}$$

where W is any $(m \times n)$ matrix such that $R(W^T) = R(J_1^T)$, or equivalently such that $W = CJ_1$ where C is a non-singular $(m \times m)$ matrix.

There is additional freedom in the choice of the projection operator when it is observed that the orthogonal projection of a vector g onto $N(J_1)$ is zero if and only if the projection onto a vector subspace F, such that $F \oplus R(J_1^T) = \mathbf{R}^n$, along $R(J_1^T)$, is zero.

Now suppose that F is given by a linear transformation of matrix A, that is $F = N(A)$ where A is an $(m \times n)$ full rank matrix and let us denote the range of A^T by $\mathbf{A}$. The projection operator P_1^A onto $N(A)$ along $R(J_1^T)$ is (Fig. 4.7)

$$P_1^A = I - W^T(AW^T)^{-1} A. \tag{4.2.28}$$

Since $\dim(\mathbf{A}) = n - m$ and $\dim(R(J_1^T)) = m$, the condition $N(A) \oplus R(J_1^T) = \mathbf{R}^n$ is equivalent to

$$N(A) \cap R(J_1^T) = \{0\}. \tag{4.2.29}$$

This last condition is equivalent to having AJ_1^T non-singular. (To prove this, let x be a vector such that $AJ_1^T x = 0$. Then $J_1^T x \in N(A) \cap R(J_1^T)$ so $J_1^T x = 0$. But J_1 is $(m \times n)$ and full rank $(m < n)$. This implies that J_1 is injective and

4 TASK REDUNDANCY

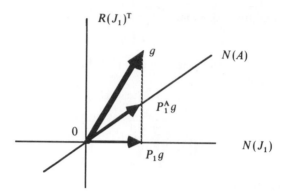

Fig. 4.7 Null space and projections

$x = 0$. AJ_1^T is a square matrix with null space $\{0\}$ so it is regular. The converse is obvious.) The non-singularity of AJ_1^T is already needed for the existence of the projection operator (4.2.28).

Defining

$$\tilde{e}(q, t) = P_1^A g_s(q, t) \tag{4.2.30}$$

gives us an n-dimensional function $\tilde{e}$ such that '$\tilde{e}(q, t) = 0$' is equivalent to '$g_s(q, t)$ is orthogonal to $N(J_1)$'.

Remark. The matrix A generally depends on q and t since the matrix AJ_1^T must be regular and J_1^T depends on q and t.

Let us now map the n-dimensional vector e_1 into $\mathbf{R}^n$. This is done simply by replacing e_1 by the n-dimensional vector $B^T e_1$ where B is a full rank $(m \times n)$ matrix. Indeed we then have

$$e_1(q, t) = 0 \Leftrightarrow B^T e_1(q, t) = 0. \tag{4.2.31}$$

Finally, let us add $\tilde{e}$, defined by (4.2.30), to $B^T e_1$ and consider the resulting vector-valued function

$$e(q, t) = B^T e_1(q, t) + \alpha(q, t) \tilde{e}(q, t) \tag{4.2.32}$$

where α is a scalar function never equal to zero (for example any non-zero real number). Conditions (1) and (2) of Lemma 4.1 are equivalent to $e(q, t) = 0$, if and only if $B^T e_1$ and $\tilde{e}$ belong to supplementary subspaces. Since $B^T e_1 \in R(B^T)$ and $\tilde{e} \in N(A)$, a necessary and sufficient condition is the non-singularity of the matrix AB^T. Like A, B may also depend on (q, t).

Let us summarize the above discussion as a Lemma.

Lemma 4.2 *Let*:

(1) $A(q, t)$ *be an* $(m \times n)$ *full rank matrix such that* $A(q, t) J_1^T(q, t)$ *is of full rank* m;

4.2 REDUNDANT TASKS

(2) $B(q, t)$ be an $(m \times n)$ *full rank matrix such that* $A(q, t)B^T(q, t)$ *is of full rank* m.

Then, '$e_1 = 0$ and g_s orthogonal to $N(J_1)$' is strictly equivalent to:

$$e = B^T e_1 + \alpha P_1^A g_s = 0 \qquad (4.2.33)$$

where $P_1^A = I_n - W^T(AW^T)^{-1}A$, $W(q, t)$ being an $(m \times n)$ matrix such that $R(W^T) = R(J_1^T)$, and $\alpha(q, t)$ is a scalar function never equal to zero.

Remarks.

1. The matrix $W^T(AW^T)^{-1}$ appearing in (4.2.28) is a generalized inverse of A.

2. A solution of equation (4.2.33) is indeed a *minimum* of the constrained problem if in addition the associated Hessian $\partial^2 H/\partial q^2$ is, at this point, positive on $N(J_1)$.

3. The scalar α was introduced in (4.2.32) for generality and because it represents an extra degree of freedom which can be used at the control level. When chosen constant, it can obviously be included in h_s.

4. Projecting g_s along $R(J_1^T)$ implies that it is necessary in practice to know J_1, or at least $R(J_1^T)$. We shall return to this point later.

4.2.2.2 General admissibility conditions

The previous lemma suggests using the new task function

$$e(q, t) = B^T(q, t)e_1(q, t) + \alpha(q, t) P_1^A(q, t) g_s(q, t) \qquad (4.2.34)$$

as a way of characterizing our constrained minimization objective.

In order to completely specify this function, once given the initial canonical redundant task function represented by e_1, the user still has to define a secondary objective by choosing a function $h_s(q, t)$ to be minimized (some existing possibilities will be described in section 4.3) and choose the matrix functions A and B, and the scalar function α.

A legitimate concern liable to influence these choices is the admissibility of the new task function e. Clearly, this admissibility must depend on certain properties of J_1, A, B, and it is therefore important to specify which conditions these matrices must satisfy.

Knowing that the central condition is that the task Jacobian matrix

$$J_T(q, t) = \frac{\partial e}{\partial q}(q, t) \qquad (4.2.35)$$

should remain non-singular along the ideal trajectory $q_r(t)$, it is possible to produce the following result.

Lemma 4.3 *Given the vector function e defined in (4.2.34) and q^*, a solution of $e(q, t) = 0$ at time t, then if*:

(1) $J_1(.)$ *is of full rank m*;

(2) $A(.)J_1^T(.)$ *and* $A(.)B^T(.)$ *are of full rank m*;

(3) $\alpha(.) \neq 0$;

(4) $x^T \dfrac{\partial^2 H}{\partial q^2}(., \lambda^*)x > 0$, *for* $x \neq 0 \in N(J_1(.))$, *with* $\lambda^* = -(A(.)J_1^T(.))^{-1}$
$\times A(.)g_s(.)$,

then

$$\det J_T(.) \neq 0$$

where $(.) = (q^*, t)$.

The proof is given on Appendix A4.1.2. Conditions (1) and (2) were already met in Lemma 4.2.

Condition (4) is intrinsic to the problem of minimization and depends on h_s. As will be seen later, it has however not to be used *explicitly*.

If the four conditions of the lemma are satisfied along an ideal trajectory $q_r(t)$ for $t \in [0, T]$, initiated at $q_r(0) = q_0$, then the task $\{e(q, t); q_0\}$ is at least 0-admissible.

Remarks

1. If q^* is a strict local minimum of $h_s(q, t)$ under the constraint $e_1(q, t) = 0$, then $(\partial^2 H/\partial q^2)(., \lambda^*)$ is *semi*-positive definite on $N(J_1(q^*, t))$. Condition (4) of Lemma 4.3 is slightly stronger.

2. This lemma provides conditions for 0-admissibility of the task $\{e(q, t), q_0\}$. Thus if, *initially*, the robot lies in a solution of the constrained problem (4.2.15) and if the task is 0-admissible on $[0, T]$, the ideal control which realizes $e(q, t) = 0$ on $[0, T]$ ensures *at each instant* that problem (4.2.15) is solved.

4.2.2.3 Some positivity conditions

Lemma 4.3 shows that regularity of $J_T(q_r(t), t)$ depends loosely on the choices made for $A(q, t)$ and $B(q, t)$. However, in practice this freedom may be exploited so as to separate more explicitly the respective contributions of the main and secondary tasks, or to simplify the implementation by reducing computing requirements; in fact, we shall see that both aspects are connected.

For example, we shall see in Chapter 6 that control algorithms can be simplified when J_T is positive definite around $q_r(t)$, in which case, it will be

4.2 REDUNDANT TASKS

shown that inverting J_T is not necessary in the synthesis of a control scheme ensuring stability (cf. Chapter 6, section 6.6.2). In this section we shall thus propose sufficient conditions on α, A, and B which lead to positivity of J_T on the ideal trajectory.

Let us define the matrix operator

$$S_{\gamma, A} = I_n + \gamma(I_n - A^+ A) \qquad (4.2.36)$$

where $\gamma(q, t)$ is a scalar positive function for any $q \in \mathbf{R}^n$ and $t \in [0, T]$. Note that this operator is the identity operator when $\gamma = 0$. Then we have the following result.

Lemma 4.4 *Let q^* denote a solution of $e(q, t) = 0$ for some $t \in [0, T]$. If:*

(1) $J_1(.)$ *is of full rank m;*

(2) $R(A^T(.)) = R(J_1^T(.))$;

(3) $B(.) = C(.)J_1(.)$; $C(.) > 0$;

(4) $\dfrac{\partial^2 H}{\partial q^2}(., \lambda^*)$ *is positive definite on $N(J_1(.))$;*

(5) $\alpha(.) > 0$;

where $(.) = (q^*, t)$, *then there exists a nondecreasing positive function $\gamma_m: \mathbf{R}^+ \to \mathbf{R}^+$ and a strictly positive real number α_M such that:*

(i) $\gamma_m(x) = 0$ *if $0 \leq x \leq \alpha_M$;*

(ii) $\gamma(.) \geq \gamma_m(|\alpha(.)|) \Rightarrow J_T(.)S_{\gamma, A}$ *is positive definite.*

The proof is given in appendix A4.1.3.

Basically, this lemma tells us that if A and B are chosen so as to satisfy conditions (2) and (3), which are stronger than the initial rank conditions of Lemma 4.2, and if γ is chosen large enough (its size depending on the size of α, also chosen positive) then we are entitled to expect the matrix $J_T S_{\gamma, A}$ to be positive at the optimum. The same result would hold in the case of a constrained maximization problem by replacing conditions (4) and (5) by:

(4') $\dfrac{\partial^2 H}{\partial q^2}(., \lambda^*)$ *is negative definite on $N(J_1(.))$;*

(5') $\alpha(.) < 0$.

This lemma also tells us that J_T may itself be rendered positive by choosing α small enough ($\alpha < \alpha_M$) since γ may then be taken equal to zero.

Let us now return to Lemma 4.3, and illustrate it with a simple example.

Example 4.5 Let us consider the system shown in Fig. 4.8.

The main task consists of 'controlling' the orientation of the second link with respect to F_0, i.e.:

$$e_1 = q_1 + q_2 - \theta(t) \qquad (4.2.37)$$

where $\theta(t)$ is the desired orientation of the second link.

The related Jacobian matrix is

$$J_1 = (1 \quad 1) \qquad (4.2.38)$$

which is always of full rank. Let us choose matrices A and B which satisfy the conditions of Lemma 4.3

$$A = J_1 \qquad (4.2.39)$$

$$B = J_1^+ = J_1^T (J_1 J_1^T)^{-1} = \tfrac{1}{2} \begin{pmatrix} 1 \\ 1 \end{pmatrix}. \qquad (4.2.40)$$

The orthogonal projection operator is

$$P_1 = \tfrac{1}{2} \begin{pmatrix} 1 & -1 \\ -1 & 1 \end{pmatrix} \qquad (4.2.41)$$

and, choosing $\alpha = 2$, a candidate task vector is

$$e = \begin{pmatrix} e_1 + g^1 - g^2 \\ e_1 - g^1 + g^2 \end{pmatrix} \qquad (4.2.42)$$

where $g_s = (g^1 \quad g^2)^T = \partial h_s / \partial q$, h_s being a secondary cost function to be minimized. Let us for example choose as a secondary goal the keeping of the height of O_2 close to a given value a. Thus

$$h_s = \tfrac{1}{2}(s_1 + s_{12} - a)^2 \qquad (4.2.43)$$

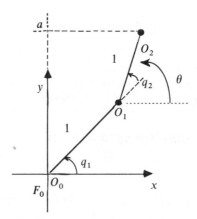

Fig. 4.8 A two-joint mechanism in a redundant problem

4.2 REDUNDANT TASKS

and
$$g^1 - g^2 = c_1(s_1 + s_{12} - a). \tag{4.2.44}$$

Then
$$J_T = \begin{pmatrix} 1 + c_1^2 - s_1^2 + c_{12}c_1 - s_{12}s_1 + as_1 & 1 + c_1 c_{12} \\ 1 - c_1^2 + s_1^2 - c_{12}c_1 + s_{12}s_1 - as_1 & 1 - c_1 c_{12} \end{pmatrix} \tag{4.2.45}$$

with determinant
$$\det(J_T) = c_1^2 - s_1^2 - s_{12}s_1 + as_1. \tag{4.2.46}$$

The following two cases may then be considered.

1. *The absolute minimum of h_s can be reached.* In this case we must have, along the ideal trajectory
$$q_1 + q_2 = \theta(t) \Rightarrow s_{12} = \sin(\theta(t)) \tag{4.2.47}$$

and
$$s_1 + s_{12} = a \Rightarrow s_1 = a - \sin(\theta(t)) \tag{4.2.48}$$

which requires
$$|a - \sin \theta(t)| \le 1. \tag{4.2.49}$$

Under this condition, we obtain
$$\det(J_T) = 1 - (\sin \theta(t) - a)^2. \tag{4.2.50}$$

A task singularity will therefore exist if, at some time t_1, we have $\sin \theta(t_1) - a = \pm 1$ (for example $\theta(t_1) = \pm \pi/2$ and $a = 0$).

Let us now come back to Lemma 4.3. Conditions (1) to (3) are satisfied. Concerning condition (4), we have, from equation (A4.8) of Appendix A4.1

$$\lambda^* = -(AJ_1^T)^{-1} A \frac{\partial h_s}{\partial q} \tag{4.2.51}$$

and therefore
$$\frac{\partial H(.,\lambda^*)}{\partial q} = P_1 g_s \tag{4.2.52}$$

where P_1 is given by (4.2.41).

$N(J_1)$ is spanned by $(1 \quad -1)$. Choosing $x = \beta(1 \quad -1)^T$, $\beta \ne 0$, and differentiating (4.2.52) we find

$$x^T \frac{\partial^2 H(.,\lambda^*)}{\partial q^2} x = \beta^2 \det(J_T) \tag{4.2.53}$$

which is always ≥ 0 under condition (4.2.49). The case where the determinant is equal to zero corresponds to the singularity and contradicts the condition (4) of Lemma 4.3.

2. *The absolute minimum of h_s cannot be reached.* When condition (4.2.49) is not satisfied (i.e. when $|a - \sin \theta(t)| > 1$), the constrained minimum of h_s is given by $c_1 = 0$. The ideal trajectory then is

$$\left\{ q_1(t) = \pm \frac{\pi}{2}; \quad q_2(t) = \theta(t) \pm \frac{\pi}{2} \right\}. \tag{4.2.54}$$

Along this trajectory, we have

$$\det(J_T) = -1 \pm (\sin \theta(t) - a) \tag{4.2.55}$$

which is always different from zero.

In fact, the singularity occurring when $|a - \sin \theta(t)| = 1$ corresponds to the switching from case 1, in which the *constrained* minimum of h_s coincides with the *absolute* one, to case 2 in which the absolute minimum ($h_s = 0$) is not reached. This problem was also encountered in the example of section 3.1.5 in Chapter 3.

In this example the primary task is redundant. However the rank of the associated Jacobian matrix J_1 is always equal to one (its maximal value). Therefore problems of admissibility of the new global task do not, in this example, result from changes in the rank of J_1. The previous discussion showed that such problems may nonetheless occur because of the *secondary* objective.

4.2.3 Extension to other systems of coordinates

Let $x(q)$ be any (local) system of generalized coordinates such that $\partial x/\partial q$ is non-singular on the domain D_x in $\mathbf{R}^n$. For example, in the specific case of joint coordinates, $x = q$ and $D_x = \mathbf{R}^n$. The basic observation which motivates this section is that inside D_x the equation $\partial H/\partial q = 0$ is equivalent to $\partial H/\partial x = 0$. A direct consequence is that all previous results can be restated exactly in the same form as before while replacing q by the new set of coordinates x (as long as the solution of the constrained minimization problem remains in D_x). The detailed rewriting of the results is left to the reader. The most noticeable effect of this change of coordinates comes from the differentiation operation which is now performed with respect to x instead of q. For example the general task function expression (4.2.34) now becomes

$$e = B^T e_1 + \alpha P_1^A g_s^x \tag{4.2.56}$$

with
$$P_1^A = I - W^T(AW^T)^{-1}A \qquad (4.2.57)$$

$$J_1^x = \frac{\partial e_1}{\partial x} \qquad (4.2.58)$$

$$R(W^T) = R((J_1^x)^T) \qquad (4.2.59)$$

$$g_s^x = \frac{\partial h_s}{\partial x} = \left(\frac{\partial x}{\partial q}\right)^{-T} g_s. \qquad (4.2.60)$$

The global task Jacobian matrix $J_T = \partial e/\partial q$ must also be replaced by $J_T^x = \partial e/\partial x$ in Lemmas 4.3 and 4.4.

The introduction of a set of generalized coordinates different from the natural set formed by the robot's joint variables is, in reality, a conceptual artifact which in some cases can be exploited to derive a simpler task function. Basically, the simplification comes from the fact that differentiating the primary task function e_1 with respect to a well-chosen set of coordinates may be simpler than differentiating it directly with respect to q. For example, this will be the case when the primary objective depends only on the location of the robot's end-effector. It is then tempting to choose for x a parametrization of the configuration of the end-effector, which itself can be advantageously replaced by the element $\bar{r}$ of SE_3 characterizing the location (position and attitude) of the end-effector. With an abuse of notation, we can write $x = \bar{r}$. Then, differentiating e_1 with respect to $\bar{r}$ will be the first step to differentiating e_1 with respect to q, according to the relationship

$$J_1 = J_1^r \frac{\partial \bar{r}}{\partial q} = J_1^r(P) \frac{\partial \bar{r}}{\partial q}(P)\Big|_F \qquad (4.2.61)$$

where $\partial \bar{r}/\partial q$ is defined from the differential of $\bar{r}$ by (cf. Chapter 1)

$$\begin{pmatrix} V(P) \\ \omega \end{pmatrix} = \frac{\partial \bar{r}}{\partial q}(P)\dot{q}\Big|_F \qquad (4.2.62)$$

with (V, ω) denoting the end-effector velocity screw and $\begin{pmatrix} V(P) \\ \omega \end{pmatrix}$ its vector representation, evaluated in a frame F at a point P linked to the end-effector.

However the choice $x = \bar{r}$ can be made only when the number of degrees of freedom of the robot is six, since $\partial x/\partial q$ has to be a square non-singular matrix. This choice is often implicitly made in *hybrid force/position* control schemes as well as in other sensor-based control schemes (see Chapter 7). In the other case, the determination of an intermediate set of generalized coordinates, more adequate than q, is not systematic, and usually $x = q$ is chosen.

4.2.4 Practical issues

As shown previously, there exists an infinite number of task functions of the form (4.2.34) or more generally (4.2.56), which can be associated with the constrained minimization problem (4.2.15). In practice, it is necessary for the design of the control to completely specify the task function which is going to be used. This means that the functions α, A and B, as well as the set of coordinates x, have to be determined. Practical constraints versus theoretical considerations will understandably influence the choice of these terms. Therefore this choice will be highly dependent on the nature of the specific task under consideration. At the present level of abstraction where we do not yet wish to specify the application, we can merely give general indications corresponding to a few important selected cases that one might encounter in practice. Some of the proposed options will be illustrated and further studied in Chapter 7 when applying the theory to sensor-based control.

4.2.4.1 Knowledge of the Jacobian matrix J_1

A first important practical issue is the knowledge that we have of the Jacobian matrix $J_1(q, t)$ and, more precisely, of the range of its transpose, $R(J_1^T)$. Indeed, minimizing a secondary cost function under the constraint $e_1 = 0$ usually requires, as seen previously, a perfect knowledge of $R(J_1^T)$ *along the ideal trajectory*. The reason is that this minimization involves projecting the gradient vector of the secondary cost onto the null space $N(A)$ of the matrix A along $R(J_1^T)$.

In practice, the perfect knowledge of $R(J_1^T)$ (or equivalently of $N(J_1)$) may be beyond our grasp because our knowledge of the primary task function e_1 may itself be partial. This will, for example, be the case when e_1 involves outputs of exteroceptive sensors, the response of which, with respect to the environment, is not precisely known; although the values taken by e_1 are measured continually, the explicit expression of e_1 is unknown and so is its partial derivative J_1. The consequence of this is that $R(J_1^T)$ will, in practice, be approximated by a subspace denoted by $\hat{R}(J_1^T)$ and the ideal projection operator P_1^A will be replaced by the operator $\hat{P}_1^A$ which projects onto $N(A)$ along $\hat{R}(J_1^T)$. Accordingly, the chosen task function will be of the form

$$e = B^T e_1 + \alpha \hat{P}_1^A g_s. \qquad (4.2.63)$$

Obviously, if $\hat{P}_1^A$ is different from P_1^A, the secondary cost will (usually) not be minimized by keeping e equal to zero (Fig. 4.9). At best, it will be possible to approximate the secondary objective by preventing the 'distance' between $\hat{R}(J_1^T)$ and $R(J_1^T)$ from being large. In some applications, usually based on the use of exteroceptive sensors, this may motivate the introduction of mechanisms providing an on-line estimation of $R(J_1^T)$ (or equivalently of $N(J_1)$

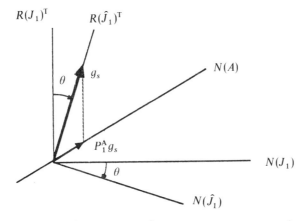

Fig. 4.9 When $\hat{R}(J_1^T) \neq R(J_1^T)$, $\hat{P}_1^A g_s = 0$ does not imply that $P_1^A g_s = 0$

along the ideal trajectory). By estimating a subspace, we of course mean estimating any matrix which spans this subspace.

Fortunately, the realization of the primary task is more robust with respect to errors made in modelling $R(J_1^T)$ than the realization of the secondary objective. Indeed, by premultiplying the right member of (4.2.63) by A^T we easily verify that $e = 0$ always implies $A^T B e_1 = 0$, which itself implies $e_1 = 0$ whenever $A^T B$ is non-singular. Therefore A and B should at least be chosen such that $A^T B$ is non-singular. This is hardly a restrictive condition. For instance, it is fulfilled in practice simply by taking $B = A$ or $(A^+)^T$, with A a full-rank $m \times n$ matrix function.

A conclusion to this discussion is that it is conceptually convenient, for design and analysis purposes, to assume that the estimated subspace $\hat{R}(J_1^T)$ is equal to the 'true' subspace $R(J_1^T)$. However, while doing so, it is important to keep in mind that all terms involving J_1 may, in practice, have to be replaced by estimates and that the regulation of the task function (4.2.63) to zero (as long as the task remains admissible) is then liable to ensure the minimization of the secondary cost function *only when the estimates are perfect along the ideal trajectory.*

4.2.4.2 Condition (4) in Lemmas 4.3 and 4.4

As already pointed out, the important condition (4), which we recall:

$$\frac{\partial^2 H}{\partial q^2}(q^*, t, \lambda^*) \text{ is positive on } N(J_1(q^*, t)),$$

is inherent to the problem of constrained minimization. Its satisfaction depends on the properties of the secondary cost function in relation to those of e_1. When it is not satisfied at some point on the ideal trajectory, the minimization problem is ill-conditioned and so is the task because J_T is singular at this point.

It may therefore be useful to test this condition in advance before implementing the task on the physical system. Let us briefly comment upon this condition when the following property of the primary task function holds:

$$N(J_1(q^*, t)) \subset N\left(\frac{\partial^2 e_{1,k}}{\partial q^2}(q^*, t)\right); \quad k = 1, \ldots, m \quad (4.2.64)$$

where the $e_{1,k}$ are the components of the vector e_1.

Then, from relation (A4.17) in appendix A4, we have:

$$x_2^T \frac{\partial^2 H}{\partial q^2}(q^*, t, \lambda^*) x_2 = x_2^T \frac{\partial^2 h_s}{\partial q^2}(q^*, t) x_2 \quad (4.2.65)$$

for all $x_2 \in N(J_1(q^*, t))$.

Therefore the condition (4) of Lemmas 4.3 and 4.4 reduces to a positivity condition bearing upon the second partial derivative of the function h_s when the primary task function is such that the property (4.2.64) is true. Then, if by any chance, h_s happens to be globally convex in the variable q, the condition is obviously satisfied.

A particularly simple case is when $e_1(q, t)$ depends linearly on the variable q. Then J_1 is a function of the single variable t, $\partial^2 e_{1,k}/\partial q^2 = 0$, and therefore $N(\partial^2 e_{1,k}/\partial q^2) = \mathbf{R}^n$. In this case, (4.2.64) is clearly true. Furthermore, from (A4.17):

$$\frac{\partial^2 H}{\partial q^2}(q^*, t, \lambda^*) = \frac{\partial^2 h_s}{\partial q^2}(q^*, t). \quad (4.2.66)$$

Let us now recall that, when using x instead of q as a set of generalized coordinates, the previous results (as well as most of the results derived in this chapter) still hold by differentiating all functions with respect to x instead of q. A practical example is provided when e_1 is the output of a sensor measuring the force resulting from the contact between the tip of a robot's end-effector and a planar surface, and with $x = \bar{r}$. The linearity of e_1 with respect to the displacements of the end-effector is a consequence of the elasticity of the contact (this assumption is discussed in Chapter 7). Then, a convex function h_s can be obtained by considering the square of some distance in SE_3 between the location of the end-effector and some predefined ideal one. This is a typical hybrid force/position problem.

4.2.4.3 A set of simple task functions

An important case, more general than the one considered in the previous section, is when $N(J_1(q^*, t))$ (where q^* satisfies the necessary conditions

4.2 REDUNDANT TASKS

associated with the constrained minimization problem) is independent of the choice of the secondary cost function h_s. This happens when, *at any time t, the null-space $N(J_1(q,t))$ is invariant over the set $\{q^{**}\}|_t$ of solutions of $e_1(q,t) = 0$.* Then:

(1) $\{q^{**}\}|_t = q_1^{**} \oplus N(J_1^*(t))$, where $N(J_1^*(t))$ denotes the invariant null space and q_1^{**} any solution of $e_1(q,t) = 0$;

(2) $\dfrac{\partial^2 e_{1,k}}{\partial q^2}(q,t)x_2 = 0$, for any $x_2 \in N(J_1^*(t))$ and $q \in \{q^{**}\}|_t$, $k = 1, \ldots, n$.

Since the solution q^* to the constrained minimization problem belongs to the set $\{q^{**}\}|_t$, it results from point (2) that the property (4.2.64) is also satisfied. This in turn implies that relation (4.2.65) is true in this case. It is possible to take advantage of this property of e_1 by choosing the task function within the following set

$$e(q,t) = B^T(t)e_1(q,t) + \alpha(q,t)[P_1^A(t)]^* g_s(q,t) \qquad (4.2.67)$$

where

$$[P_1^A(t)]^* = P_1^A(q^{**}, t) \qquad (4.2.68)$$

is the operator which projects onto $N(A(t))$ along $R(J_1^*(t)^T)$. In practice $[P_1^A(t)]^*$ is computed as follows:

$$P_1^A(t) = I_n - W^T(t)(A(t)W^T(t))^{-1}A(t) \qquad (4.2.69)$$

where $A(t)$ is chosen such that $A(t)W(t)$ is non-singular, and $W(t)$ such that $N(W(t)) = N(J_1^*(t))$. Of course, $B(t)$ is chosen such that $A(t)B^T(t)$ is non-singular.

With respect to the class of functions (4.2.34), the set (4.2.67) presents two obvious advantages.

1. The global function $e(q,t)$ is simpler because $[P_1^A(t)]^*$ and $B(t)$ do not depend on the variable (q).

2. Only the knowledge of the subspace $N(J_1^*(t))$, which is invariant on the set of solutions of $e_1(q,t) = 0$ is required. In the case of a function (4.2.34), it is necessary to know the null space of J_1 everywhere. In practice, this may prove to be a very restrictive condition (cf. section 4.2.4.1).

Furthermore, as $[P_1^A(t)]^*$ coincides with $P_1^A(q,t)$ on the set $\{q^{**}\}|_t$ of solutions of e_1, one can show that both Lemmas 4.3 and 4.4 continue to hold when a task function of the form (4.2.67) instead of (4.2.34) is considered, and when e_1 satisfies the conditions given at the beginning of this section. The proofs are similar to those given in Appendix A4 and are left to the interested reader. The most noticeable differences are that $(\partial^2 H/\partial q^2)(q^*, t, \lambda^*)$ has now to be replaced by $(\partial^2 h_s/\partial q^2)(q^*, t)$ in the expression A4.23 of the task

Jacobian $J_T(q^*, t)$, and that the condition (4) in Lemmas 4.3 and 4.4 can be written equivalently:

$$\frac{\partial^2 h_s}{\partial q^2}(q^*, t) \text{ is positive on } N(J_1^*, (t)).$$

As a consequence, the conditions for 0-admissibility of the task function are the same.

An additional simplification arises when $N(J_1^*)$ is constant for all $t \in [0, T]$. Then, constant matrices A and B can be utilized. The importance of this case will appear more clearly in Chapter 7 where primary tasks based on the use of exteroceptive sensors are considered. Then the natural change of coordinates will be $x = \bar{r}$ and a nice property of e_1 will be the invariance of $N(\partial e_1^*/\partial \bar{r}(P)|_F)$ over the set of $e_1 = 0$, $\partial e_1^*/\partial \bar{r}$ being evaluated at a point P and a frame F tied to the end-effector.

4.2.4.4 Choice of $A(q, t)$ and $B(q, t)$

Let us now describe two particularly interesting possibilities for the choice of A and B in practice.

Preliminary remark. For the sake of simplicity, it is assumed from now on that the task function is chosen within the set (4.2.34). However, when the conditions described in the previous paragraph are satisfied and when the task function is instead chosen within the set (4.2.67), all the following conditions bearing upon A and B (and W) are required only on the set $\{q^{**}\}|_t$ of solutions of $e_1(q, t) = 0$, instead of being required everywhere.

Choice 1

$$\left. \begin{array}{l} A = (I_m \quad 0) \\ B = A(= (A^+)^T) \\ \alpha \neq 0 \end{array} \right\} \quad (4.2.70)$$

Let $W(q, t)$ denote an $m \times n$ matrix, known to the user, and such that $R(W^T) = R(J_1^T)$. Recall that this equality between subspaces is equivalent to the existence of a non-singular matrix C such that $W = CJ_1$. It is also equivalent to having the following two conditions simultaneously satisfied:

$$\left. \begin{array}{l} J_1 W^T \text{ non-singular} \\ (I_n - W^+ W)J_1^T = 0 \end{array} \right\} \quad (4.2.71)$$

4.2 REDUNDANT TASKS

or
$$\left.\begin{array}{c} J_1 W^T \text{ non-singular} \\ (I_n - J_1^+ J_1) W^T = 0 \end{array}\right\}. \quad (4.2.72)$$

A particular case is of course $W = J_1$, in which case J_1 must be perfectly known.

With the previous choice of A and B, and with $P_1^A = I_n - W^T(AW^T)^{-1}A$, it is easy to verify that the task function (4.2.63) can also be written

$$e = \begin{pmatrix} e_1 \\ e_2 \end{pmatrix} \quad (4.2.73)$$

with

$$e_2 = \alpha R_1^T g_s \quad (4.2.74)$$

$$R_1 = \begin{pmatrix} -W_1^{-1} W_2 \\ I_{n-m} \end{pmatrix} \quad (4.2.75)$$

$$W = (W_1 \; W_2) \quad (4.2.76)$$

where $\dim W_1 = m \times m$, $\dim W_2 = m \times (n - m)$.

As could be expected from the previous analysis the column vectors of R_1 form a particular basis of $N(J_1)$ when W_1 is non-singular. We also see, from (4.2.73), that this particular choice of A and B corresponds to adding $(n - m)$ independent constraints (the components of e_2) to the primary task function e_1. Clearly these constraints represent the contribution of the secondary objective to the new global task. Remember that this solution, along with the task function (4.2.73)–(4.2.76), has already been proposed, following a more conventional derivation, in the introductory section 4.2.2.1. As was pointed out then a first drawback of this solution is that it requires W_1 to be non-singular. The non-singularity of W_1 is also a necessary condition for AJ_1^T to be non-singular, a condition already met in several previous lemmas. However this condition is also too strong because W_1 may become singular while W (or equivalently J_1) remains of full rank. In this case it would clearly be possible to have AJ_1^T non-singular by choosing A differently, for example $A = J_1$. In fact, the same type of difficulty would appear whenever attempting to exhibit a basis of $N(J_1)$ that remains a basis when J_1 continuously changes. It is known that besides the case when $\dim N(J_1) = 1$, there is no systematic and 'nice' way of doing this. The difficulty can only be avoided by choosing A so that the projection onto $N(A)$ along $R(J_1^T)$ does not require the specification of a basis of $N(J_1)$. Choice 2 will illustrate this possibility.

Another potential drawback is that choice 1 does not usually confer interesting properties, such as positivity, to the task Jacobian J_T. A consequence of this, at the control level, is that it will usually be necessary to explicitly compute this matrix in order to form the control. The reason is explained in Chapter 6 in relation to an important positivity condition which should be satisfied for ensuring stability and robustness of the control. The necessity of computing J_T can in turn raise difficulties. Indeed this computation can be complex because it involves differentiating $e_1(q,t)$ and $h_s(q,t)$ twice. It may also in some cases prove to be impossible when the explicit expression of the primary task function is not known. This will typically be the case when the task consists of monitoring exteroceptive sensor outputs.

Choice 2

$$\left.\begin{array}{c} A = W \\ B = (W^+)^T \\ \alpha > 0 \end{array}\right\} \qquad (4.2.77)$$

where $W(q,t)$ is a matrix function, *known* to the user, and such that, in this case, the two following conditions are simultaneously satisfied

$$R(W^T) = R(J_1^T) \qquad (4.2.78)$$

$$J_1 W^T > 0. \qquad (4.2.79)$$

Condition (4.2.78) was already present in choice 1. One can easily verify that the pair of conditions (4.2.78) and (4.2.79) is also equivalent to

$$(I_n - W^+ W)J_1^T = 0 \qquad (4.2.80)$$

$$J_1 W^T > 0. \qquad (4.2.81)$$

With this choice of A and B, and with $P_1^A = (I_n - W^+ W)\ (= P_1)$, the task function (4.2.34) becomes

$$e = W^+ e_1 + \alpha(I_n - W^+ W)g_s. \qquad (4.2.82)$$

A first potential advantage of using this task function rather than the one resulting from choice 1 is that possible singularities of W_1 (or of any given $(m \times m)$ block matrix extracted from W) do not create task singularities when W (or J_1) remain full rank matrices.

In addition, the conditions (2), (3) and (5) of Lemma 4.4, the satisfaction of which depends on the choice of α, A and B, are satisfied in the case of choice 2. Indeed condition (5) is the same as choosing α positive. Condition (2) is the same as (4.2.78) with $A = W$. From (4.2.80) we have $(W^+)^T = CJ_1$ with $C = (J_1 W^T)^{-1}$. Since the matrix $J_1 W^T$ is positive, according to (4.2.81), its inverse is also positive and so is C. This proves that condition (3) is satisfied. Therefore, we are within the domain of application of Lemma 4.4 and this

4.2 REDUNDANT TASKS

lemma tells us that the task Jacobian matrix J_T, associated with the task function (4.2.82), post-multiplied by the matrix $S_{y,w}$, is likely to be positive definite. As already pointed out, this property can in turn be utilized in order to simplify the control expression because the complete calculation of J_T can then be avoided (more details about this will be found in Chapter 6). The good features of the task functions (4.2.82) that we have just emphasized make these functions particularly well suited, among the class (4.2.34) of possible task functions, for the practical treatment of many redundant tasks. We shall illustrate this in Chapter 7. For instance, we shall see how classical hybrid force/position control strategies and related concepts can be derived from (4.2.82).

4.2.5 A differential motion point of view

Much of the work reported in the literature on redundant robots does not take the complete problem of control into account because the question of redundancy is mainly examined from a differential motion point of view. The solution, generally proposed under the name of *resolved motion rate control*, consists of finding a joint velocity vector in the form

$$\dot{q} = \left(\frac{\partial x}{\partial q}\right)^+ \dot{x} - \left(I_n - \left(\frac{\partial x}{\partial q}\right)^+ \frac{\partial x}{\partial q}\right)\frac{\partial h_s}{\partial q} \qquad (4.2.83)$$

where $x = f(q)$ is the direct kinematic transformation, $\dot{x}$ is a predetermined velocity of the end-effector in the Cartesian space and $h_s(q)$ is a secondary cost function to be minimized.

For control purposes, this solution is usually transformed into a 'velocity control' scheme. This is achieved, for example, by rewriting (4.2.83) in the following form

$$\dot{q}_d = -\left(\left(\frac{\partial x}{\partial q}\right)^+ \dot{x}_d + \left(I_n - \left(\frac{\partial x}{\partial q}\right)^+ \frac{\partial x}{\partial q}\right)\frac{\partial h_s}{\partial q}\right) \qquad (4.2.84)$$

with

$$\dot{x}_d = -k(x - x_r); \quad k > 0 \qquad (4.2.85)$$

where $x_r(t)$ is usually the desired position of the end-effector in the Cartesian space at time t. In this case, $\dot{x}_d$ has the meaning of a 'desired velocity', while the computed vector $\dot{q}_d$ has the meaning of a 'desired joint velocity', the components of which become the inputs of the actuators' built-in control loops.

The connections between the task function approach developed in this book and conventional velocity control schemes will be detailed in the next chapter. However, we need, at this point, to slightly anticipate the discussion of this matter in order to formally show how the classical differential motion approach of redundancy relates to the present approach. From our point of view, equations (4.2.84) and (4.2.85) are a particular case of a more general relation which can be written

$$\dot{q}_d = -\mu e \qquad (4.2.86)$$

where μ is a positive scalar gain and e a task function associated with the robot task, such that the task Jacobian matrix $J_T = \partial e/\partial q$ is *positive* along the ideal trajectory. This relation clearly indicates how a 'desired velocity' vector $\dot{q}_d$ can be computed simply from the measurements of the values taken by the task function. Relations even more general than (4.2.86) will be derived in Chapter 5.

By comparing (4.2.84) and (4.2.85) with (4.2.86) it is easy to verify that the task function which is implicitly involved in the differential motion approach of redundancy belongs to the subclass of functions (4.2.82) and that it is obtained by making the following assignments:

$$e_1(q, t) = x(q) - x_r(t) \qquad (4.2.87)$$

$$W(q, t) = J_1(q, t) = \frac{\partial x}{\partial q}(q) \qquad (4.2.88)$$

$$\alpha = \frac{1}{k}. \qquad (4.2.89)$$

Then the gain μ is equal to k. A first conclusion is that the task function approach potentially contains the classical differential motion approach of redundancy. Indeed not only can relation (4.2.86) be applied to other redundant tasks, based for example on the use of sensors (a different function e_1 is then to be considered), but also the task function concept is not attached to any particular control strategy such as velocity control.

In order to complete the parallel between the two approaches, a simple analysis, performed at the differential motion level as it is often done, can be applied to the task function (4.2.82). To this end, let us consider the relation (4.2.86) with e given by (4.2.82). Let us also assume that it is possible via an ideal control scheme to have the vector of joint velocities of the robot exactly follow the desired velocity $\dot{q}_d$, i.e. $\dot{q} = \dot{q}_d$. Then we have

$$\dot{q} = -\mu e, \quad \mu > 0 \qquad (4.2.90)$$

with

$$e = W^+ e_1 + \alpha(I_n - W^+ W)g_s. \qquad (4.2.91)$$

Let us recall that the matrix function W is such that the following two properties hold:

$$(I_n - W^+ W)J_1^T = 0 \qquad (4.2.92)$$

$$J_1 W^T > 0 \qquad (4.2.93)$$

4.2 REDUNDANT TASKS

and that the task Jacobian matrix J_T is positive in the neighbourhood of the ideal trajectory (assuming that all the conditions of Lemma 4.4 are satisfied and that α is chosen positive and small enough). Therefore there exists some positive real number such that

$$x^T J_T x > \beta \|x\|^2 \tag{4.2.94}$$

for $x \neq 0$.

For the sake of simplicity α is chosen constant and the functions e_1, h_s and W do not depend on the independent time variable. As a consequence, e does not itself depend on the time variable and we have

$$\dot{e} = J_T \dot{q}. \tag{4.2.95}$$

Premultiplying (4.2.95) by e^T and taking (4.2.90) into account leads to:

$$\frac{1}{2}\frac{d}{dt}\|e\|^2 = -\mu e^T J_T e \tag{4.2.96}$$

which, combined with (4.2.94), gives

$$\frac{1}{2}\frac{d}{dt}\|e\|^2 < -\mu\beta\|e\|^2. \tag{4.2.97}$$

It follows from this inequality that e tends to zero. This in turn implies (Lemma 4.2) that e_1 and $P_1 g_s$ also tend to zero, and that the constrained minimization objective is asymptotically reached.

A second possible analysis, which does not explicitly rely on the positivity property of the global task Jacobian matrix J_T, is as follows. Let us consider the case when J_1 is known and when

$$W = J_1. \tag{4.2.98}$$

Then, left-multiplying (4.2.90) by J_1 and replacing e by the right-hand side of (4.2.90), while taking (4.2.98) into account, leads to

$$J_1 \dot{q} = -\mu e_1. \tag{4.2.99}$$

Premultiplying (4.2.99) by e_1^T, and since $\dot{e}_1 = J_1 \dot{q}$, we obtain

$$\frac{1}{2}\frac{d}{dt}\|e_1\|^2 = -\mu\|e_1\|^2 \tag{4.2.100}$$

which shows that e_1 tends to zero. The primary task is thus asymptotically achieved.

Let us now premultiply (4.2.90) by g_s^T and replace e by the right-hand side of (4.2.91) while taking again (4.2.98) into account. We obtain

$$g_s^T \dot{q} = -\mu g_s^T J_1^+ e_1 - \mu\alpha g_s^T P_1 g_s. \tag{4.2.101}$$

If g_s and J_1^+ are bounded, the first term on the right-hand side of (4.2.101) asymptotically vanishes because e_1 tends to zero. Therefore, since $\dot{h}_s = g_s^T \dot{q}$, we have asymptotically

$$\dot{h}_s \approx -\mu \alpha g_s^T P_1 g_s. \qquad (4.2.102)$$

Now, since P_1, as the matrix of a projection operator, is symmetric and idempotent, it is positive (although not strictly positive). Relation (4.2.102) thus indicates that the time derivative of h_s stays negative (h_s decreases) until $P_1 g_s$ becomes equal to zero, i.e. until the constrained minimum is reached.

This kind of differential motion analysis is useful because it gives an intuitive understanding of the global task function e. However, in no sense should it be considered as a clean proof of good control behaviour. Indeed, it is based on several simplifying and restrictive assumptions, and overall the problem of realizing the underlying ideal control scheme remains unsolved. A more general, realistic and mathematically sound analysis will be proposed in Chapter 6 when taking the control expression and the robot's dynamics into account.

4.3 EXAMPLES OF SECONDARY TASKS

4.3.1 Introduction

It has been shown previously how it is possible to modify an underdimensioned task (i.e. not admissible) into a possibly admissible one by adding a supplementary goal to the initial problem. The general form of the new task is

$$e = B^T e_1 + \alpha P_1^A g_s \qquad (4.3.1)$$

with P_1^A a projection operator onto $N(A)$ along $R(J_1^T)$ given in (4.2.28). All terms are functions of q and t.

In the general case, the explicit calculation of the function e, and especially of $J_T = \partial e / \partial q$ is not straightforward in practice. However, it will be shown in Chapter 6 that two situations may occur when deriving the control associated with e:

(1) J_T is not *a priori* positive, and it is necessary to have at least an approximate knowledge of it;
(2) J_T is known to be positive (as with the particular choices of A and B given in the previous section), and it is not absolutely necessary to compute it.

For this reason we shall not discuss the issues of the complexity of e in this section, and we shall focus on some practical examples of supplementary task functions. In all cases, these supplementary tasks will be derived in the form of minimization problems, even if the original objective is given as a set of

4.3 EXAMPLES OF SECONDARY TASKS

supplementary constraints, so as to exploit in the control the freedom allowed by the choices of A and B.

Traditionally, three classes of secondary tasks may be distinguished.

1. *Tasks based upon internal trajectory criteria*, i.e. which do not depend on the environment. In this case, $h_s(q, t)$ depends only on the internal position, q, and is often time-independent. The most frequently encountered goals are *singularity avoidance* and *removal from joint limitations*.

2. *Tasks based upon external trajectory criteria*. Here, the goal expressed by $h_s(q, t)$ depends on external information or on the environment of the robot. One of the main goals here is to avoid obstacles or to stay close to a desired trajectory (which might be generated in real time).

3. *Tasks based upon cost functions connected with energy aspects*. Basically this last class takes the magnitude of drive torques into account, or more general criteria such as the global kinetic energy, or the norm of the joint velocity vector. However, such cost functions are not directly included in the present analysis, for they depend not only on q and t, but also on $\dot{q}$ and $\ddot{q}$. Some transformations or approximations will thus be needed.

These three classes of task will be analysed successively in the present section.

4.3.2 Internal cost functions

4.3.2.1 Singularities of e_1

Let us return to the function $e_1(q, t)$. It is not admissible by itself due to its dimension, and thus the notion of singularity for this function in the sense of Chapter 3 is not defined. The construction of the new task e we gave in section 4.2 is based on the assumption that J_1 is of full rank. Therefore problems may arise when forming the new task if no specific care is taken to ensure that J_1 stays of full rank all along the ideal trajectory. As a consequence of definition D4.6 we may put forward the following definition:

Definition D4.7 *A point q^* is a singular point at time t^* of the function $e_1(q, t)$ if $e_1(q^*, t^*) = 0$ and $J_1(q^*, t^*)$ is not of full rank.*

If we now consider an ideal trajectory for the 'primary task' $\{e_1(q, t); q_0\}$, i.e. a trajectory $\{q_r(t)\}$ such that $e_1(q_r(t), t) = 0$ for $t \in [0, T]$, it is clear that J_1 will be of full rank along this trajectory if it does not contain singular points of e_1. However, the existence of singular points on an ideal trajectory is not as dramatic as when the task is not redundant, because the ideal trajectory is no longer unique. We still may find another trajectory which does not pass through a singular point. The question which arises then is the following: is it possible to take advantage of redundancy to avoid singular points of the primary task? A necessary condition is obviously the existence of at least one ideal trajectory along which J_1 is of full rank. The use of a well-chosen secondary cost function h_s to be minimized is a way of finding such a trajectory, but is has no *a priori* guarantee of success.

In some cases, the property of 'avoidability' can be easily exhibited, as in the following example.

Example 4.6 Let us consider the three-bar mechanism of Fig. 4.10, with the time-independent function

$$e_1(q) = (x(q) - x_0 \quad y(q) - y_0)^T \tag{4.3.2}$$

where

$$x(q) = c_1 + c_{12} + c_{123} \tag{4.3.3}$$

$$y(q) = s_1 + s_{12} + s_{123}. \tag{4.3.4}$$

The Jacobian is:

$$J_1 = \begin{pmatrix} -s_1 - s_{12} - s_{123} & -s_{12} - s_{123} & -s_{123} \\ c_1 + c_{12} + c_{123} & c_{12} + c_{123} & c_{123} \end{pmatrix}. \tag{4.3.5}$$

1. Let us consider the point $\{x_0 = 1; y_0 = 0\}$, and the joint position vector

$$q^* = (0 \quad 0 \quad \pi)^T. \tag{4.3.6}$$

It is a singular point, because $e_1(q^*) = 0$ and because the rank of

$$J_1(q^*) = \begin{pmatrix} 0 & 0 & 0 \\ 1 & 0 & -1 \end{pmatrix} \tag{4.3.7}$$

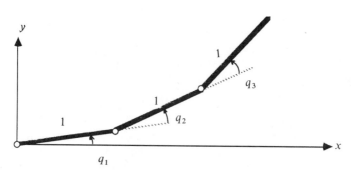

Fig. 4.10 Three-bar mechanism

4.3 EXAMPLES OF SECONDARY TASKS

is equal to one. This singular point is avoidable, because we have $e_1(q) = 0$ for any q belonging to the subspace defined by

$$\left. \begin{array}{l} q_3 = q_1 - \pi \\ q_2 = -q_1 \end{array} \right\}. \qquad (4.3.8)$$

2. If we now choose as a goal the point $\{x_0 = 3;\ y_0 = 0\}$, the *unique* solution of $e_1(q) = 0$ is obviously

$$q^* = (0\ 0\ 0)^T \qquad (4.3.9)$$

with the rank of

$$J_1(q^*) = \begin{pmatrix} 0 & 0 & 0 \\ 3 & 2 & 1 \end{pmatrix}, \qquad (4.3.10)$$

equal to one. Because of the uniqueness of the solution, the task is not redundant and this singular point is unavoidable.

Remark. When the singularities are of kinematic origin, it is generally easy to find which singularities are avoidable. For general or more complex redundant tasks e_1, the analysis may not be so simple.

4.3.2.2 Avoidance of singular points of the primary task

Redundancy can be exploited to keep the robot's ideal trajectory far from avoidable singularities. A method consists of using a cost function h_s, which is able to point out a possible tendency to rank deficiency of J_1.

Several functions, the values of which are related to the rank of J_1, have been proposed. One of them is the sum of the squares of the principal minors of J_1. In Example 4.6, this reduces to the norm of the cross-product of the two lines of J_1, which spans $N(J_1(q))$ out of singular points.

A general approach is obtained by studying the m 'singular values', σ_i, of J_1, in the classical sense of this notion. Indeed, the σ_i are the square roots of the eigenvalues of $J_1 J_1^T$, and the number of *non-zero* singular values is equal to the rank of J_1. From this comes the idea of choosing

$$h_s(q) = \prod_{i=1}^{m} \sigma_i = \sqrt{\det(J_1 J_1^T)} \qquad (4.3.11)$$

and trying to *maximize* h_s, which may be called the 'generalized manipulability function' by extension of the usual one. In the particular case when we are interested in avoiding kinematic singularities, we may, for instance, try to maximize

$$h_s(q) = \sqrt{\det(J_N(q) J_N^T(q))} \qquad (4.3.12)$$

where J_N is the basic Jacobian matrix associated with link N (cf. Chapter 2). This function is known as the *manipulability*.

The advantage of this approach lies in its generality, as an analytical knowledge of singular domains is not needed. However, the generalized manipulability functions defined in this way have two main drawbacks:

1. Recalling that we have to derive $g_s = \partial h_s/\partial q$ analytically, and, if needed, $\partial P_1^A g_s/\partial q$ (in order to compute $\partial e/\partial q$ in the control), the total number of computations with h_s defined in (4.3.11) may be very high.

2. When $e_1(q, t)$ includes a representation of the orientation with respect to a given frame, it has been shown in Chapter 3 that the associated singularities lead to unboundedness of the Jacobian. Then, the attempt to maximize h_s given by (4.3.11) is not recommended. Another function which may instead be considered is the *dispersion* of the singular values, $d = (\sigma_{\min}/\sigma_{\max})(J_1)$. This ratio is also the inverse of the *condition number*. The maximization of d takes care of the singularities of J_1 through $\sigma_{\min}$ and of its size through $\sigma_{\max}$.

In practice, it is sometimes possible to know the singular domains explicitly. Then simpler manipulability-like functions may be derived. Returning to Example 4.6, we easily find:

$$\|j_1 \times j_2\|^2 = \det J_1 J_1^T = \sum (\text{principal minors})^2$$
$$= s_3^2 + (s_3 + s_{23})^2 + (s_2 + s_{23})^2 \qquad (4.3.13)$$

which vanishes if and only if

$$s_2 = s_3 = 0. \qquad (4.3.14)$$

When starting at the minimum and being sure that it is possible to stay close to it, several functions may be formed, among which are

$$h_s(q, t) = \lambda_1 s_2^2 + \lambda_2 s_3^2; \quad \lambda_i > 0 \qquad 2 \quad (4.3.15)$$

to be maximized; and

$$h_s(q, t) = \tfrac{1}{2}(\lambda_1 (q_2 - q_2^*)^2 + \lambda_2 (q_3 - q_3^*)^2), \quad \sin q_2^* \neq 0, \quad \sin q_3^* \neq 0 \qquad (4.3.16)$$

to be minimized.

4.3.2.3 Avoidance of joint limitations

Generally (except when infinite rotations are allowed, or when some joints are coupled), $q(t)$ has to lie inside a hypercube of $\mathbf{R}^n$:

$$q_i^{\min} < q_i < q_i^{\max}, \quad i = 1, \ldots, n. \qquad (4.3.17)$$

4.3 EXAMPLES OF SECONDARY TASKS

It is rather simple to form functions, the minimization of which will tend to avoid joint limitations. For example

$$h_s(q) = \sum_{i=1}^{n} \frac{(q_i - \frac{1}{2}(q_i^{\min} + q_i^{\max}))^2}{(q_i^{\min} - q_i^{\max})^2}. \quad (4.3.18)$$

However, we have to take care that removing the robot from joint limits does not allow too close an approach to the singularities (and inversely that avoiding singular domains in the sense of rank deficiency of J_1 does not send the joints close to their bounds). This situation is encountered, for example, with h_s given by (4.3.18) when joint limits are located symmetrically around a value which may correspond to a singular point. In Example 4.4, this would mean

$$q_i^{\min} = -q_i^{\max}, \quad i = 2, 3. \quad (4.3.19)$$

As this case frequently occurs in practice, it is thus recommended that a function be used which combines both objectives, e.g. simultaneous avoidance of joint limitations and singular domains.

Remark. Let us recall that in all these methods the minimization of h_s is done under *constraints*. Therefore, in the general case, nothing ensures that singularities or joint limitations are effectively avoided.

4.3.2.4 Cyclic tasks

In practical applications, a common requirement is the *cyclic* behaviour of a robot. This intuitively means that, when the end-effector of a manipulator has to come back to its starting location at the end of a workcycle, the corresponding joint variables should be the same as the initial ones. In other words, a closed path in the workspace should ideally result in a closed path in the joint space.

More generally, we may distinguish four classes of tasks with some properties of 'periodicity'; they are:

(T1): *periodic tasks*, such that $e(q, t) = e(q, t + \tau)$, for any q and t; $\tau \neq 0$ being fixed; for example

$$e(q, t) = x(q) - \sin(t);$$

(T2): *closed-in-time tasks*, such that $e(q, t^*) = e(q, t^* + \tau)$, for any q; $\tau \neq 0$ and t^* being fixed; for example

$$e(q, t) = x(q) + t(t - \tau); \quad t^* = 0;$$

(T3): *periodic tasks for some q^**, such that $e(q^*, t) = e(q^*, t + \tau)$ for any t; $\tau \neq 0$ being fixed; for example

$$e(q, t) = \sin(t) + \cos(qt); \quad \tau = 2\pi; \quad q^* \in \mathbf{Z};$$

(T4): *closed-in-time tasks*, such that for some q^* $e(q^*, t^*) = e(q^*, t^* + \tau)$, t^* and $\tau \neq 0$ being fixed; for example

$$e(q, t) = t(qt + 1); \quad q^* = -\frac{1}{\tau + 2t^*}.$$

The inclusion graph is:

$$(T1) \subset (T2) \subset (T4)$$

$$(T1) \subset (T3).$$

The most frequent case in practice is (T1), but the analysis below is valid for the wider class (T4). We shall assume for simplicity but without loss of generality that $t^* = 0$ and $\tau = T$. Finally, in order to be consistent with the analysis of Chapter 3, we may propose the following definition.

Definition D4.8 *A task function $e(q, t)$ is called cyclic if it is 0-admissible and if the ideal trajectory is such that*

$$q_r(0) = q_r(T). \tag{4.3.20}$$

Remarks.

1. A cyclic task is closed-in-time at $q_r(0)$. The converse is not true.

2. In practice, if infinite joint rotations are possible, (4.3.20) may sometimes be weakened to $q_r(0) = q_r(T) \bmod(2\pi)$.

In other words, due to the admissibility conditions of Chapter 3, a task function e is cyclic for some $t^* \leq T$ if it is 0-admissible (i.e. $q_r(t)$ is C^2 and does not pass through singularities) and if the projection of the connected set $C_{0,T}$ on the set of joint variables is closed.

Let us now come back to the problem of redundancy. A practical problem is to find a global task function of the form (4.3.1) i.e. $e = B^T e_1 + \alpha P_1^A g_s$, such that the associated task is certainly cyclic in the sense of the above definition, given the initial function e_1 belonging to (T1) or (T4). Let us then assume that e_1 belongs to (T4) with $q^* = q_r(0)$. The remaining choices concern W, A, B, α and the secondary cost function h_s, to be minimized. Usually, it is possible to choose matrix functions $W(q, t)$, $A(q, t)$ and $B(q, t)$ which share with e_1 the same periodicity property (for example $A = W = J_1$, $B^T = J_1^+$). Let us assume that this choice has been made. Then a necessary condition for e to belong to (T4) with $q^* = q_r(0)$ is that $P_1^A g_s$ also belongs to (T4). Thus

$$(P_1^A g_s)(q_r(0), 0) = (P_1^A g_s)(q_r(0), T). \tag{4.3.21}$$

Since $q_r(0)$ is on the ideal trajectory, we have $P_1^A g_s(q_r(0), 0) = 0$; furthermore, with the choices made for A and W, we have $P_1^A(q_r(0), 0) = P_1^A(q_r(T), 0)$.

4.3 EXAMPLES OF SECONDARY TASKS

Therefore, the condition to be satisfied by the gradient of the cost function to be minimized is

$$P_1^A(q_r(0), 0)g_s(q_r(0), T) = 0. \quad (4.3.22)$$

If this condition is not satisfied, e is not closed-in-time at $q_r(0)$ and is not cyclic. Note that (4.3.22) is satisfied when h_s does not depend on the time variable.

Remark. In the particular case where $m = n - 1$, the dimension of the subspace orthogonal to $N(A)$ is 1, and the condition (4.3.22) reduces to the collinearity of $g_s(q_r(0), 0)$ and $g_s(q_r(0), T)$.

This necessary condition only indicates in practice that, when e_1, W, A, B, belong to (T4) with $q^* = q_r(0)$, only secondary goals with the property (4.3.22) should be considered in order to preserve a cyclic behaviour of the robot. Checking the cyclic character of the obtained task has then to be done in the same way as for admissibility.

4.3.3 External cost functions for the problem of obstacle avoidance

4.3.3.1 Introduction

A problem sometimes encountered is: given a desired trajectory $q_l(t)$, $t \in [0, T]$, try to follow it as closely as possible while performing $e_1(q, t) = 0$. An obvious candidate is then

$$h_s(q, t) = \tfrac{1}{2}(q - q_l(t))^T W(q - q_l(t)). \quad (4.3.23)$$

This cost function may be called 'external' in the sense that generally $q_l(t)$ is a trajectory which may have been *taught*, or heuristically determined to satisfy some criteria of positioning within the environment. The method mainly applies to obstacle avoidance problems when the location of obstacles is known in advance. There are however two drawbacks in using a function like (4.3.23) to solve this kind of problem.

1. Minimizing h_s under the constraint $e_1 = 0$ does not imply that the minimum $h_s(q^*, t)$ is *small*. The resulting trajectory may thus be significantly different from $q_l(t)$ and there is nothing to ensure that obstacle avoidance is actually performed.

2. When $q_l(t)$ exists, it is generally *one* possible trajectory among an infinite number of trajectories for which the obstacles are avoided. Furthermore, it is determined in relation to the primary task function e_1. For the same environment, and a different e_1, a new trajectory $q_l(t)$ may have to be considered. But there is no systematic method for designing this trajectory and no guarantee that the problem can even be solved in this way.

If we now analyse the problem of obstacle avoidance in its generality, two situations have to be considered.

1. The avoidance of obstacles must be achieved with certainty. Since it is not possible to properly define an obstacle avoidance goal as a primary task e_1, this excludes the use of the framework of redundancy. For example, when in a known environment, the robot has to move to a known objective position without touching the objects, the problem is known as the one of finding a collision-free path; usually, this level is performed 'off-line' because of the high computational load required. Generally this step results in a desired trajectory $q_d(t)$ to be tracked *exactly*, which shows that the use of a function of kind (4.3.23) must be excluded.

2. The avoidance of obstacles may be considered as a secondary goal. It is for example the case when only slight and local modifications of the trajectory are needed to take into account modelling errors or approximations. The problem then reduces to finding a secondary cost, the minimization of which will be equivalent to increasing the distance between the robot and the obstacles. Some examples of such 'potential-like' functions will be proposed in the following.

It may be noted that this approach may sometimes lead to a global collision-free trajectory solving problem 1, but without any *a priori* guarantee of finding it. We can also distinguish two kinds of methods for solving this problem of real-time local avoidance:

(a) Use of local sensors, which measure the proximity of obstacles. This approach will be presented in Chapter 7.

(b) Simultaneous use of three-dimensional models of the robot and of the environment, with measured joint positions. In the following we shall discuss two methods based on this last approach.

4.3.3.2 Minimal distance method

1. *Gradient of a distance cost function.* In a first step, every rigid body constituting the robot and the obstacles is embedded in a larger convex volume. These volumes fully contain the actual solids and are generally simple analytical primitives, such as spheres or ellipsoids. A non ambiguous *distance* measurement between robot and obstacles may then be defined as the *smallest* distance, d, between the surfaces associated with the bounding volumes of the robots and those of the obstacles. At each time instant, this distance can be interpreted as the distance between two points R and S with the following properties:

4.3 EXAMPLES OF SECONDARY TASKS

(a) The points R and S respectively belong to the surfaces bounding the robot and the set of obstacles;

(b)
$$\|RS\| = d; \tag{4.3.24}$$

(c) the tangent planes to the surfaces in R and S are parallel.
Therefore
$$RS = dn_r = -dn_s \tag{4.3.25}$$

where n_r and n_s are the unitary vectors represented in Fig. 4.11.

Remark. Although it is beyond the scope of this book, it should be noted that, using adequate hierarchical geometric databases, it is possible to derive simple and recursive algorithms for finding the two points R and S. However, the computing cost may be high if the number of primitives is large.

Having determined the points R and S, let us consider the following cost function
$$h_s(q, t) = \lambda d(q, t)^{-k} \tag{4.3.26}$$

with
$$\lambda > 0$$
$$k \in N^*$$
$$d(q, t) = \|RS\|(q, t). \tag{4.3.27}$$

In order to form the task function (4.3.1), we need to calculate
$$\frac{\partial h_s}{\partial q}(q, t) = -k\lambda d(q, t)^{-(k+1)} \frac{\partial d}{\partial q}(q, t). \tag{4.3.28}$$

This in turn requires the calculation of $(\partial d/\partial q)(q, t)$. To this end let us define R′ and S′, two points sliding on the surfaces of the robot and the obstacle

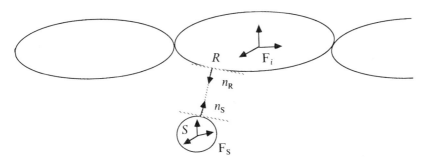

Fig. 4.11 Robot and obstacles

respectively, and which coincide at each instant with R and S. Then

$$d(q, t) = \langle R'S', n_r \rangle. \tag{4.3.29}$$

Let F_S and F_i be two frames attached respectively to the obstacle containing S and to the link number i containing R. Differentiating (4.3.29) yields

$$\dot{d} = \langle (V_{S'/F_S} - V_{R'/F_S}), n_r \rangle \tag{4.3.30}$$

where $V_{P/F}$ denotes the velocity of P in frame F.
As R' and S' move on the surfaces, we have

$$\langle V_{R'/F_i}, n_r \rangle = \langle V_{S'/F_S}, n_r \rangle = 0. \tag{4.3.31}$$

Using the last equality in (4.3.30) yields

$$\dot{d} = -\langle V_{R'/F_S}, n_r \rangle. \tag{4.3.32}$$

On the other hand

$$V_{R'/F_S} = V_{R'/F_i} - V_{R/F_i} + V_{R/F_S}. \tag{4.3.33}$$

As R is motionless in F_i, we have

$$V_{R'/F_i} = V_{R'/F_S} - V_{R/F_S}. \tag{4.3.34}$$

Thus, using (4.3.31) and (4.3.34) in (4.3.32)

$$\dot{d} = -\langle V_{R/F_S}, n_r \rangle. \tag{4.3.35}$$

Recalling that F_0 is the fixed frame linked to the base of the robot, we have

$$V_{R/F_S} = V_{R/F_0} - V_{S/F_0} - \omega_{F_S/F_0} \times SR \tag{4.3.36}$$

which leads to

$$\dot{d} = -\langle n_r, V_{R/F_0} \rangle + \langle n_r, V_{S/F_0} \rangle \tag{4.3.37}$$

since n_r is collinear with SR and because of the properties of the mixed product.

Let J_L^R be the $(3 \times n)$ translation Jacobian matrix associated with point R (cf. Chapter 2), and $\{x_R, x_S\}$ the respective (3×1) matrices of coordinates of R and S in F_0. Then, with all vector coordinates expressed in the basis of F_0, we have

$$V_{R/F_0} = J_L^R \dot{q} \tag{4.3.38}$$

and

$$n_r(q, t) = \frac{1}{d(q, t)} (x_S(t) - x_R(q)). \tag{4.3.39}$$

Recalling that

$$\dot{d} = \frac{\partial d}{\partial q} \dot{q} + \frac{\partial d}{\partial t} \tag{4.3.40}$$

4.3 EXAMPLES OF SECONDARY TASKS

and using (4.3.37)–(4.3.40), we obtain

$$\frac{\partial d}{\partial q}(q,t) = -\frac{1}{d(q,t)}(J_L^R)^T(q)(x_S(t) - x_R(q)) \quad (4.3.41)$$

and finally, using (4.3.41) in (4.3.28)

$$\frac{\partial h_s}{\partial q}(q,t) = \lambda k d(q,t)^{-(k+2)}(J_L^R)^T(q)(x_S(t) - x_R(q)). \quad (4.3.42)$$

Remark. Recalling that

$$V_{R/F_0} = V_{O_i/F_0} + \omega_{F_i/F_0} \times O_i R \quad (4.3.43)$$

where O_i is the origin of the standard frame attached to link i, we may write

$$\langle n_r, V_{R/F_0} \rangle = \langle n_r, V_{O_i/F_0} \rangle + \langle n_r \times RO_i, \omega_{F_i/F_0} \rangle. \quad (4.3.44)$$

Using (4.3.39), this leads to

$$\frac{\partial h_s}{\partial q} = \lambda k d^{-(k+2)} J_i^T \begin{pmatrix} x_S - x_R \\ As(RO_i)(x_S - x_R) \end{pmatrix} \quad (4.3.45)$$

where $J_i = \begin{pmatrix} J_L^i \\ J_R^i \end{pmatrix}$ is the basic Jacobian matrix associated with link i and $As(RO_i)$ is the matrix associated with the cross-product ($\times RO_i$), all terms of (4.3.45) being expressed in F_0. Relations (4.3.42) and (4.3.45) are equivalent.

2. *Discussion.* The interpretation of (4.3.42) in terms of potential functions is obvious. If the point S is associated with a repulsive potential function of Newtonian type, the associated gradient field thus generates at point R a repulsive force

$$F = -\frac{\lambda}{d^2} n_R. \quad (4.3.46)$$

Therefore, with the previous notation, and F_0 being the coordinates of F in the frame F_0:

$$F_0 = \lambda d^{-2}(x_S - x_R). \quad (4.3.47)$$

The principle of virtual power allows us to find the joint torque vector needed for getting the desired force F at point R (cf. Chapter 7):

$$\Gamma = (J_L^R)^T F_0. \quad (4.3.48)$$

By combining (4.3.47) and (4.3.48), and choosing $k = 1$, we find that $\partial h_s/\partial q$ given by (4.3.42) is equal to Γ.

Although the physical interpretation is simple, an important drawback of this approach lies in the fact that only *one* distance is taken into account at each time; due to the choice of the global minimum, this distance is obtained for a pair of points {R, S} which may jump from one body to another, leading to discontinuities in the derivatives.

To illustrate this type of problem, let us take once more Example 4.4 of the three-bar mechanism, now with simple obstacles shown in Fig. 4.12, and let us choose as in case 1 of Example 4.6 the task e_1:

$$e_1 = \begin{pmatrix} x(q) - 1 \\ y(q) \end{pmatrix} \tag{4.3.49}$$

where $x(q)$ and $y(q)$ are given by (4.3.3) and (4.3.4). In this simple case, we can transform the constrained minimization problem into a lower-dimensional unconstrained one, using explicitly the constraint $e_1 = 0$ (i.e. O_3 is fixed as shown in Fig. 4.12). Then, 'minimizing $h_s(q_1, q_2, q_3)$ under $e_1 = 0$' becomes 'minimizing a new cost function $h_s^c(q_1)$' which depends only on q_1. Considering for simplicity that obstacle avoidance is desired only for link 2, the previous method leads to minimization (with the choice $k = 1$) of

$$h_s^c(q) = \max \left\{ \frac{1}{d_1}; \frac{1}{d_2} \right\} \tag{4.3.50}$$

with

$$d_1 = l - s_1 \tag{4.3.51}$$

$$d_2 = l' + s_1. \tag{4.3.52}$$

The resulting constrained cost function is graphed in Fig. (4.13), and it is obvious that h_s does not satisfy the conditions of section (4.2.2), as it is not C^1 at the minimum.

Methods given in the next subsection will prove less sensitive to this kind of problem.

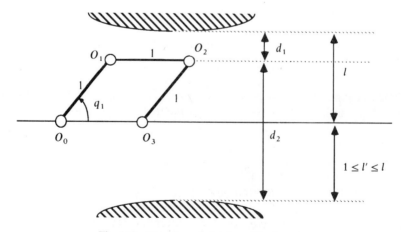

Fig. 4.12 Three-bar mechanism with obstacles

4.3 EXAMPLES OF SECONDARY TASKS

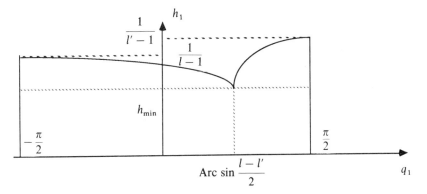

Fig. 4.13 Constrained cost function

4.3.3.3 Approximation methods

In order to avoid having to find points R and S at each instant, we may select in advance a set of fixed points on the robot, on the obstacles, or on both, and try to minimize a function including several distances between such points. We consider two cases:

1. *Fixed Points on the Robot Only.* Let $\{R_j\}$ be a set of points on the bounded convex volumes of the robot. With each R_j is associated the closest point S_j on the surface of a convex volume bounding one of the obstacles (Fig. 4.14).

Let

$$d_j = \|R_j S_j\|. \tag{4.3.53}$$

If we choose

$$h_s(q, t) = \sum_j d_j^{-k}, \tag{4.3.54}$$

we have

$$\frac{\partial h_s}{\partial q} = -k \sum_j d_j^{-(k+1)} \frac{\partial d_j}{\partial q}. \tag{4.3.55}$$

Let S'_j be the point which slides onto the obstacle while coinciding at each time with S_j. Thus

$$d_j = \langle S'_j R_j, n_{S_j} \rangle \tag{4.3.56}$$

and

$$\dot{d}_j = \langle n_{S_j}, (V_{R_j/F_s} - V_{S'_j/F_s}) \rangle \tag{4.3.57}$$

where F_s is a frame linked to the obstacle.

As S'_j slides on the surface of the obstacle, we have

$$\langle V_{S'_j/F_s}, n_{S_j} \rangle = 0 \tag{4.3.58}$$

and

$$\dot{d}_j = \langle n_{S_j}, V_{R_j/F_s} \rangle. \tag{4.3.59}$$

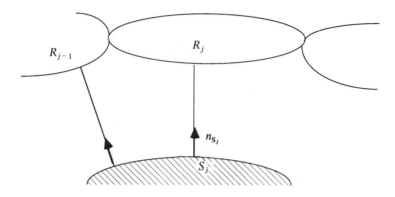

Fig. 4.14 Obstacles and fixed points on the robot

Therefore, by proceeding in the same way as we did to obtain (4.3.41) we find

$$\frac{\partial d_j}{\partial q} = -d_j^{-1}(J_L^{R_j})^T(x_{S_j} - x_{R_j}). \tag{4.3.60}$$

Remark. As in the previous method, if several points R_j belong to the same link L_i of the robot, we may replace (4.3.60) by

$$\frac{\partial d_j}{\partial q} = -d_j^{-1} J_i^T \begin{pmatrix} x_{S_j} - x_{R_j} \\ As(R_j O_i)(x_{S_j} - x_{R_j}) \end{pmatrix} \tag{4.3.61}$$

2. *Fixed Points on Robot and Obstacles.* Let $\{S_l\}$ be a set of points on the surfaces of the bounding volumes of the obstacles, and $\{R_j\}$ a corresponding set of points on the robot. An associated cost function is

$$h_s(q, t) = \sum_j \sum_l d_{jl}^{-k} \tag{4.3.62}$$

where

$$d_{jl} = \|R_j S_l\|. \tag{4.3.63}$$

Therefore

$$\frac{\partial h_s}{\partial q} = -k \sum_j \sum_l d_{jl}^{-(k+1)} \frac{\partial d_{jl}}{\partial q}. \tag{4.3.64}$$

Let

$$u_{jl} = \frac{R_j S_l}{\|R_j S_l\|}. \tag{4.3.65}$$

Then

$$d_{jl} = \langle R_j S_l, u_{jl} \rangle \tag{4.3.66}$$

and

$$\dot{d}_{jl} = -\langle u_{jl}, V_{R_j/F_{S_l}} \rangle \tag{4.3.67}$$

4.3 EXAMPLES OF SECONDARY TASKS

where F_{S_l} is a frame linked to the obstacle containing S_l. Knowing that

$$V_{R_j/F_{S_l}} = V_{R_j/F_0} - V_{S_l/F_0} - \omega_{F_{S_l}/F_0} \times S_l R_j \qquad (4.3.68)$$

we find

$$\dot{d}_{jl} = -\langle u_{jl}, V_{R_j/F_0}\rangle + \langle u_{jl}, V_{S_l/F_0}\rangle. \qquad (4.3.69)$$

Thus

$$\frac{\partial d_{jl}}{\partial q} = -d_{jl}^{-1}(J_L^{R_j})^T(x_{S_l} - x_{R_j}) \qquad (4.3.70)$$

and

$$\frac{\partial h_s}{\partial q} = -k \sum_j \sum_l d_{jl}^{-(k+2)}(J_L^{R_j})^T(x_{S_l} - x_{F_j}). \qquad (4.3.71)$$

Note that, as in (4.3.61), it is possible to use basic Jacobian matrices for points R_j belonging to the same link L_i.

This method is the simplest of the three that have been presented, since no computation of the minimal distance is needed. Furthermore, the problems of discontinuities previously evoked do not appear. However, its efficiency depends greatly on the number of points selected on the robot and on the obstacles.

A final remark concerns all these methods of obstacle avoidance. The cost function which is used tends to remove the robot from the obstacles as far away as it is allowed by the task e_1. This may lead to moving the robot towards joint limits or singularities (see for example Fig. 4.12 when $l = l'$). This is why it is again recommended that cost functions h_s be used which combine the obstacle avoidance goal with joint limits and removal from singularities. As usual, nothing ensures that obstacles, joint limits and singularities are actually avoided.

4.3.4 Energy-based cost functions

4.3.4.1 Joint velocity limitations

As $h_s(q, t)$ depends only on time and joint positions, it is not directly possible to take into account joint velocities or accelerations. Let us thus consider instead the following cost function

$$h_s(q, t) = \tfrac{1}{2}(q - y)^T W(q - y) \qquad (4.3.72)$$

where $y(t)$ is a filtered function of $q(t)$, obtained, for example, from

$$\left.\begin{array}{l}\dot{y} + \alpha y = \alpha q(t) \\ y(0) = q(0); \alpha > 0\end{array}\right\}. \qquad (4.3.73)$$

Obviously, $(q(t) - y(t))$ is an image of the rate of variation of $q(t)$ and may be considered as a kind of approximation to $\dot{q}(t)$. Indeed, differentiating

(4.3.73) and setting α and $\dot{q}(t)$ constant shows that in that case the steady state solution is $\alpha(q(t) - y(t)) = \dot{y}(t) = \dot{q}(t)$.

In (4.3.72), W is a symmetric positive definite matrix. Two possible choices are the following:

1. $W = M(q)$, where $M(q)$ is the inertia matrix given in Chapter 2. Then, $1/2\dot{q}^T M(q)\dot{q}$ is the overall kinetic energy, K, and the minimization of K is a way of saving energy. In practice, in order to avoid the differentiation of $M(q)$, an approximated expression of K is obtained by using $M(y(t))$ instead of $M(q)$.

2. $W = I$. It may be desired to minimize the norm of the joint velocity vector, for example, for safety reasons.

When W is constant, the computation of the q derivatives of h_s is quite simple. Indeed we then have

$$\frac{\partial h_s}{\partial q}(q, t) = W(q - y(t)) \tag{4.3.74}$$

which by (4.3.73) takes at $q(t)$ the value

$$\frac{\partial h_s}{\partial q}(q(t), t) = \frac{1}{\alpha} W \dot{y}(t) \tag{4.3.75}$$

and

$$\frac{\partial^2 h_s}{\partial q^2} = W. \tag{4.3.76}$$

Remarks.

1. The parameter α may be tuned so as to limit the velocity of $y(t)$ independently of the robot velocity (recall that $\|\partial e/\partial t\|$ has to be bounded for the task to be admissible). By choosing

$$\alpha = \alpha(t) = \frac{\lambda_1}{1 + \lambda_2(\lambda_3^2 + \|q(t) - y(t)\|^2)^{1/2}}, \quad \lambda_i > 0 \tag{4.3.77}$$

we get

$$0 < \alpha(t) \leq \frac{\lambda_1}{1 + \lambda_2 \lambda_3} \tag{4.3.78}$$

and

$$\|\dot{y}(t)\| < \frac{\lambda_1}{\lambda_2}. \tag{4.3.79}$$

2. As seen in Chapter 3, cost functions such as (4.3.72) may be used for improving the admissibility by providing a way of passing through singularities. However in that case, h_s is used in a different way: it is added to a

4.3 EXAMPLES OF SECONDARY TASKS

primary cost function which characterizes the initial objective. The new objective being the minimization of the sum of the two cost functions, the new task function is the gradient of the sum. This is no longer a constrained minimization problem. More generally, $y(t)$ may then be chosen in different ways. It may, for example, be the vector of joint coordinates of another robot to be tracked (cases of trajectory learning or telemanipulation).

3. In practice the control and the measurements are realized in discrete-time. Computing $y(t)$ according to $y(kt) = q(kt - \Delta)$, where k is the discrete-time index and Δ a given delay, may give good results. However, this choice does not provide us with any theoretical guarantee of velocity boundedness. A safer method consists of implementing the non-linear filter $y(kt) = y((k-1)t) + \alpha(q(kt) - y((k-1)t))$ with

$$\alpha = \begin{cases} \dfrac{V_{\max}}{\|q(kt) - y((k-1)t)\|}, & \text{when } \|q(kt) - y((k-1)t)\| > V_{\max}; \\ 1 & \text{otherwise.} \end{cases}$$

Then $\|y(kt) - y((k-1)t)\|$ is bounded by $V_{\max}$ which can be interpreted as the desired upper bound of the joint velocity.

4.3.4.2 Limitation of torques

Limitation of drive torques offers some advantages, such as a reduction in energy consumption or avoidance of overheating of D.C. motors. However, more generally, it may be desired that the exerted torque be at each time as close to a given value Γ^* as possible. This value is not necessarily equal to zero. For example, in the case of D.C. motors, it is known that the allowed torque Γ_i is approximately limited at each joint i according to

$$\Gamma_i^{\min}(t) = -a_i - b_i \dot{q}_i(t) \leq \Gamma_i \leq a_i - b_i \dot{q}_i(t) = \Gamma_i^{\max}(t), \quad a_i, b_i > 0. \tag{4.3.80}$$

In order to preserve at each instant the greatest capacity for braking or acceleration, a reasonable objective torque is

$$\Gamma^*(t) = \tfrac{1}{2}(\Gamma^{\min} + \Gamma^{\max})(t) = -B\dot{q}(t) \tag{4.3.81}$$

where $B = \text{diag}(b_1, \ldots, b_n)$. Recalling (cf. Chapter 2) that the overall dynamic model is

$$\Gamma = M(q)\ddot{q} + N(q, \dot{q}) \tag{4.3.82}$$

an *ad hoc* idea for staying close to the desired ideal torque is to extend the method used for velocities, and choose to minimize

$$h_s(q, t) = \tfrac{1}{2} \|\hat{M}(y(t))w(t) + \hat{N}(y(t), z(t), t) + \alpha B(q - y(t))\|^2 \tag{4.3.83}$$

where $\hat{M}(.)$, $\hat{N}(.)$ are approximated functions for $M(.)$ and $N(.)$, and

$$\dot{y} + \alpha y = \alpha q(t) \tag{4.3.84}$$

$$\dot{z} + \beta z = \beta \dot{y} \tag{4.3.85}$$

$$\dot{w} + \gamma w = \gamma \dot{z}. \tag{4.3.86}$$

z may be considered as an image of $\dot{y}$, and therefore of the velocity; and w as an image of $\ddot{y}$, and therefore of the acceleration. By choosing α as previously, and β and γ equal to constant values, $\dot{y}$, z and w are bounded.

Note however that, unless rough approximations are used in (4.3.83), the computation cost of this method may be high. Further, $w(t)$ is not the actual acceleration but at best a smoothed and phase-shifted approximation to it. Independently of the main task requirements, it thus may be expected to follow only roughly the objective torque $\Gamma^*(t)$.

4.4 MONITORING OF THE TRANSIENT PHASE OF THE SECONDARY COST MINIMIZATION

The central issue of this chapter is the constitution of a new non-redundant task function

$$e = B^T e_1 + \alpha P_1^A g_s, \tag{4.4.1}$$

starting from the basic constrained minimization problem

$$\min h_s(q, t); \quad e_1(q, t) = 0, \quad t \in [0, T].$$

As seen in Chapter 3, an ideal situation is when the robot starts on the ideal trajectory, i.e. when

$$e(q(0), 0) = 0. \tag{4.4.2}$$

This means that the initial position, $q(0) = q_0$, should be such that the main task is realized:

$$e_1(q_0, 0) = 0, \tag{4.4.3}$$

and the constrained minimization is performed:

$$P_1^A(q_0, 0) g_s(q_0, 0) = 0. \tag{4.4.4}$$

In practice, it is generally possible to satisfy condition (4.4.3), by an adequate choice of e_1. But there is generally no reason for the robot to minimize the chosen secondary cost function from the beginning. In that case there exists an initial phase during which the value of $h_s(q(t), t)$ has to be *different* from the optimum, $h_s(q_r(t), t)$ (Fig. 4.15).

This transient phase, necessary for h_s to converge to its constrained minimum, should be:

4.4 MONITORING OF THE TRANSIENT PHASE

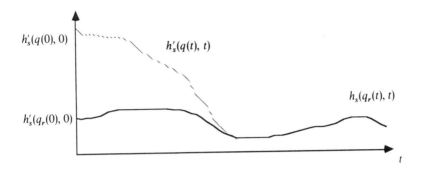

Fig. 4.15 Transient period on h_s

(1) not too fast, in order to avoid motions with high velocities;

(2) not too slow, in order to realize the secondary task after a given amount of time;

(3) as independent as possible from control schemes, disturbances, etc., so as to ensure in a robust way the repetitivity of the movement.

This leads to the idea of specifying the characteristics of the transient motion within the task definition itself. A method of realizing this is to replace h_s by the modified cost function

$$h'_s(q, t) = h_s(q, t) + \tfrac{1}{2}\beta(t) \| q - q'_0 \|^2 \qquad (4.4.5)$$

where

$$\frac{\partial h'_s}{\partial q}(q'_0, 0) = 0 \qquad (4.4.6)$$

and $\beta(t)$ is a C^2 scalar positive function such that

$$\lim_{t \to +\infty} \beta(t) = 0. \qquad (4.4.7)$$

The associated new task function is then

$$e' = B^T e_1 + \alpha P_1^A(g_s + \beta(q - q'_0)) \qquad (4.4.8)$$

and

$$\frac{\partial^2 h'_s}{\partial q^2}(q_0, 0) = \frac{\partial^2 h_s}{\partial q^2}(q_0, 0) + \beta(0)I. \qquad (4.4.9)$$

By choosing $\beta(0)$ large enough, this matrix can clearly be made positive. Then, because of (4.4.6), $h'_s(q, t)$ is minimized without constraints at time $t = 0$. Furthermore, if $e_1(q_0, 0) = 0$, we also have $e'(q_0, 0) = 0$. The initial Lagrange multipliers associated with the problem $\{\min h'_s; e_1 = 0\}$ are thus equal to zero, and the Hessian of the problem

$$\frac{\partial^2 H}{\partial q^2}(q_0, 0, \lambda^*(0)) = \frac{\partial^2 h_s}{\partial q^2}(q_0, 0) + \beta(0)I \quad (4.4.10)$$

is also positive. This ensures that condition (4) of Lemma 4.3 is satisfied, which in turn can be used to prove the admissibility of the task. Having determined $\beta(0)$, the vector q'_0 is obtained from (4.4.5) and (4.4.6):

$$q'_0 = q_0 + \frac{1}{\beta(0)} g_s(q_0, 0). \quad (4.4.11)$$

Finally note that the rate of convergence of $\beta(t)$ toward zero determines the duration of the transient phase, after which the modification of the task has only a negligible (asymptotically null) effect on the realization of the original task. An example of a function β which is C^2 at the origin is

$$\beta(t) = \frac{\beta(0)}{\lambda_1 - \lambda_2}(\lambda_1 e^{-\lambda_2 t} - \lambda_2 e^{-\lambda_1 t}); \quad 0 \leq \lambda_2 < \lambda_1. \quad (4.4.12)$$

4.5 BIBLIOGRAPHIC NOTE

With the exception of a few recent works, the study of redundant systems has rarely been performed at the task level (we exclude here the problem of off-line determination of collision-free trajectories in a complex environment). One of the basic common methods, which seems to have been broached in Liegeois (1977), is close to the differential motion approach presented in section 4.2.5. An example of constrained minimization technique is given in Gabay and Luenberger (1976). Early works in this direction may be found in Fournier (1980), Baillieul (1984), Klein and Ching-Hsiang (1983), Klein (1984). Generalized inverses are an old mathematical invention (Ben Israel and Greville 1974; Boullion and Odell 1971; Campbell and Meyer 1979; Klema and Laub 1980), and the contribution to the problem of redundancy in robotics usually consists of finding the right function to be minimized: manipulability (Nakamura and Hanafusa 1984; Yoshikawa 1983; Yoshikawa 1985a, b); joint torques (Hollerbach and Suh 1985; Suh and Hollerbach 1987); 'distances' to obstacles (Khatib 1980, 1985b; Freund and Hoyer 1984, 1985; Kircanski and Vukobratovic 1984; Maciejewski 1984). Some of the techniques evoked in section 4.3 for the computation of the distances between convex objects are from Faverjon (1984). The use of sensors in a redundant manipulator is presented in Boulic (1986), Espiau and Boulic (1985b), Espiau (1986a). A modified potential field is used in Krogh (1984). Cyclic paths have mainly been studied by Wampler (1987), but from the point of view of inverse functions. The concept of 'extended Jacobian' (Baillieul 1986, 1987) has

APPENDIX A4.1 PROOFS OF LEMMAS OF CHAPTER 4

something in common with the first choice given in paragraph 4.2.4.4. Minimization of the number of arithmetic operations involved in the computation of pseudo-inverse matrices is presented in Chevallereau and Khalil (1987).

The general idea of analyzing a redundant system with the *task* point of view is from Samson (1987a).

APPENDIX A4.1 PROOFS OF LEMMAS OF CHAPTER 4

A4.1.1 Proof of Lemma 4.1

Conditions (4.2.16) and (4.2.20) being the same, the proof focuses on conditions (4.2.17) and (4.2.21).

1. *A solution q^* of the constrained minimization problem satisfies condition* (4.2.21). Let $F_1(q^*, t)$ be an $(n \times (n - m))$ matrix, the column vectors of which form a basis of $N(J_1(q^*, t))$. Then, (4.2.17) gives

$$F_1^T \frac{\partial H}{\partial q} = F_1^T g_s + F_1^T J_1^T \lambda = 0. \tag{A4.1}$$

From the definition of F_1, we have

$$F_1^T J_1^T = (J_1 F_1)^T = 0. \tag{A4.2}$$

Then

$$F_1^T g_s = 0. \tag{A4.3}$$

Therefore $g_s(q^*, t)$ is orthogonal to the vectors of the basis, i.e. to $N(J_1)$.

2. *If q^* satisfies condition* (4.2.20) *and* (4.2.21), *it also satisfies condition* (4.2.17). Since $g_s(q^*, t)$ is orthogonal to $N(J_1(q^*, t))$, its orthogonal projection on $N(J_1(q^*, t))$ is zero, and therefore

$$(I - J_1^T(J_1 J_1^T)^{-1} J_1)g_s = 0. \tag{A4.4}$$

By choosing the Lagrange multipliers according to

$$\lambda^* = -(J_1 J_1^T)^{-1} J_1 g_s \tag{A4.5}$$

relation (A4.4) can also be written

$$g_s + J_1^T \lambda^* = 0 \tag{A4.6}$$

which is condition (4.2.17).

Note that it is not necessary to use an orthogonal projection. Indeed, if $A(q^*, t)$ is a $(m \times n)$ matrix such that AJ_1^T is non-singular, then (A4.4) is equivalent to

$$(I - J_1^T(AJ_1^T)^{-1} A)g_s = 0. \tag{A4.7}$$

This relation means that the 'oblique' projection of g_s onto $N(A)$ along $R(J_1^T)$ is also zero.

From (A4.7), we verify that condition (4.2.17) is also satisfied by using the following set of Lagrange multipliers

$$\lambda^* = -(AJ_1^T)^{-1} A g_s. \tag{A4.8}$$

A4.1.2 Proof of Lemma 4.3

Let us define

$$e_2(q, t) = P_1^A(q, t) g_s(q, t). \tag{A4.9}$$

Thus, from (4.2.33)

$$J_T(q, t) = \frac{\partial e}{\partial q}(q, t) = B^T \frac{\partial e_1}{\partial q} + \sum_{k=1}^{m} e_1^k \frac{\partial b_k^T}{\partial q} + \alpha \frac{\partial e_2}{\partial q} \tag{A4.10}$$

where the b_k^T are rows of B^T, and the e_1^k the components of e_1. If q^* is a solution of $e(q, t) = 0$, then we have (Lemma 4.2) $e_1(q^*, t) = 0$, and therefore

$$J_T(q^*, t) = B^T J_1(.) + \alpha \frac{\partial e_2}{\partial q}(.), \quad (.) = (q^*, t). \tag{A4.11}$$

Let us compute $(\partial e_2/\partial q)(.)$. From Lemma 4.2

$$e_2(q(.)) = 0 \tag{A4.12}$$

and since $e_2 \in N(A)$, we also have

$$A e_2 = 0; \quad \text{for any } q, t. \tag{A4.13}$$

Differentiating $A e_2$ with respect to the variable q, and using the two previous relations, we obtain

$$A(.) \frac{\partial e_2}{\partial q}(.) = 0. \tag{A4.14}$$

Now, starting from

$$H(q, t, \lambda^*) = h_s(q, t) + \lambda^{*T} e_1(q, t) \tag{A4.15}$$

we define

$$D_1 H(q, t, \lambda^*) = \frac{\partial H}{\partial q}(q, t, \lambda^*) = g_s + J_1^T \lambda^*. \tag{A4.16}$$

We also define a Hessian by differentiating (A4.15) a second time with respect to the variable q

$$\frac{\partial^2 H}{\partial q^2}(q, t, \lambda^*) = \frac{\partial g_s}{\partial q} + \sum_{k=1}^{m} \lambda_k^* \frac{\partial^2 e_{1,k}}{\partial q^2}. \tag{A4.17}$$

Let us then consider the following vector function

$$\lambda = -(AJ_1^T)^{-1} A g_s. \tag{A4.18}$$

APPENDIX A4.1 PROOFS OF LEMMAS OF CHAPTER 4

Note that $\lambda(q, t) = \lambda^*$ for $q = q^*$. From (A4.8) and (A4.9), we also have

$$e_2(q, t) = g_s + J_1^T \lambda. \tag{A4.19}$$

Differentiating (A4.19) with respect to the variable q and using (A4.17), we obtain

$$\frac{\partial e_2}{\partial q}(.) = \frac{\partial^2 H(., \lambda^*)}{\partial q^2} + J_1^T(.) \frac{\partial \lambda}{\partial q}(.). \tag{A4.20}$$

Now, since $(\partial e_2/\partial q)(.)$ belongs to $N(A(.))$ (relation (A4.14)), we have

$$\frac{\partial e_2}{\partial q}(.) = P_1^A(.) \frac{\partial e_2}{\partial q}(.). \tag{A4.21}$$

Application of the projection operator P_1^A to (A4.20) therefore yields

$$\frac{\partial e_2}{\partial q}(.) = P_1^A(.) \frac{\partial^2 H}{\partial q^2}(., \lambda^*). \tag{A4.22}$$

Thus, from (A4.11)

$$J_T(.) = B^T(.) J_1(.) + \alpha(.) P_1^A(.) \frac{\partial^2 H}{\partial q^2}(., \lambda^*). \tag{A4.23}$$

To complete the proof, let us check the non-singularity of $J_T(.)$. Let x be a vector of $N(J_T(.))$, which can be decomposed into the sum of two vectors:

$$x = x_1 + x_2 \tag{A4.24}$$

with

$$x_1 = A^T(.)(J_1(.)A^T(.))^{-1} J_1(.) x \tag{A4.25}$$

$$x_2 = (P_1^A(.))^T x. \tag{A4.26}$$

$(P_1^A(.))^T$ is a projection operator onto $N(J_1(.))$, and x_1 is the projection of x onto $R(A^T(.))$ along $N(J_1(.))$. Multiplying (A4.23) by $A(.)$, and applying the result to x yields

$$A(.) J_T(.) x = A(.) B^T(.) J_1(.) x = 0. \tag{A4.27}$$

Since $A(.) B^T(.)$ is non-singular, we get

$$J_1(.) x = 0 \tag{A4.28}$$

and since $J_1(.) x = J_1(.) x_1$, x_1 must necessarily belong to $N(J_1(.))$. From the way x_1 was defined (relation (A4.25)), the only possibility is

$$x_1 = 0 \tag{A4.29}$$

and x reduces to x_2 which belongs to $N(J_1(.))$. Left-multiplying (A4.23) by x^T and right-multiplying by x leads to

$$x_2^T \frac{\partial^2 H}{\partial q^2}(., \lambda^*) x_2 = 0. \tag{A4.30}$$

Because of condition (4) in the lemma, the only possibility is

$$x_2 = 0. \tag{A4.31}$$

Therefore, $x = 0$ is the only element of $N(J_T(.))$, and the lemma is proved.

A4.1.3 Proof of Lemma 4.4

From (A4.23) we have

$$J_T(.) = B^T(.)J_1(.) + \alpha(.)P_1^A(.)\frac{\partial^2 H}{\partial q^2}(., \lambda^*). \tag{A4.32}$$

Because of condition (2) of the lemma, we also have

$$P_1^A(.) = P_1(.) = I - A^+(.)A(.), \tag{A4.33}$$

$P_1(.)$ denoting the orthogonal projection operator onto $N(J_1(.))$. Therefore, from the definition (4.2.36) of $S_{\gamma, A}$

$$S_{\gamma, A}(.) = I_n + \gamma P_1(.). \tag{A4.34}$$

Post-multiplying $J_T(.)$ by $S_{\gamma, A}(.)$ and taking condition (3) into account, we obtain

$$J_T(.)S_{\gamma, A}(.) = J_1^T C^T J_1 + \alpha P_1 \frac{\partial^2 H}{\partial q^2} + \alpha\gamma P_1 \frac{\partial^2 H}{\partial q^2} P_1 \bigg|_{(.); \lambda = \lambda^*} \tag{A4.35}$$

where C is a positive $(m \times m)$ matrix.

Let x be any vector of $\mathbf{R}^n$ which we decompose according to

$$x = x_1 + x_2 \tag{A4.36}$$

$$x_2 = P_1(.)x. \tag{A4.37}$$

Then, by construction, $x_2 \in N(J_1(.))$ and $x_1 \in R(J_1^T(.))$. Relation (A4.35) yields

$$x^T J_T(.)S_{\gamma, A}(.)x = x_1^T J_1^T C^T J_1 x_1 + \alpha x_2^T \frac{\partial^2 H}{\partial q^2} x_1 + \alpha(\gamma + 1)x_2^T \frac{\partial^2 H}{\partial q^2} x_2. \tag{A4.38}$$

Now, since $J_1^T C J_1$ is positive on $R(J_1^T(.))$, there exists a positive real number k_1 such that

$$x_1^T J_1^T C J_1 x_1|_{(.)} \geq k_1 \|x_1\|^2. \tag{A4.39}$$

From condition 4, there also exists a positive real number k_2 such that

$$x_2^T \frac{\partial^2 H}{\partial q^2}(., \lambda^*)x_2 \geq k_2 \|x_2\|^2. \tag{A4.40}$$

APPENDIX A4.1 PROOFS OF LEMMAS OF CHAPTER 4

If k_3 denotes the matrix norm of $(\partial^2 H/\partial q^2)(.,\lambda^*)$ associated with the Euclidean product, we also have

$$x_2^T \frac{\partial^2 H}{\partial q^2}(.,\lambda^*)x_1 \leq k_3 \|x_1\| \|x_2\|. \tag{A4.41}$$

Therefore, by (A4.38)–(A4.41), and from the positivity of $\alpha(.)$ (condition 5), we have

$$x^T J_T(.)S_{\gamma,A}(.)x \geq k_1 \|x_1\|^2 + \alpha(\gamma+1)k_2\|x_2\|^2 - \alpha k_3 \|x_1\|\|x_2\| \tag{A4.42}$$

which can also be written

$$x^T J_T(.)S_{\gamma,A}(.)x \geq k_1\left(\|x_1\| - \frac{\alpha k_3}{2k_1}\|x_2\|\right)^2 + \alpha\left((\gamma+1)k_2 - \frac{k_3^2}{2k_1}\right)\|x_2\|^2. \tag{A4.43}$$

Therefore, if $x \neq 0$, a sufficient condition for having

$$x^T J_T(.)S_{\gamma,A}(.)x > 0 \tag{A4.44}$$

is

$$\gamma \geq \gamma_m(\alpha(.)) \tag{A4.45}$$

where

$$\gamma_m(\alpha(.)) = \max\left(\frac{\alpha(.)k_3}{4k_1 k_2} - 1 + \varepsilon, 0\right), \tag{A4.46}$$

ε being a small positive real number ($\varepsilon \ll 1$).

$\gamma_m(\alpha(.))$ is a non-decreasing positive function and result (ii) of the lemma is established. From (A4.46), we see that when $\alpha(.) < \alpha_M$, with

$$\alpha_M = \frac{4k_1 k_2}{k_3}(1-\varepsilon) \tag{A4.47}$$

then $\gamma_m(\alpha(.)) = 0$. This establishes result (i) of the lemma.

5
CONTROL: A GENERAL APPROACH

5.1 GENERAL BACKGROUND

5.1.1 Introduction

As we have seen throughout the previous chapters, a robot task can be defined by a task function and an initial condition:

$$\{e(q, t); q_0\} \qquad (5.1.1)$$

with, on an ideal trajectory:

$$e(q_r(t), t) = 0, \qquad t \in [0, T] \qquad (5.1.2)$$

and the aim of any control scheme is to achieve effectively $e(q(t),t) = 0$, through the available control variables of the robot. Obviously, because of the need to compensate for errors and disturbances, an efficient control scheme will necessarily be in *closed loop*. Further, good *stability* properties are basic requirements in any control system; in the case of robots, it will appear that *robustness* (i.e. the feature of insensitivity to modelling errors and disturbances) is the basic concept to be used in analysing such stability properties and deriving efficient control schemes. This concept will underlie the whole control analysis.

We are concerned with *on-line* control problems for which the structure of the task function $e(q, t)$ has already been fixed. The *higher order control decisions* leading to the choice of $e(q, t)$ are not treated in this chapter. For example, the analysis covers the problem of tracking a trajectory provided by a C^2 reference model:

$$\ddot{q}_r(t) = u_r(t); \qquad (5.1.3)$$

and the use of force and proximity sensors inside a servo-loop. The analysis does *not* cover the generation of trajectories, for example computation of $u_r(t)$ in (5.1.3), or of a C^2 trajectory $q_r(t)$ passing close to predefined points; and it does not cover the use of high level vision sensors. A stage of pattern recognition will for example influence the choice of the task function $e(q, t)$, or the decision of starting a new task, but will not interfere with the on-line regulation of e. However, if a visual sensor is used as a real-time position or velocity sensor, it falls within the scope of the present approach.

This chapter will be devoted to the design of possible control schemes. Robustness and stability issues will be presented in Chapter 6. In the first

5.1 GENERAL BACKGROUND

section, we will review various errors and disturbances introduced by different components of a robotics system. This will allow us to state the basic hypothesis used in what follows. Of course, one of the most important subsystems is the set of actuators. This is why a special subsection will be devoted to the analysis of the actuators from the control point of view, in the particular (and frequent) case of D.C. motors.

The next section will start from the idea that a very large set of possible control schemes belongs to one of two classes defined by analogy with gradient and Newton minimization methods. These two subclasses will in turn be gathered into a single one, which will constitute our general control formulation. Finally, we shall show how this approach includes the existing classical control schemes.

5.1.2 Subsystems, unknowns and uncertainties

In order to analyse the structure of robot models, we may distinguish several open-loop subsystems (i.e. excluding the control unit) in a robot manipulator (Fig. 5.1).

5.1.2.1 *The actuators*

The function of actuators is to generate the forces and torques needed to move the robot joints. Three kinds of actuators are used in robotics: electric, hydraulic and pneumatic drives. The control input is classically a *current* or a *voltage* and the output variable is a current (torque) or a fluid pressure.

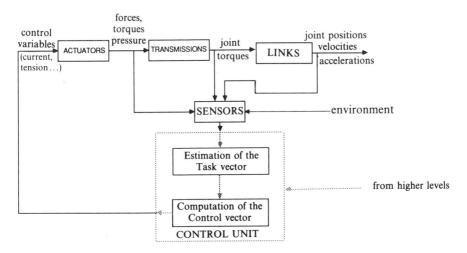

Fig. 5.1 Subsystems in a robot manipulator

5 CONTROL: A GENERAL APPROACH

The difficulties induced by the actuators mainly arise from non-linearities, for instance intrinsic non-linearities of hydraulic drives, limits and saturation of D.C. motors. We shall concentrate here on classical D.C. motor drives. Their simplified equations are given, at each joint i, by

$$\tau_i^m = m_i \ddot{q}_i^m + \beta_i \dot{q}_i^m + \tau_i^r \qquad (5.1.4)$$

$$\tau_i^m = k_i i_i \qquad (5.1.5)$$

$$u_i = r_i i_i + l_i \frac{di_i}{dt} + k_i \dot{q}_i^m, \qquad |i_i| < i_i^{Max} \qquad (5.1.6)$$

where we find the following variables and parameters (Fig. (5.2)):

(1) τ_i^m: torque on the motor axis;

(2) τ_i^r: resistance torque due to the load and other disturbances reduced on the motor axis;

(3) m_i: inertia of the rotating part, including the transmission inertia reduced on the motor axis;

(4) $q_i^m, \dot{q}_i^m, \ddot{q}_i^m$: angular position, velocity and acceleration of the motor axis;

(5) β_i: internal damping (viscous friction) parameter; it is often set equal to zero because the major part of viscous friction is generally located in the transmissions;

(6) k_i: torque coefficient, generally assumed to be equal to the back-e.m.f. coefficient, by conservation of power;

(7) i_i: current inside the motor;

(8) u_i: voltage;

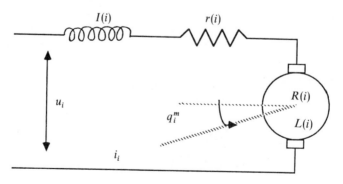

Fig. 5.2 A simple d.c. motor

(9) $r_i = R(i) + r(i)$: overall resistance;

(10) $l_i = L(i) + l(i)$: overall inductance.

Usually, due to the structure of power units, inner loops are added to this basic scheme. The effects of these loops from the control point of view will be discussed in section 5.1.3.

5.1.2.2 The transmissions

The function of transmissions is in general to realize both an adaptation of mechanical impedance and to transmit the torques generated by the actuators to the joints. When conventional D.C. motors are used, their axis velocities are generally much higher than the required joint velocities, while their nominal torques are too low. Transmissions are then designed to increase the available torque on the joints, and reduce the angular velocity by a given ratio called the *transmission ratio*. Transmission devices usually encountered in robots, and associated disturbances, are of three main kinds.

1. *Gears.* Conventional gear trains are generally rigid and reversible mechanical systems. They are responsible for most of the *dry friction* in a robot manipulator. Transmissions in high precision robots sometimes include systems for the compensation of backlash, which may introduce some *elasticity*. Another well-known transmission device is the 'harmonic drive', with which small transmission ratios (less than 1/300) are allowed, but sometimes with a rather high internal inertia and a loss of rigidity. Two other transmission systems, highly rigid and presenting large velocity-reduction capacities, are used for transforming a rotation into a translation: they are pinion-rack and screw-nut transmissions. However, these systems are not mechanically reversible, and we shall not consider them here.

2. *Cables and belts.* Cables are often used when the available space for transmissions inside the links is very limited. They are used for example in remote manipulators. Generally friction level is lower than with gears, but on the other hand some mechanical elasticity may occur. Belts, rubbers and chains are largely used in industrial robots, although they present various non-linearities and drawbacks: friction, backlash, elasticity, hysteresis.

3. *Cranks and levers.* These are sometimes used with reduction devices. These transmissions have two drawbacks: they usually reduce the range of possible joint positions; and additionally they introduce mechanical loops into the system which complicate the dynamic robot model.

Transmissions may therefore be responsible for a large proportion of the internal non-linear disturbances in a robotic system. Furthermore, these disturbances are often non-stationary (i.e. vary with time), and 'reasonable' models (i.e. useful and relevant) are rarely available. The main consequences of this situation are: a loss of precision; the generation of vibrations leading to mechanical fatigue, or to stability problems; and the advent of stick-slip phenomena.

Remark. It should be emphasized that the transmission subsystem does not always exist: a recent trend is the use of 'direct drive' actuators, realized by directly mounting high torque D.C. motors on the joints. This technique allows very high instantaneous torques to be obtained when needed, and friction to be drastically reduced. Its concomitant drawback is the increase in inertia and mass due to the on-board motors, and the possibly large variations of the load on the motor axis.

A simplified model of the effect of transmissions, which neglects flexibility and backlash is, at joint i:

$$p_i \tau_i^r = \tau_i + h_i(\dot{q}_i) \qquad (5.1.7)$$

$$\dot{q}_i^m = p_i \dot{q}_i \qquad (5.1.8)$$

where the parameters and variables are:

(1) $1/p_i$: transmission ratio;

(2) τ_i: torque at joint i;

(3) $h_i(\dot{q}_i)$: general friction term;

(4) $\dot{q}_i$: velocity of joint i.

5.1.2.3 The mechanical manipulator (arm)

Of course, the function of this subsystem is to perform physical displacements, or motions, of selected rigid bodies (for example the end-effector) inside the three-dimensional workspace of the robot, in relation to the assigned task. The input variables are the joint drive torques, and the outputs are the joint accelerations. As seen in Chapter 2, the dynamic model of a rigid robot has the general form:

$$[\tau_i] = \tau = M_a(q)\ddot{q} + N_a(q, \dot{q}) \qquad (5.1.9)$$

where $M_a(q)$ is the kinetic energy (inertia) matrix, and $N_a(q, \dot{q})$ includes Coriolis, centrifugal, gravity and friction terms, all computed without taking transmissions and actuators into account. The subscript a indicates that this equation is the model of the bare arm.

In practice, the parameters of model (5.1.9) may be evaluated by using CAD facilities or identification algorithms (cf. section 5.3.3). The values so obtained are only approximations and may sometimes be wrong: this is the

5.1 GENERAL BACKGROUND

case, for example, when the inertia and mass characteristics of the payload are unknown.

The structure of the model itself is generally also approximated. Underlying assumptions are that the links are perfectly rigid and that the reference frame is Galilean. No real manipulator meets these specifications exactly. Nevertheless we shall assume, in the coming analysis, that equation 5.1.9 provides an accurate description of the dynamics of the part of the physical system described as the arm.

5.1.2.4 *The sensors*

A sensor is a device which provides a measurement of a desired physical quantity. The output is generally an electrical signal which may be subject to noise and extraneous disturbances. The main kinds of sensors with which we are concerned are as follows.

1. *Internal sensors.* These sensors measure joint positions and velocities directly. Measurement of angular (or linear) positions is generally fairly easy, using potentiometers, resolvers, digital encoders, differential transformers, etc. Independently of classical problems of intrinsic accuracy, repeatability, or noise level, the *location* of the sensor itself is important in robotics: for example, measurement of the angular position of a motor axis is often easier than that of the corresponding joint axis, but the latter is better for compensation of transmission disturbances. Classically, the measurement of velocity is less accurate than that of the position. A device frequently used is the tachometer (which may provide noisy and undulatory signals), but velocity estimation or computation from encoder signals are also encountered.

Remark. Direct joint force/torque sensors also exist, although they are not frequently used, and we will not consider them here.

2. *External sensors (proximity, force, distance ...).* These sensors measure *interactions* between themselves and the *environment*. In taking these measurements, some uncertainties may appear, as detailed in Chapter 7. For example, optical proximity sensors provide mixed information on distance, optical properties of the sensor, albedo, and local orientation of the objects, etc. A difficulty, when using such sensors, is to extract and utilize the part of the information which is useful to the application.

Finally it is to be emphasized that the measurement of the task velocity $\dot{e}(t)$ is generally less reliable, and more difficult to obtain than the measurement of $\dot{q}(t)$, except in the indirect form:

$$\dot{e}(q(t), t) = \frac{\partial e}{\partial q}(q(t), t)\dot{q}(t) + \frac{\partial e}{\partial t}(q(t), t). \qquad (5.1.10)$$

5.1.2.5 Complete dynamic models

We have presented the basic models associated with each subsystem. Combination of these models leads to complete dynamic representations of the system.

5.1.2.5.1 Model in joint space. Combining equations (5.1.4)–(5.1.9) yields

$$\Gamma = M(q)\ddot{q} + N(q, \dot{q}, t) \qquad (5.1.11)$$

where

$$\Gamma_i = p_i \tau_i^m \qquad (5.1.12)$$

(this is the effect of the drive motor torque transmitted at joint i);

$$M(q) = M_a(q) + \begin{pmatrix} p_1^2 m_1 & & 0 \\ & \ddots & \\ 0 & & p_n^2 m_n \end{pmatrix} \qquad (5.1.13)$$

which is again a symmetric positive definite matrix; and

$$N(q, \dot{q}) = N_a(q, \dot{q}) + \begin{pmatrix} h_1(\dot{q}_1) \\ \vdots \\ h_n(\dot{q}_n) \end{pmatrix} + \begin{pmatrix} p_1^2 \beta_1 \dot{q}_1 & & 0 \\ & \ddots & \\ 0 & & p_n^2 \beta_n \dot{q}_n \end{pmatrix}. \qquad (5.1.14)$$

Equation (5.1.11) is the complete dynamic model in joint space which will be used from now on. Note that gyroscopic effects of transmissions and actuators are neglected. In practice, the control variables are not exactly the Γ_i, but more precisely the desired associated currents inside the motor (see section 5.1.3):

$$i_i = \frac{\Gamma_i}{p_i k_i}, \qquad i = 1, \ldots, n. \qquad (5.1.15)$$

Remark. It is well known that small transmission ratios reduce the effects of robot inertia variations. Let us suppose for simplicity that $M_a(q)$ is diagonal, with terms $M_i(q)$. Then (5.1.9) yields for each link:

$$\tau_i = M_i(q)\ddot{q}_i + N_i(q, \dot{q}). \qquad (5.1.16)$$

Using (5.1.7) and (5.1.8) leads to

$$\tau_i^r = \frac{1}{p_i^2} M_i(q)\ddot{q}_i^m + \frac{1}{p_i}(N_i(q, \dot{q}) + h_i(\dot{q}_i)) \qquad (5.1.17)$$

and, by (5.1.4)

$$\tau_i^m = \left(m_i + \frac{1}{p_i^2} M_i(q) \right) \ddot{q}_i^m + \frac{1}{p_i}(N_i(q, \dot{q}) + h_i(\dot{q}_i)) + \beta \dot{q}_i^m. \qquad (5.1.18)$$

5.1 GENERAL BACKGROUND

At the motor level, the robot inertia variations are thus multiplied by the square of the transmission ratio, while the effects of Coriolis and centrifugal terms, which appear in $N_i(q, \dot{q})$ in the form $\dot{q}^T B_i(q)\dot{q}$, are reduced by the order of p_i^3.

5.1.2.5.2 Model in task space.
Differentiating (5.1.10) with respect to time gives

$$\ddot{e} = \frac{\partial e}{\partial q}(q, t)\ddot{q} + f(q, \dot{q}, t) \tag{5.1.19}$$

where

$$f(q, \dot{q}, t) = \begin{pmatrix} \dot{q}^T W_1(q, t)\dot{q} \\ \vdots \\ \dot{q}^T W_n(q, t)\dot{q} \end{pmatrix} + 2\frac{\partial^2 e}{\partial q \partial t}(q, t)\dot{q} + \frac{\partial^2 e}{\partial t^2}(q, t) \tag{5.1.20}$$

in which the matrix $W_i(q, t)$ is the partial derivative of the ith column of $(\partial e^T/\partial q)(q, t)$ with respect to q, i.e. the Hessian of the ith coordinate of e.

Combining (5.1.11) and (5.1.19), and writing, as in the previous chapter

$$\frac{\partial e}{\partial q} = J_T(q, t) \tag{5.1.21}$$

we get

$$\ddot{e} + J_T(q, t)M^{-1}(q)N(q, \dot{q}, t) - f(q, \dot{q}, t) = J_T(q, t)M^{-1}(q)\Gamma. \tag{5.1.22}$$

This is the model of the dynamics of the task vector e which will be used in the control analysis.

We have seen throughout this subsection that various errors, disturbances and approximations exist within the models leading to the expression (5.1.22). The underlying assumptions that led to this equation are restated below.

(1) The actuators are D.C. motors with a small electrical response time, and without limits, such that any desired instantaneous current i_i is immediately obtained (see section 5.1.3).

(2) The transmissions are ideal, with no backlash or 'stiction'.

(3) The structure is perfectly rigid.

It will also be assumed that the control is continuous, and that there is no computation delay.

These classical assumptions are needed for the general analysis; however, as they are not fully satisfied in practice, we shall offer some comments on the consequences of violating them in Chapter 6. Since modelling errors, noises, and disturbances cannot be completely avoided or rejected, it is understandable that the performance of any control algorithm is highly dependent on its ability to tolerate such errors. From this arises the need for *robustness*, which will be discussed mainly in the next chapter.

5.1.3 Conventional control of D.C. motors

Before studying the problem of controlling the whole system, it is fruitful to analyse stability properties in the case of a single joint driven by a D.C. motor. The aim of this subsection is thus, starting from a very conventional description of D.C. motor control techniques, to show how existing inner loops in power units (current or velocity feedback) may contribute to modify the performance of the system (accuracy and stability). Another objective of the present study is to specify the relationships that exist between the joint torques, the state variables and the user's control variables.

5.1.3.1 Open-loop models

We start from equations (5.1.4) to (5.1.6), which we rewrite, to simplify the notation:

$$\gamma = m\ddot{\theta} + \beta\dot{\theta} + \gamma_r \quad (5.1.23)$$

$$\gamma = ki \quad (5.1.24)$$

$$u = ri + L\frac{di}{dt} + k\dot{\theta} \quad (5.1.25)$$

where i is the current, θ is the angular position of the motor axis, γ_r the external resistive torque, and γ the motor torque; m combines all the inertias, expressed on the motor axis; β is the internal viscous coefficient, r and L respectively denote the resistance and the inductance, and k is the back-e.m.f. parameter, equal to the torque coefficient. The related block diagram is given in Fig. 5.3, where p is the Laplace variable.

The internal friction coefficient β is small. Setting it equal to zero is a realistic assumption and does not qualitatively influence the results which will be given below. With zero initial conditions, and $\gamma_r = 0$, the open-loop transfer functions of the system, derived from expressions (5.1.23)–(5.1.25), are

$$H_v(p) = \frac{\dot{\theta}(p)}{u(p)} = \frac{1}{D(p)} \quad (5.1.26)$$

and

$$H_p(p) = \frac{\theta(p)}{u(p)} = \frac{H_v(p)}{p} \quad (5.1.27)$$

with

$$D(p) = k(1 + \tau_m p + \tau_e \tau_m p^2) \quad (5.1.28)$$

5.1 GENERAL BACKGROUND

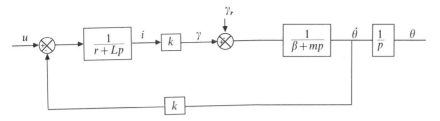

Fig. 5.3 Motor block diagram

where we have defined the *mechanical* time constant as

$$\tau_m = \frac{mr}{k^2} \tag{5.1.29}$$

and the *electrical* time constant as

$$\tau_e = \frac{L}{r}. \tag{5.1.30}$$

Since τ_e is usually much smaller than τ_m, $H_v(p)$ may be considered as the product of two first-order transfer functions (Fig. 5.4):

$$H_v(p) \approx \frac{1}{k(1 + \tau_e p)(1 + \tau_m p)}. \tag{5.1.31}$$

It is also possible to describe the system of Fig. 5.4 in terms of state variables, in the classical linear form:

$$\dot{X} = AX + BU \tag{5.1.32}$$

by choosing

$$X^T = (\dot{\theta} \quad \ddot{\theta}) \tag{5.1.33}$$

$$A = \frac{1}{\tau_e \tau_m} \begin{pmatrix} 0 & \tau_e \tau_m \\ -1 & -\tau_m \end{pmatrix} \tag{5.1.34}$$

and

$$B = \begin{pmatrix} 0 \\ \frac{1}{k \tau_e \tau_m} \end{pmatrix} \tag{5.1.35}$$

when the output variable is only the velocity. If we are interested in both position and velocity, a trivial extension of the state vector in the form of

$$X^T = (\theta \quad \dot{\theta} \quad \ddot{\theta}) \tag{5.1.36}$$

can be chosen.

5 CONTROL: A GENERAL APPROACH

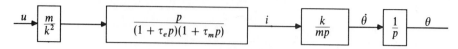

Fig. 5.4 Simplified motor block diagram

In both cases, the system is controllable using a state feedback:

$$u = -G^T X. \tag{5.1.37}$$

Arbitrary pole placement requires an acceleration/velocity feedback in the first case and an acceleration/velocity/position feedback in the second.

5.1.3.2 Closing the loop

Intuitively, the useful control variables of a motor are roughly the voltage u for the velocity $\dot{\theta}$ and the current i for the torque γ. However, these two variables are not often individually controlled in the commercially available motor control units; the practical value to be controlled is only u, while only safety functions take i into account (in order to avoid saturation or overheating.) Let us thus first examine the most frequent control scheme, without current feedback, which will be called 'angular velocity control' in what follows.

5.1.3.2.1 Angular velocity control

1. *Velocity Feedback Only.* This loop, the most traditional one, is generally implemented using a tachometer mounted on the motor axis, with (Fig. 5.5):

$$u = g_v(\mu \dot{\theta}_d - \dot{\theta}) \tag{5.1.38}$$

where g_v is the control gain, and the coefficient μ allows a unit stationary gain between $\dot{\theta}$ and $\dot{\theta}_d$ to be obtained. Its value is then

$$\mu = 1 + \frac{k}{g_v}. \tag{5.1.39}$$

Using (5.1.26), (5.1.28), and (5.1.38), the transfer function in closed loop is

$$\frac{\dot{\theta}(p)}{\dot{\theta}_d(p)} = \frac{g_v \mu}{D(p) + g_v} \tag{5.1.40}$$

which is stable for any positive g_v.

2. *Velocity and Position Feedback.* Most D.C. motor controllers are based on the aforementioned velocity loop. This is due to the fact that, outside of robotic applications, angular velocity is often the desired output variable.

5.1 GENERAL BACKGROUND

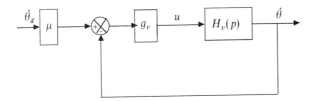

Fig. 5.5 Angular velocity loop

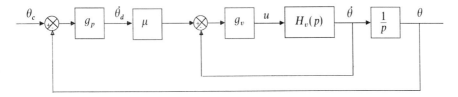

Fig. 5.6 Position/velocity feedback

When position servoing is needed, a classical approach is to add a supplementary loop to the previous scheme, in which the desired velocity $\dot{\theta}_d$ is made proportional to the position error (Fig. 5.6):

$$\dot{\theta}_d = g_p(\theta_c - \theta) \tag{5.1.41}$$

where θ_c is the desired angular position of the motor axis.

Using (5.1.41) in (5.1.38), it appears that this method is fully equivalent to implementing a proportional-derivative feedback:

$$u = -\mu g_p g_v(\theta - \theta_c) - g_v \dot{\theta}. \tag{5.1.42}$$

However, the *explicit* separation of velocity and position loops is frequently done in robotics, and $\dot{\theta}_d$ is often considered as *the only control variable allowed to the user* because it is the traditional 'input' of angular velocity controlled motors. Section 5.2 will return to this common approach in a more general framework.

The resulting closed-loop transfer function, using (5.1.40) and (5.1.41) is

$$\frac{\theta(p)}{\theta_c(p)} = \frac{g_p g_v \mu}{p(D(p) + g_v) + g_p g_v \mu}. \tag{5.1.43}$$

Applying the Routh criterion (cf. Chapter 0) to this third-order transfer function gives as stability conditions:

$$g_v > 0 \tag{5.1.44}$$

and

$$0 < g_p < \frac{1}{\tau_e}. \tag{5.1.45}$$

3. *Integral, Position and Velocity Feedback.* In order to limit the steady-state error, arising in particular from the previously neglected disturbance torque γ_r, integral feedback may be used. The feedback (5.1.42) then becomes

$$u = -g_p g_v (\theta - \theta_c) - g_v \dot{\theta} - g_i \int (\theta - \theta_c) dt. \qquad (5.1.46)$$

The new closed-loop transfer function is then

$$\frac{\theta(p)}{\theta_c(p)} = \frac{p g_p g_v \mu + g_i}{p^2 (D(p) + g_v) + p g_p g_v \mu + g_i} \qquad (5.1.47)$$

and the Routh criterion gives the following supplementary condition to (5.1.44) and (5.1.45):

$$g_i < \frac{(g_v + k)^2 g_p}{k \tau_m} (1 - g_p \tau_e). \qquad (5.1.48)$$

4. *Conclusions on Velocity Control.* From the previous simple analysis, it appears that the performance and robustness of velocity control are limited by the bounds existing on the allowed control gains. For instance, recalling that the sum of the poles, $\sum$, in a polynomial of type $\sum_{i=0}^{n} a_j p^j$ is $-\frac{a_{n-1}}{a_n}$, we have

$$\sum = -\frac{1}{\tau_e} = -\frac{r}{L} \qquad (5.1.49)$$

which is the limit of the smallest possible negative real part of the poles of the closed-loop system. This shows that a better stability margin and faster closed-loop dynamics can theoretically be obtained by reducing τ_e. Unfortunately, this is not always possible because of the existence of a practical lower limit on L.

This fact, added to gain limitations, gives an idea of the drawbacks of this kind of control. In reality, the limitations of this control come from the fact that the theoretical conditions for arbitrary pole placement are not satisfied (cf. 5.1.39): a complete state feedback would have required an acceleration loop, which is not present here. A way of overcoming this difficulty is to include a *current* feedback, as explained in the following.

5.1.3.2.2 Adding a current loop. Intuitively, it is easy to understand that a current feedback may be an alternative to an acceleration feedback: torque and acceleration are linked by a basic mechanical relationship, and current and torque are approximately proportional. If current measurements are more readily available than an acceleration measurement, it is logical to utilize them in a supplementary loop. The general form of such a loop is

5.1 GENERAL BACKGROUND

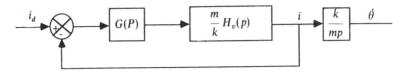

Fig. 5.7 Basic current loop

(cf. fig. 5.7):
$$u = G(p)(i_d - i) \qquad (5.1.50)$$

where i_d is a 'desired' current and $G(p)$ is a rational polynomial.
From (5.1.26), and knowing from (5.1.23) that

$$\frac{\dot\theta(p)}{i(p)} = \frac{k}{mp} = B(p) \qquad (5.1.51)$$

the closed-loop transfer function is

$$\frac{i(p)}{i_d(p)} = \frac{1}{1 + \dfrac{D(p)B(p)}{G(p)}}. \qquad (5.1.52)$$

Usually, $G(p)(i_d - i)$ is either a proportional or a proportional-integral term. In this last case, it is possible to obtain for (5.1.52) a simple first-order transfer function. The important fact is that, in both cases, there is little theoretical limitation on the size of the gains in $G(p)$. This is why it may be assumed that $G(p)$ is always chosen such that the response time of (5.1.52) is fast with respect to other dynamics of the system. In other words, the current loop can be set so as to have

$$i \approx i_d \qquad (5.1.53)$$

and the control variable becomes the current itself. From (5.1.24), this means that *the joint torque may now be considered as a control variable also*. This hypothesis, which underlies many control schemes, becomes justifiable with a current loop.

Let us now examine the effects of a current loop on velocity and position control with the assumption (5.1.53).

1. *Velocity control.* As i_d is the new control input, we may set

$$i_d = k_v(\dot\theta_d - \dot\theta) \qquad (5.1.54)$$

where $\dot\theta_d$ is the desired velocity. We have then simply

$$\frac{\dot\theta(p)}{\dot\theta_d(p)} \approx \frac{kk_v}{kk_v + mp} \qquad (5.1.55)$$

instead of (5.1.40).

Remark. Note that in this case, the commonly used assumption of proportionality between torque and velocity error becomes approximately true.

2. *Position control.* With $\dot{\theta}_d = (k_p/k_d)(\theta_c - \theta)$, the feedback is now

$$i_d = -k_p(\theta - \theta_c) - k_v \dot{\theta} \tag{5.1.56}$$

and the transfer function which replaces (5.1.43) is

$$\frac{\theta(p)}{\theta_c(p)} \approx \frac{k_p}{\frac{m}{k}p^2 + k_v p + k_p} \tag{5.1.57}$$

which is stable for any k_p, k_v positive. The sum of the poles of the closed-loop system is now proportional to the gain k_v the choice of which is free.

Remark. Qualitatively, the same types of results may be obtained in considering the complete transfer function (5.1.52) instead of the approximation (5.1.53).

5.1.3.3 Concluding remarks

The advantages of including a current loop inside the controller of a D.C. motor were emphasized. In this case the current intensity (and hence the torque) becomes a true control variable, yielding improved stability and robustness properties. With this loop, the motor behaves like a double integrator.

5.2 A GENERAL CONTROL STRUCTURE

5.2.1 Introduction

We now return to the problem of controlling the whole robot. A first point is the choice of the control variable. We shall assume that the power units of all D.C. motors include, as previously discussed, a current loop fast enough to allow us to consider the vector of joint torques, Γ, as a control variable. Then, there is no *a priori* limitations on the gains of the control loops to be designed, and model (5.1.11) is sufficient to represent the internal behaviour of the robot.

A second requirement, which will be assumed to be satisfied in the following is that the task $\{e(q, t); q_0\}$ is ρ-admissible during the expected task duration $[0, T]$. With this assumption, the control problem, which is defined as the regulation of the error vector $e(t) = e(q(t), t)$ is well-posed. Also, the properties of admissibility required for the task function ensure the existence

5.2 GENERAL CONTROL STRUCTURE

of a local diffeomorphism between (e, t) and (q, t), as discussed in Chapter 3. This allows the passage from joint space to task space, i.e. *the system dynamics to be fully expressed in the output space.*

We shall thus begin for the analysis from model (5.1.22), where Γ is the effective control vector. We make the natural choice

$$Y = \begin{pmatrix} e \\ \dot{e} \end{pmatrix} \tag{5.2.1}$$

as the new state vector. Then the dynamic state equation is

$$\dot{Y} = F(Y, \Gamma, t) = \begin{pmatrix} 0 & I \\ 0 & 0 \end{pmatrix} Y + \begin{pmatrix} 0 \\ J_T(q, t) M^{-1}(q) \end{pmatrix} \Gamma$$

$$- \begin{pmatrix} 0 \\ J_T(q, t) M^{-1}(q) N(q, \dot{q}, t) - f(q, \dot{q}, t) \end{pmatrix}. \tag{5.2.2}$$

It should be emphasized that the original state vector, $X = \begin{pmatrix} q \\ \dot{q} \end{pmatrix}$ must, when employed in (5.2.2), be implicitly considered as a function of Y.

The control of a non-linear system such as (5.2.2) may be treated in two stages:

(1) linearization of the system;

(2) control of the new system obtained after linearization.

In the present case, the first step may be formally realized simply by setting

$$\Gamma = M J_T^{-1}(u_e - f) + N \tag{5.2.3}$$

where u_e is the new control vector. Then (5.2.2) becomes

$$\dot{Y} = \begin{pmatrix} 0 & I \\ 0 & 0 \end{pmatrix} Y + \begin{pmatrix} 0 \\ I \end{pmatrix} u_e. \tag{5.2.4}$$

Note that equation (5.2.3) combines *two* steps of linearization, namely:

(1) an *intrinsic* one, corresponding to the linearization and decoupling of the original robot's dynamic equations (5.1.11), where only M and N are involved at this level;

(2) an *extrinsic* one, corresponding to the linearization in the task space, in which only J_T and f are involved.

This splitting will later be related to the main classes of control schemes.

The system is now linear. If the objective is to keep the state vector equal to zero, the classical theory of linear systems shows that this system is fully controllable (i.e. with arbitary pole placement) with a state feedback:

$$u_e = -GY = -(G_1 \quad G_2) \begin{pmatrix} e \\ \dot{e} \end{pmatrix}. \quad (5.2.5)$$

The closed-loop system is stable if and only if the eigenvalues of

$$\begin{pmatrix} 0 & I \\ -G_1 & -G_2 \end{pmatrix}$$

have negative real parts. For example, a natural approach consists in choosing the G_i as diagonal positive matrices. This choice ensures both stability and decoupling of the closed-loop system. However, at this stage the important thing to be kept in mind is that, to control the system fully, *both* position and velocity feedback are needed. Further, and as mentioned in section 5.1.3.2, if we now wish to compensate for steady-state errors (due to, for example, unmodelled disturbances) an integral feedback is required; this may easily be demonstrated by increasing the size of the state vector.

Obviously, the class of control defined by equations (5.2.3) and (5.2.5) is not the only possible one among state feedback controls. However, this approach of control is very natural because it takes us back to a linear framework. Relation (5.2.5) also indicates that robot control schemes originating from linear control theory necessarily have a *minimal proportional-derivative structure*, despite different appellations. We shall characterize this common form further. Of course, all robot control schemes are not exactly alike and do not show similar performance, but true structural differences cannot usually be guessed from the names given to these schemes. We shall try, in the following, to illustrate this by proposing a new classification.

More precisely, we shall define two classes of control schemes, nuclei of which are named after well-known optimization techniques: gradient and Newton methods. The reason for establishing a parallel between some control schemes and classical minimization methods is that it may help to evaluate rapidly the kind of performance that can be expected from a control scheme and to discover the role of some sensitive terms in the control expression. The differences between the two approaches are related to the *class* of the data used in control synthesis, in the sense that only *first-order* data is used in *gradient*-type schemes, and *second-order* data are needed in *Newton*-type control schemes.

5.2.2 Gradient versus Newton

In this subsection we compare the two methods from the point of view of numerical analysis only, by considering the problem of minimizing a scalar cost function. It will be shown later that the extension to control problem may be done if we assume that the admissible task function e is itself the

5.2 GENERAL CONTROL STRUCTURE

gradient of a cost function. Then, staying at an extremum of this function will be strictly equivalent to keeping the task error $e(t)$ equal to zero.

5.2.2.1 Gradient method

We start from the unconstrained time-independent problem:

$$\min h(q), \qquad q \in \mathbf{R}^n. \tag{5.2.6}$$

A classical continuous 'gradient' strategy for finding a minimum of h consists of making $q(t)$ vary according to the evolution equation

$$\dot{q} + \mu Q g = 0 \tag{5.2.7}$$

where μ is a positive scalar, Q is a constant positive matrix and

$$g = \frac{\partial h^T}{\partial q}(q). \tag{5.2.8}$$

Premultiplying (5.2.7) by g^T, we get

$$\frac{d}{dt}(h(q)) = -\mu g^T Q g \leq 0. \tag{5.2.9}$$

Thus h decreases with time as long as $g \neq 0$, and remains constant when $g = 0$, i.e. when a stationary point q^* has been reached. If the Hessian

$$H(q) = \frac{\partial^2 h}{\partial q^2}(q) \tag{5.2.10}$$

is positive at q^*, then q^* is a minimum. A common and simple choice for Q is the identity matrix I. In this case, q moves in a direction opposite to the gradient g.

The main advantage of this method is that, from (5.2.9), it always ensures that h decreases to an area where its value is smaller than $h(q(0))$. Moreover, if q becomes close to a minimum q^* of h, then q converges to q^*. A well-known shortcoming of the gradient method is that the convergence rate may be slow if the shape of h around h^* is prolate: this is the *valley* problem. A consequence is that a small disturbance $V(t)$ in (5.2.7) such as:

$$\dot{q} + \mu Q g = V(t) \tag{5.2.11}$$

may produce large variations in q and keep it away from q^*.

5.2.2.2 Newton method

A means of overcoming the valley problem consists in 'spherizing' the isofunction surfaces by using the inverse of the Hessian matrix $H(q)$ instead of

Q in (5.2.7):

$$\dot{q} + \mu H^{-1}(q)g = 0 \qquad (5.2.12)$$

which, due to $\dot{g} = H\dot{q}$ is equivalent to setting

$$\dot{g} + \mu g = 0. \qquad (5.2.13)$$

The gradient g thus exponentially and uniformly converges to zero, for any initial value $q(0)$. In the neighbourhood of q^*, the convergence rate is isotropic and the sensitivity to disturbances is reduced. The improvement is all the more sensitive when the condition number of the Hessian matrix is large near the minimum. A shortcoming of this method is that the convergence towards a minimum is ensured only when $H(q)$ is positive along the trajectory of q. For example, if $H(q(0))$ is negative, $h(q)$ increases.

5.2.2.3 Time-dependent case

Let us now consider the problem of minimizing $h(q, t)$. Then

$$\dot{h}(q, t) = \frac{d}{dt}(h(q, t)) = g^T(q, t)\dot{q} + \frac{\partial h}{\partial t}(q, t) \qquad (5.2.14)$$

where

$$g(q, t) = \frac{\partial h}{\partial q}(q, t) \qquad (5.2.15)$$

and the gradient method (5.2.7) leads to

$$\dot{h}(q, t) = -\mu g^T Q g + \frac{\partial h}{\partial t}(q, t) \qquad (5.2.16)$$

instead of (5.2.9).

Obviously, if $\partial h/\partial t \neq 0$, convergence to the minimum of h is not ensured. Moreover, if $\partial h/\partial t$ is rapidly varying, with a small value of μ, q may move away from the minimum.

Another method consists of having the equation (5.2.13) always satisfied so as to ensure the exponential convergence of g to zero. From (5.2.15)

$$\dot{g}(q, t) = \frac{\partial g}{\partial t}(q, t) = H(q, t)\dot{q} + \frac{\partial g}{\partial t}(q, t) \qquad (5.2.17)$$

where $H(q, t)$ is the Hessian matrix given by (5.2.10). Then (5.2.13) yields

$$\dot{q} + \mu H^{-1}(q, t)g + H^{-1}(q, t)\frac{\partial g}{\partial t}(q, t) = 0. \qquad (5.2.18)$$

This is a modified Newton method which ensures convergence to the minimum $h^*(q, t)$ as long as $H(q, t)$ is positive.

Thus in the time-dependent case, perfect tracking of $h^*(q, t)$ certainly requires the use of $H(q, t)$, and a Newton method is needed. However, a drawback to the tracking capabilities of this method lies in the possibility that H may become singular. Let us suppose that $h(q, t)$ varies with time according to the course shown in Fig. 5.8.

Between times t_2 and t_3, $H(q^*, t)$ becomes singular, and the method is numerically unstable. By comparison, a gradient method always tends to reach a minimum of $h(q, t)$, as a ball subjected only to gravity chooses the steepest slope. This motion may even start before time t_3, as q does not exactly coincide with q^*. Moreover, a disturbance acting on the evolution equation of q may again quicken the motion of q towards a new minimum.

We have seen that, in order to minimize a time-varying function accurately, the Newton method is desirable. However, this method is more 'fragile' from the numerical point of view than the gradient method. This difference in robustness is connected with the variations of convexity of $h(q, t)$ in the neighbourhood of the minimum $q^*(t)$. The favourable situation is obviously when h warps smoothly with time, remaining convex in a large enough neighbourhood of $q^*(t)$. In that case, the control problem is not ambiguous: the aim is to drive q to the minimum $q^*(t)$, and to hold it at this point. This task is equivalent to the *regulation* of the gradient $g(q, t)$ around zero. We again find that, *when a task is well-conditioned, it can be expressed in the form of a regulation problem during the whole task execution time.*

5.2.2.4 Convexification

In practice, it is often not so easy to verify the convexity of $h(q, t)$ around $q^*(t)$ everywhere that is required. If $h(q, t)$ is a C^2 function, points where $h(q, t)$ fails to be convex correspond to points where the Hessian is singular, so that the Jacobian of the gradient is singular. When this happens the gradient is not an admissible function. To avoid this problem, we may apply a

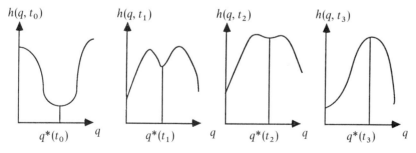

Fig. 5.8 Evolution of a cost function

method similar to that used in section 3.3.4 of Chapter 3, which consists of modifying the task itself.

Suppose that $H(q, t)$ is uniformly bounded, but not necessarily positive. We may then define

$$h'(q, t) = h(q, t) + \tfrac{1}{2}\lambda(t)\|q - y(t)\|^2 \qquad (5.2.19)$$

where $\lambda(t) > 0$ for every t, $y(t)$ is a n-dimensional real function, and $\|.\|$ denotes the Euclidean norm. Then

$$H'(q, t) = \frac{\partial^2 h'}{\partial q^2}(q, t) = \frac{\partial^2 h}{\partial q^2}(q, t) + \lambda(t)I. \qquad (5.2.20)$$

$\lambda(t)$ has to be chosen large enough so as to ensure the positivity of H'. It should, at the same time, be chosen not too large so as to modify the initial task as little as possible whenever H is known to be positive. It remains also to determine $y(t)$ so as to keep the global contribution of the supplementary cost function as small as possible. We may see by an example that (5.2.19) may be interpreted as a way of combining the advantages of both Newton and gradient methods.

Suppose that $h(q)$ is time-independent, and that

$$h'(q, t) = h(q) + \tfrac{1}{2}\lambda\|q - y(t)\|^2 \qquad (5.2.21)$$

where λ is an upper bound of the module of the most negative eigenvalue of $H(q)$. This ensures the definite positivity (and hence the regularity) of

$$H'(q) = \frac{\partial^2 h'}{\partial q^2}(q, t) = H(q) + \lambda I. \qquad (5.2.22)$$

On the other hand, as in section 4.3.4.1 of Chapter 4, let us define $y(t)$ as a filtered value of $q(t)$, through

$$\dot{y}(t) + \alpha y(t) = \alpha q(t), \qquad \alpha > 0 \qquad (5.2.23)$$

$$y(0) = q(0) = q_0. \qquad (5.2.24)$$

It may be noticed that $y(t)$ stays close to $q(t)$ especially when α is large and that $y(t) \to q^*$ when $q(t) \to q^*$. Let us now examine the effect of this modification of h on the two minimization methods.

5.2.2.4.1 Gradient method in the minimization of $h'(q, t)$. Equation (5.2.7), when applied to $h'(q)$ yields

$$\dot{q} + \mu Q g' = 0 \qquad (5.2.25)$$

where

$$g'(q, t) = \frac{\partial h'}{\partial q}(q) = g(q) + \lambda(q - y(t)). \qquad (5.2.26)$$

5.2 GENERAL CONTROL STRUCTURE

Thus
$$\dot{q} + \mu Q(g + \lambda(q - y)) = 0 \tag{5.2.27}$$

with $\dot{q}(0)$ not changed, owing to (5.2.24).

Left-multiplying (5.2.27) by $(\dot{q} - \dot{y})^T Q^{-1}$ leads to

$$\dot{q}^T(Q^{-1}\dot{q} + \mu(\lambda(q - y) + g)) - \dot{y}^T Q^{-1}(\dot{q} + \mu Q g) - \mu\lambda\dot{y}^T(q - y) = 0. \tag{5.2.28}$$

Using (5.2.23) and (5.2.27) yields

$$\dot{y}^T Q^{-1}(\dot{q} + \mu Q g) = -\alpha\lambda\mu(q - y)^T(q - y). \tag{5.2.29}$$

By substituting in (5.2.28), we get

$$\mu\dot{q}^T g + \lambda\mu(\dot{q} - \dot{y})^T(q - y) = -\dot{q}^T Q^{-1}\dot{q} - \alpha\lambda\mu(q - y)^T(q - y) \tag{5.2.30}$$

that is to say

$$\frac{d}{dt}(\mu h + \tfrac{1}{2}\lambda\mu(q - y)^T(q - y)) = -\dot{q}^T Q^{-1}\dot{q} - \alpha\lambda\mu(q - y)^T(q - y) \tag{5.2.31}$$

which is always ≤ 0. Therefore, $h' = h + \tfrac{1}{2}\lambda(q - y)^T(q - y)$ decreases, and $\dot{q}, q - y$, and $g(q)$, by (5.2.17), tend to zero. Since $h'(q(0)) = h(q(0))$, and h' decreases with time, if q converges, it stops at a minimum q^* of h', which is a stationary point of h, with

$$h'(q^*, \infty) = h(q^*) \leq h(q(0)) \tag{5.2.32}$$

(the equality holds when $q^* = q(0)$).

We have thus verified that the gradient method, when applied to $h'(q, t)$, gives results similar to those obtained with $h(q)$.

5.2.2.4.2 Newton method in the minimization of $h'(q, t)$. Equation (5.2.13) becomes

$$\dot{g}' + \mu g' = 0 \tag{5.2.33}$$

where $g'(q, t)$ is given by (5.2.26). As $\dot{g}' = (H + \lambda I)\dot{q} - \lambda\dot{y}$, (5.2.33) yields

$$(H + \lambda I)\dot{q} + \lambda\mu(q - y) + \mu g - \lambda\dot{y} = 0. \tag{5.2.34}$$

Using (5.2.22), (5.2.23), and (5.2.26) leads to

$$\dot{q} + H'^{-1}(\alpha g + (\mu - \alpha)g') = 0 \tag{5.2.35}$$

where $H'(q)$ is always positive, which overcomes the drawbacks of the original Newton method.

Let us finally consider two cases:

(1) $\alpha = \mu$. Then (5.2.35) becomes

$$\dot{q} + \mu H'^{-1} g = 0. \tag{5.2.36}$$

This equation shows that, in this case, applying the Newton method to $h'(q,t)$ is strictly equivalent to applying a composite Newton–gradient method to $h(q)$.

(2) $\alpha \neq \mu$. Left-multiplying (5.2.35) by g^T yields

$$\dot{h}(q) = -\alpha g^T(q) H'^{-1}(q) g(q) - (\mu - \alpha) g^T(q) H'^{-1}(q) g'(q,t). \tag{5.2.37}$$

Now, by (5.2.33), $g'(q,t)$ tends asymptotically to zero; so, roughly

$$\dot{h}(q) \approx -\alpha g^T(q) H'^{-1}(q) g(q) \tag{5.2.38}$$

which shows that q tends to a minimum of $h(q)$.

5.2.3 Control schemes based on Newton and gradient approaches

It is possible to design *control* schemes by analogy with the two *minimization* methods presented above. This classification appears naturally when considering the approach of control presented in the introduction. Let us therefore return to equation (5.2.3) of feedback linearization. As evoked in section 5.2.1, it may be split into two successive levels:

(1) linearization of the robot dynamics through

$$\Gamma = M u_q + N; \tag{5.2.39}$$

(2) linearization in the task space, by

$$u_q = J_T^{-1} u_e - J_T^{-1} f. \tag{5.2.40}$$

Independently of the fact that more or less approximate models may be used in (5.2.39) and (5.2.40), the explicit *presence* or *absence* of the second step of feedback linearization is a way of characterizing the class of a control scheme.

In the ideal case where steps (1) and (2) are both implemented, the evolution equation becomes, using (5.1.22)

$$\ddot{e} = u_e \tag{5.2.41}$$

We shall see in what follows that it is then possible, through the choice of u_e, to make the closed-loop system behave as if a Newton minimization procedure were being performed.

In the case where only (1) is used, then $u_e = u_q$, and the resulting evolution equation is

$$\ddot{q} = u_e. \tag{5.2.42}$$

5.2 GENERAL CONTROL STRUCTURE

Expression (5.2.42) associated with an adequate control u_e will be shown to constitute an extension of the gradient method.

To clarify and pursue the analogy between control classes and minimization methods further, we must assume that the task function $e(q, t)$ is the *gradient* of a cost function to be minimized

$$e(q, t) = \frac{\partial h}{\partial q}(q, t). \qquad (5.2.43)$$

The previous analysis of gradient and Newton techniques has led to the derivation of joint trajectories which were defined by first-order differential equations. In the framework of control, due to the nature of the system to be controlled, we now have to extend these methods to the second order.

5.2.3.1 Newton-like control

Let us start from equation (5.2.41), with a proportional-derivative control of the form

$$u_e = -kG(\mu De + \dot{e}) \qquad (5.2.44)$$

in which k and μ are two positive scalars and G, D are two symmetric positive definite matrices. Choosing four tuning terms k, μ, G and D clearly introduces some redundancy in the control design. However, this choice will be useful for stating the results of the stability analysis.

Then, the equation of the closed-loop system is

$$\ddot{e} + kG(\dot{e} + \mu De) = 0. \qquad (5.2.45)$$

It is easy to show, from (5.2.45), that e exponentially converges to zero. Assigning to e the behaviour given by (5.2.45) thus theoretically allows the value of the cost function $h(q, t)$ to be kept at its minimum during the task execution time. If we now make the scalar gain k tend to $+\infty$, the solutions of equation (5.2.45) have the same asymptotic behaviour as the one given by

$$\dot{e} + \mu De = 0. \qquad (5.2.46)$$

Remark. When k is large enough, equation (5.2.46) may be understood as the slow part of the system (5.2.45) with 'singular perturbation'.

Equation (5.2.46) means that, asymptotically, the evolution of the system is approximately that given by a Newton minimization method, as in equation (5.2.13).

Using equations (5.1.22) and (5.2.45), and recalling that

$$\dot{e} = J_T \dot{q} + \frac{\partial e}{\partial t}, \qquad (5.2.47)$$

where $J_T = (\partial e/\partial q)(q, t)$ is the Hessian matrix associated with the cost function $h(q, t)$, we find that the ideal control leading to the closed-loop system equation (5.2.45) is

$$\Gamma = -kM(q)J_T^{-1}(q,t)G(\mu D e(q,t) + J_T(q,t)\dot{q} + \frac{\partial e}{\partial t}(q,t))$$
$$+ N(q, \dot{q}, t) - M(q)J_T^{-1}(q,t)f(q,\dot{q},t) \qquad (5.2.48)$$

where f is given by (5.1.20).

Implementing this ideal control requires a complete knowledge of $M(q)$, $N(q, \dot{q}, t)$, $J_T(q, t)$ and $f(q, \dot{q}, t)$, and perfect measurements of q and $\dot{q}$. In practice, this control will be approximated by

$$\Gamma = -k(\hat{q},\hat{\dot{q}},t)\hat{M}(\hat{q})\hat{J}_T^{-1}(\hat{q},t)G(\mu D \hat{e}(\hat{q},t) + \hat{J}_T(\hat{q},t)\hat{\dot{q}} + \frac{\partial \hat{e}}{\partial t}(\hat{q},t))$$
$$+ \hat{N}(\hat{q},\hat{\dot{q}},t) - \hat{M}(\hat{q})\hat{J}_T^{-1}(\hat{q},t)\hat{f}(\hat{q},\hat{\dot{q}},t) \qquad (5.2.49)$$

where:

(1) $\hat{q}(t), \hat{\dot{q}}(t)$ are measurements or estimates at time t of true joint positions and velocities;

(2) $\hat{M}$ and $\hat{N}$ are functions representing estimates of the true inertia matrix M and of the term N;

(3) $\partial \hat{e}/\partial t$, $\hat{J}_T$ and $\hat{f}$ are functions representing estimates of $\partial e/\partial t$, J_T and f;

(4) μ, G, and D are defined as previously, while k is a scalar gain, and is a function of the state and the time (see Chapter 6);

(5) $\hat{e}(\hat{q}, t)$ is an estimate of the true error $e(q(t), t)$.

The proportional-derivative nature of this control expression appears in the feedback term $-k\hat{M}\hat{J}_T^{-1}G(\mu D\hat{e} + \hat{J}_T^{-1}\hat{\dot{q}} + \partial \hat{e}/\partial t)$ which is needed to obtain good robustness properties.

We may now propose the following definition.

Definition D5.1. *Any control scheme of the form*

$$\Gamma = -k(\hat{q},\hat{\dot{q}},t)\hat{M}(\hat{q})\hat{J}_T^{-1}(\hat{q},t)G(\mu D \hat{e}(\hat{q},t) + \hat{J}_T(\hat{q},t)\hat{\dot{q}} + \frac{\partial \hat{e}}{\partial t}(\hat{q},t))$$
$$+ \hat{N}(\hat{q},\hat{\dot{q}},t) - \hat{M}(\hat{q})\hat{J}_T^{-1}(\hat{q},t)\hat{f}(\hat{q},\hat{\dot{q}},t) \qquad (5.2.50)$$

where the task function is the gradient of a cost function $h(q, t)$:

$$e(q,t) = \frac{\partial h}{\partial q}(q,t) \qquad (5.2.51)$$

is called Newton-type control **(NTC)**.

5.2 GENERAL CONTROL STRUCTURE

This control, which can be utilized to minimize $h(q, t)$ in the same way as a Newton minimization method, uses first- and second-derivative data ($\partial h/\partial q$ and $\partial e/\partial q$). Intuitively, we may guess that this kind of control will have properties similar to Newton methods, as seen in section 5.2.2:

(1) *advantages*:
   (a) ability to track the ideal trajectory well;
   (b) spherization of the valley around the ideal trajectory, leading to greater insensitivity to measurement noise;

(2) *drawbacks*:
   (a) the complexity of the control scheme is greater than in the forthcoming gradient method;
   (b) control is defined only inside the regions where J_T is regular, and moreover convergence to a minimum of h only occurs where $J_T = H$ is strictly positive (or strictly negative for a maximization problem). This last condition is obviously linked to the problem of admissibility.

5.2.3.2 Gradient-like control

Analogously to Newton-like control (cf. equation 5.2.45), an extension of gradient techniques consists of trying to make the system evolve according to

$$\ddot{q} + kG\left(\dot{q} + \mu D \frac{\partial h}{\partial q}\right) = 0 \qquad (5.2.52)$$

where $k, \mu > 0$ and matrices G, D are chosen positive diagonal. From (5.2.42), this implies using the auxiliary control

$$u_e = -kG(\dot{q} + \mu De). \qquad (5.2.53)$$

The connection with a gradient minimization method is easily seen if we suppose that $h(q, t)$ depends weakly on t, i.e. $\partial h/\partial t \approx 0$; then, if we left-multiply (5.2.52) by $\dot{q}^T(GD)^{-1}$, we find

$$\frac{d}{dt}(\tfrac{1}{2}\dot{q}^T(GD)^{-1}\dot{q} + k\mu h) = -k\dot{q}^T D^{-1}\dot{q} \leq 0. \qquad (5.2.54)$$

This implies

$$\dot{q} \longrightarrow 0 \qquad (5.2.55)$$

$$h \longrightarrow \text{constant}. \qquad (5.2.56)$$

Let us define

$$\eta = \dot{q} + \mu D \frac{\partial h}{\partial q}. \qquad (5.2.57)$$

By differentiating (5.2.57) and using (5.2.52), we obtain the following evolution equation

$$\dot{\eta} = -kG\eta + \mu DH\dot{q} \qquad (5.2.58)$$

where H is again the Hessian matrix $\partial^2 h/\partial q^2 = \partial e/\partial q = J_T$.

As $\dot{q} \to 0$, $\mu DH\dot{q}$ tends to zero, and, from (5.2.58), η rapidly becomes smaller when the gain k is large. In this last case, equation (5.2.57) indicates that the asymptotic evolution of the system is approximately given by

$$\dot{q} + \mu D \frac{\partial h}{\partial q} = 0. \qquad (5.2.59)$$

The comparison of this equation with the expression (5.2.7), justifies the appellation 'gradient-like control'.

The control Γ, associated with this method, is obtained in using (5.2.53) in (5.2.39) with $u_q = u_e$:

$$\Gamma = -kM(q)G(\mu De(q,t) + \dot{q}) + N(q,\dot{q},t). \qquad (5.2.60)$$

We may analyse the contributions of the main variables of this expression in the following way:

(1) $M(q)$ is the inertia matrix; including $M(q)$ in Γ allows us to *decouple* the system (cf. section 5.3.1);

(2) $N(q, \dot{q}, t)$ is an anticipation term, the aim of which is to compensate for various modelled non-linearities (gravity, centrifugal/Coriolis torques, etc.);

(3) $-kG(\mu De(q,t) + \dot{q})$ is a kind of position/velocity feedback, since the task variable $e(q,t)$ depends on the position of the robot.

As in the Newton case, all the terms of (5.2.60) have to be replaced in practice by the related models (approximations or estimations), which will be denoted using carets.

We are now ready to put forward:

Definition D5.2. *Any control scheme of the form*

$$\Gamma = -k(\hat{q}, \hat{\dot{q}}, t)\,\hat{M}(q)\,G(\mu D\hat{e}(\hat{q},t) + \hat{\dot{q}}) + \hat{N}(\hat{q}, \hat{\dot{q}}, t) \qquad (5.2.61)$$

where the task function is the gradient of a cost function $h(q,t)$:

$$e(q,t) = \frac{\partial h}{\partial q}(q,t) \qquad (5.2.62)$$

is called gradient-type control (GTC).

By using control (5.2.61) the cost function $h(q,t)$ tends to be minimized in the same way as a gradient method would perform the minimization, i.e.

5.2 GENERAL CONTROL STRUCTURE

using only *first-derivative* information $((\partial h/\partial q)(q,t))$. Intuitively, we may guess that this kind of control will present similar properties to gradient minimization methods, as seen in section 5.2.2:

(1) *advantages*:
   (a) simplicity of the control expression;
   (b) insensitivity with respect to positivity properties of the Hessian matrix, $H = \partial e/\partial q$;

(2) *drawbacks*:
   (a) inability to track perfectly the ideal trajectory when $(\partial e/\partial t)(q,t) \neq 0$, i.e. to keep $h(q,t)$ at its minimum;
   (b) sensitivity to 'valley' effects, i.e. risk of slow convergence rate due to gain limitations, and sensitivity to noise measurements.

In the second aspect of point (1) lies the basic difference between gradient- and Newton-type control schemes: while the useful working domain of NTC exactly coincides with the task admissibility domain, the GTC may be computed and used in a larger domain. However, the behaviour of the controlled system, during the time where admissibility constraints are not satisfied, is quite unpredictable.

This again shows the importance of the problem of task definition: fine NTC schemes may be used only when the ideal trajectory is defined without ambiguity; GTC schemes are less accurate, and more flexible when the task definition presents some incoherence. For example, if we return to the example of Fig. 5.8, it appears that, when approaching the time where the minimum becomes a maximum (i.e. where the ideal trajectory is not defined), NTC schemes become numerically unstable, while GTC schemes allow us to overcome the difficulty and to find (in an unknown way) another minimum of $h(q,t)$.

From this, we may expect that NTC will be numerically less robust than GTC, when the task is not properly defined. In Chapter 6 a deeper analysis will be performed, on a common basis, i.e. by assuming full task admissibility during $[0, T]$. It was shown in 5.2.2.4 that this condition was not so restrictive in practice, because it is often possible to modify the task so as to improve its admissibility. The method consists in convexifying the cost function to be minimized by adding a suitable convex function to the initial one.

5.2.4 A general control scheme

When the task function $e(q,t)$ is not the gradient of a scalar cost function, the notion of GTC or NTC no longer applies. Moreover, it is possible to design control schemes which lie between GTC and NTC. Formally, the GTC scheme may even be deduced from NTC by setting $\hat{J}_T = I$, $\partial \hat{e}/\partial t = \hat{f} = 0$.

However, keeping the error $e(t)$ as close to zero as possible remains an objective in all cases. The combination of GTC and NTC schemes into a single, more general control structure is then possible. In fact, equations (5.2.50) and (5.2.61) are particular cases of the general control scheme

$$\Gamma = - k(\hat{q}, \hat{\dot{q}}, t) \hat{M}(\hat{q}, t) J_q^{-1}(\hat{q}, t) G(\mu D \hat{e}(\hat{q}, t)$$
$$+ J_q(\hat{q}, t)\hat{\dot{q}} + J_t(\hat{q}, t)) + \hat{l}(\hat{q}, \hat{\dot{q}}, t). \quad (5.2.63)$$

Applying this control expression to the robot finally gives the complete block diagram of Fig. 5.9, where it is assumed that a current loop exists inside each D.C. motor power drive.

In expression (5.2.63), J_q and J_t represent respectively the user's choices made in the control scheme for J_T and $\partial e/\partial t$. The *position/velocity* feedback appears in the term $- k\hat{M}J_q^{-1}G(\mu D \hat{e} + J_q\hat{\dot{q}} + J_t)$, while $\hat{l}(\hat{q}, \hat{\dot{q}}, t)$ collects other components that may be used in the control: compensation of N (Coriolis, centrifugal, gravity forces, etc.) and f, integral correction, etc.

Obviously, GTC and NTC schemes may be derived from the above general expression by setting:

$$J_q(\hat{q}, t) = I \quad (5.2.64)$$

$$J_t(\hat{q}, t) = 0 \quad (5.2.65)$$

$$\hat{l}(\hat{q}, \hat{\dot{q}}, t) = \hat{N}(\hat{q}, \hat{\dot{q}}, t) \quad (5.2.66)$$

for gradient-type control; and

$$J_q(\hat{q}, t) = \hat{J}_T(\hat{q}, t) \quad (5.2.67)$$

$$J_t(\hat{q}, t) = \frac{\partial \hat{e}}{\partial t}(\hat{q}, t) \quad (5.2.68)$$

$$\hat{l}(\hat{q}, \hat{\dot{q}}, t) = \hat{N}(\hat{q}, \hat{\dot{q}}, t) - \hat{M}(\hat{q}) \hat{J}_T^{-1}(\hat{q}, t) \hat{f}(\hat{q}, \hat{\dot{q}}, t) \quad (5.2.69)$$

for Newton-type control.

As shown in the next section, most existing robot control schemes are particular derivations of the general form (5.2.63).

5.3 APPLICATION TO PROMINENT CONTROL SCHEMES

Preliminary remark. This section is devoted to the presentation of some known control techniques as particular cases of the general form (5.2.63). To simplify the descriptions, we shall suppose that the measurements (or estimates) of the state, $(q, \dot{q})$, and the error, e, are perfect. This means: $\hat{q} = q$; $\hat{\dot{q}} = \dot{q}$; $\hat{e} = e$. However, we have to remember that in practice this ideal situation never occurs.

5.3 APPLICATION TO PROMINENT CONTROL SCHEMES

5.3.1 Control schemes based on linearization and decoupling

5.3.1.1 Introduction

This approach, well known in non-linear systems theory, finds a natural application in robot control. As stated in section 5.2.1, the principle consists of two steps:

(1) transform the original system into a linearized one;

(2) control the new system using classical tools of linear control theory.

In non-linear systems, the first step is the most difficult one, and a solution does not always exist in the general case. However, in the case of rigid robots represented by equations of type (5.1.11) and when all joints are controlled, we saw in deriving the general control approach that application of this method becomes theoretically quite trivial. Starting from

$$\Gamma = M(q)\ddot{q} + N(q, \dot{q}, t) \tag{5.3.1}$$

the ideal decoupling control vector is obviously:

$$\Gamma = M(q)u_q + N(q, \dot{q}, t) \tag{5.3.2}$$

which leads to the linear decoupled behaviour

$$\ddot{q} = u_q. \tag{5.3.3}$$

At this stage we have now merely to control n invariant linear systems which are simply double integrators.

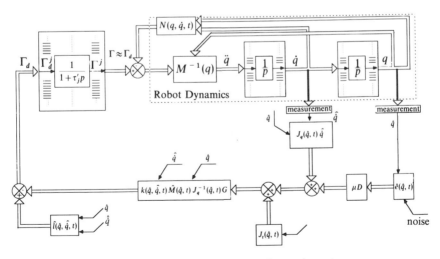

Fig. 5.9 Block diagram of the general control structure

In task space, we saw that by rewriting model (5.1.22) in the form
$$\Gamma = M(q)J_T^{-1}(q)\ddot{e} + N(q, \dot{q}, t) - M(q)J_T^{-1}(q)f(q, \dot{q}, t) \quad (5.3.4)$$
the ideal decoupling control was
$$\Gamma = M(q)J_T^{-1}(q)(u_e - f(q, \dot{q}, t)) + N(q, \dot{q}, t) \quad (5.3.5)$$
which yields the n invariant linear decoupled equations in *task* space:
$$\ddot{e} = u_e. \quad (5.3.6)$$
Note that, by combining (5.3.2) and (5.3.5), u_q and u_e are related by the equation
$$u_e = J_T^{-1}(q)u_q + f(q, \dot{q}, t). \quad (5.3.7)$$

Any classical linear control method (except minimal-time techniques with bounds on u_e, which lead to bang-bang control) applied to equation (5.3.6) leads to a feedback control of the form
$$u_e = -kG(\mu De + \dot{e}) \quad (5.3.8)$$
where $k > 0$, $\mu > 0$, and G and D are diagonal positive matrices. From (5.3.6) and (5.3.8), the resulting evolution equation of the error is
$$\ddot{e} + kG(\mu De + \dot{e}) = 0 \quad (5.3.9)$$
which is exactly the expression produced by an ideal NTC scheme. The related control torque vector is obtained by using (5.3.7) in (5.3.5):
$$\Gamma = -kMJ_T^{-1}G\left(\mu De + J_T\dot{q} + \frac{\partial e}{\partial t}\right) - MJ_T^{-1}f + N \quad (5.3.10)$$
which is a particular case of the general expression (5.2.63), obtained when
$$J_q(q, t) = J_T(q, t) \quad (5.3.11)$$
$$J_t(q, t) = \frac{\partial e}{\partial t}(q, t) \quad (5.3.12)$$
$$l(q, \dot{q}, t) = N(q, \dot{q}, t) - M(q)J_T^{-1}(q)f(q, \dot{q}, t) \quad (5.3.13)$$

All schemes of linearization and decoupling control may thus be considered as belonging to the NTC class.

5.3.1.2 Examples of linearizing and decoupling control schemes

The ideal decoupling control (5.3.10) is generally proposed for two kinds of tasks.

5.3.1.2.1 Control in joint space.
This technique is encountered under the names of 'computed torque method', 'inverse problem technique' and

5.3 APPLICATION TO PROMINENT CONTROL SCHEMES

'dynamic control in joint space'. In all cases, the objective is the tracking of trajectories specified in the joint space and the involved task function is

$$e(q) = q - q_r(t). \tag{5.3.14}$$

From (5.3.10), we find that the control expression is

$$\Gamma = -kM(q)G(\mu D(q - q_r(t)) + (\dot{q} - \dot{q}_r(t))) + M(q)\ddot{q}_r(t) + N(q, \dot{q}, t). \tag{5.3.15}$$

5.3.1.2.2 Control in operational space. The operational space formulation is simply a usual way of expressing the robot dynamics at the end-effector level. Let $T_6 = \begin{pmatrix} \dot{x} \\ \omega \end{pmatrix}$ denote the velocity screw of the last body of a six degree-of-freedom robot. Let F be the screw of internal forces expressed at the end-effector level in the same frame. The principle of virtual power allows us to write

$$\langle \Gamma, \dot{q} \rangle = T_{60} \bullet F. \tag{5.3.16}$$

Knowing that

$$T_{60} = J(q)\dot{q} \tag{5.3.17}$$

we have

$$\Gamma = J^T F. \tag{5.3.18}$$

Note that equation (5.3.18) expresses only the formal transformation from joint torques to effector forces. It has to be carefully used when external or contact forces exist (cf. Chapter 7).

Assuming the regularity of J everywhere where it is needed and differentiating (5.3.17) yields

$$\ddot{q} = J^{-1}(q)(\dot{T}_6 - \dot{J}(q, \dot{q})). \tag{5.3.19}$$

Using the original robot dynamic equation (5.3.1), and the expressions (5.3.18) and (5.3.19), allows us to finally express the robot dynamics at the effector level through

$$F = M'(q)\dot{T}_{60} + R(q, \dot{q}, t) \tag{5.3.20}$$

where M' is the symmetric positive definite matrix

$$M'(q) = J^{-T}M(q)J^{-1}(q), \tag{5.3.21}$$

often called the kinetic energy matrix in the operational space, and

$$R(q, \dot{q}, t) = J^{-T}(q)N(q, \dot{q}, t) - M'(q)\dot{J}(q, \dot{q})\dot{q}. \tag{5.3.22}$$

Expression (5.3.20) is known as an operational space formulation of the dynamics. Owing to its structure, it is possible to apply to (5.3.20) a linearization and decoupling technique, by the control

$$F = M'(q)u_e + R(q, \dot{q}, t) \tag{5.3.23}$$

implemented at the joint level using (5.3.18), i.e.

$$\Gamma = M(q)J^{-1}(q)u_e + N(q, \dot{q}, t) - M(q)J^{-1}(q)\dot{J}(q, \dot{q})\dot{q}. \quad (5.3.24)$$

The resulting behaviour is:

$$\dot{T}_{60} = u_e \quad (5.3.25)$$

where $\dot{T}_{60}$ is the acceleration of the end-effector.

In the usual operational space context, the task to be performed is the tracking of a trajectory in position of the end-effector frame F_6. A difficulty arises because the three-component angular velocity vector in T_{60} is not directly the time-derivative of a three-dimensional representation of the orientation (cf. Chapter 1). Two approaches may then be used.

1. *The 'resolved acceleration scheme'*. This method first uses the linearization and decoupling stage given by (5.3.23) and then computes the control by using $\sin \Theta \, u$, defined in Chapter 1 (section 1.2.4.3), as an 'orientation error'. Given a reference trajectory of F_6, $\bar{r}_r(t) = \begin{pmatrix} x_r(t) \\ r_r(t) \end{pmatrix}$, the classically used control expression is:

$$u_e = -K_1 \begin{pmatrix} x - x_r \\ \sin \Theta \, u \end{pmatrix} - K_2 \left(\begin{pmatrix} \dot{x} \\ \omega \end{pmatrix} - \begin{pmatrix} \dot{x}_r \\ \omega_r \end{pmatrix} \right) + \begin{pmatrix} \ddot{x}_r \\ \dot{\omega}_r \end{pmatrix} \quad (5.3.26)$$

where K_1 and K_2 are gain matrices and ω_r and $\dot{\omega}_r$ are the angular velocity and acceleration vectors related to r_r. It is easy to verify that the equation of the closed-loop system given by (5.3.25) and (5.3.26) is linear in the Cartesian position error $x(q) - x_r(t)$, while it is not in the orientation error. Despite this difficulty, local properties of stability of this system may be proved. One can also verify that the control (5.3.24)–(5.3.26) is a special case of the general expression (5.2.63) obtained by setting

$$e(q, t) = \begin{pmatrix} x(q, t) - x_r(q, t) \\ \sin \Theta \, u(q, t) \end{pmatrix} \quad (5.3.27)$$

$$J_q(q, t) = J(q) \quad (5.3.28)$$

$$J_t(q, t) = \begin{pmatrix} \dot{x}_r \\ \omega_r \end{pmatrix}(t). \quad (5.3.29)$$

The contribution of $\begin{pmatrix} \ddot{x}_r \\ \dot{\omega}_r \end{pmatrix}(t)$ is included in $l(q, \dot{q}, t)$.

Remarks

- In (5.3.26), the attitude error $\sin \Theta \, u$ may be computed through:

$$\sin \Theta \, u = \tfrac{1}{2}(s \times s_r + n \times n_r + a \times a_r)$$

5.3 APPLICATION TO PROMINENT CONTROL SCHEMES

where s, n, and a are the column vectors of the actual rotation matrix R, and s_r, n_r, and a_r are the column vectors of the desired reference one, R_r.

- The choice $J_q = J$ (5.3.28) neglects the contribution of the parametrization. Recalling that $d(PL)/dr = \frac{1}{2}(tr(R)Id - R)$ (eqn. (1.2.67)), we see that this choice of J_q is close to the true value of J_T only for small values of the attitude error. However, a justification of this choice from the point of view of stability will be given in section 6.6.1.3.

2. *The use of a parametrization of Euler type.* This methods falls immediately into our usual framework: given for example a parametrization RPY of the rotations, a task vector then is

$$e(q, t) = \begin{pmatrix} x(q) - x_r(t) \\ RPY(q) - RPY_r(t) \end{pmatrix}. \tag{5.3.30}$$

The related Jacobian matrix may be computed in the form

$$\frac{\partial e}{\partial q} = J_T = \begin{pmatrix} I_3 \\ \dfrac{d(RPY)}{dr} \end{pmatrix} J \tag{5.3.31}$$

where the derivative $d(RPY)/dr$ is given in Chapter 1, section 1.2.5.4.

Then the ideal linearization and decoupling scheme in task space may be applied and equations (5.3.10)–(5.3.13) hold. This was shown to be a true NTC scheme, contained in the general form (5.2.63).

5.3.1.3 Conclusions on linearization and decoupling control schemes

Theoretically, the major advantages of this class of control lie in the following two points:

1. From equation (5.3.9), it is possible to perform perfect execution of the task, i.e. to maintain the error $e(q(t))$ at zero (or close to zero when measurements are noisy).

2. This perfect execution may be achieved with the sole condition of gain *positivity*, without the need to have a large gain k. This important aspect will be discussed in Chapter 6. The drawback of this approach is the counterpart of perfect decoupling. Once it is assumed that $M(q)$, $N(q, \dot{q}, t)$, $f(q, \dot{q}, t)$, $J_T(q, t)$, $J_T^{-1}(q, t)$, and $J_t(q, t)$ are perfectly known, the related expressions have to be computed in real time (i.e. with high sampling rate), which needs powerful computing facilities or dedicated architectures. Furthermore, for the reasons given in section 5.2.1, the assumption of perfect knowledge of the model is unrealistic (for example, the contribution of the payload to the whole model may be important, and its mass/inertia characteristics are often

unknown). Also, there exist tasks for which it is not possible to know $J_T(q, t)$ and $J_t(q, t)$ exactly (see Chapter 7). In practice the exact linearization of the system is never possible, and only approximate solutions of the form

$$\Gamma = -k(\hat{q}, \dot{\hat{q}}, t)\hat{M}(\hat{q})\hat{J}_T^{-1}(\hat{q}, t)G(\mu D\hat{e} + \hat{J}_T(\hat{q}, t)\dot{\hat{q}} + \frac{\partial \hat{e}}{\partial t}(\hat{q}, t))$$

$$+ \hat{N}(\hat{q}, \dot{\hat{q}}, t) - \hat{M}(\hat{q})\hat{J}_T^{-1}(\hat{q}, t)\hat{f}(\hat{q}, \dot{\hat{q}}, t) \qquad (5.3.32)$$

can be implemented.

Finally, concerning this type of control, one might wonder whether the efforts made in establishing and computing a refined model of the system are always reasonable and fruitful. As will be shown in Chapter 6, the answer is not straightforward, since, from the control point of view, there is a balance to be found between refining the model and making the control insensitive to modelling errors.

5.3.2 Velocity control

5.3.2.1 Introduction

This class of control is at the origin of robot control history. As recalled in section (5.1.3), this is mainly due to the fact that classical control of the D.C. motors used in the first robots consisted in setting their 'desired' rotation velocity. Starting from this point of view, the only variable to be considered then is $\dot{q}_d$, the vector of the *desired* joint velocities. The underlying assumption to this approach is that the servo-loops located at the motor level are able to ensure $\dot{q}(t) = \dot{q}_d(t)$. This assumption is easily justified when the robot's joints are not dynamically coupled (which is not the general case), and when $\dot{q}_d(t)$ varies slowly, which may not be true when $\dot{q}_d$ is computed as a function of the actual robot state (usually its position). Therefore, most of the time, a true justification of this assumption requires a stability analysis of the complete system dynamics. Such an analysis will be undertaken in Chapter 6 in a more general framework.

Usually $\dot{q}_d$ is determined exclusively from geometric and variational analysis; this means in particular that only the *kinematics* of the robot are taken into account, and that its *dynamic* characteristics (inertia and mass) are not considered. At a conceptual level, this greatly simplifies the control synthesis. Its implementation is also simplified by the existence of individual and simple motor servoloops. Whether this solution ensures a satisfactory performance of the system is another matter, aspects of which will be examined in the analysis of Chapter 6.

When considering this class of control methods, the following assumptions are implicitly made.

5.3 APPLICATION TO PROMINENT CONTROL SCHEMES

1. The control objective can be expressed under the form of a 'desired' velocity vector $\dot{q}_d$. Note that, from the remarks made in section 5.2.1, stable positioning of the system will need, somewhere in the control loop, the existence of a *position* feedback, which will appear indirectly in the computation of $\dot{q}_d$.

2. The local D.C. motor controllers implement a velocity loop for the tracking of $\dot{q}_d$. At each joint i, the control is

$$\Gamma_i = -k\delta_i(\dot{q}_i - \dot{q}_d^i). \qquad (5.3.33)$$

As pointed out in section (5.1.3) this loop is combined with a current loop. Gathering the Γ_i within a single vector Γ, we obtain

$$\Gamma = -k\Delta(\dot{q} - \dot{q}_d) \qquad (5.3.34)$$

with Δ the diagonal matrix of the δ_i. Starting from these assumptions, we may distinguish *a priori* two classes of velocity-control methods, according to the way $\dot{q}_d$ is computed.

5.3.2.2 Direct methods

5.3.2.2.1 Gradient and velocity control (GVC). Let us return again to the problem of minimizing the scalar cost function $h(q, t)$. An ideal gradient method would consist of having the actual joint velocity $\dot{q}$ coincide with a desired one given by:

$$\dot{q}_d = -\mu D \frac{\partial h}{\partial q} \qquad (5.3.35)$$

where D is a symmetric positive definite matrix, and $\mu > 0$. Since a possible task function for this minimization objective is

$$e(q, t) = \frac{\partial h}{\partial q}(q, t) \qquad (5.3.36)$$

we obtain

$$\dot{q}_d = -\mu D e \qquad (5.3.37)$$

which, using (5.3.34) leads to the following control

$$\Gamma = -k\Delta(\mu D e + \dot{q}). \qquad (5.3.38)$$

This expression is a particular case of the general control structure (5.2.63). More precisely, it is a simple version of the GTC scheme, obtained by choosing

$$J_q = I; \quad J_t = 0 \qquad (5.3.39)$$

for the GTC class; and

$$G = I; \quad \hat{M} = \Delta; \quad \hat{l} = 0 \qquad (5.3.40)$$

as the simplest possible choice of these terms.

Therefore, one can expect that this approach has the shortcomings and, in the best cases, the qualities of GTC schemes.

5.3.2.2.2 Newton and velocity control (NVC). Recalling that the desired behaviour in Newton minimization methods is given by

$$\dot{e} = -\mu De \qquad (5.3.41)$$

and that

$$\dot{q} = -J_T^{-1}\left(-\dot{e} + \frac{\partial e}{\partial t}\right) \qquad (5.3.42)$$

it is natural to choose as a desired velocity

$$\dot{q}_d = -J_T^{-1}\left(\mu De + \frac{\partial e}{\partial t}\right). \qquad (5.3.43)$$

Using (5.3.34), the related control is

$$\Gamma = -k\Delta\left(J_T^{-1}\left(\mu De + \frac{\partial e}{\partial t}\right) + \dot{q}\right). \qquad (5.3.44)$$

Again, this expression is a special case of equation (5.2.63). It is also a simple version of the NTC scheme where:

$$J_q = J_T; \quad J_t = \frac{\partial e}{\partial t} \qquad (5.3.45)$$

as in the NTC general scheme; and

$$G = I; \quad \hat{M} = \Delta; \quad \hat{l} = 0 \qquad (5.3.46)$$

as the simplest possible choice of these terms.

5.3.2.2.3 Discussion. These two methods may thus be classed respectively as GTC and NTC schemes with crude model approximations: $\hat{M} = \Delta$; $\hat{N} = \hat{f} = 0$. These options are intuitively justified when velocities and coupling effects are small; for example when low transmission ratios are used (cf. equation (5.1.18)). Note finally that, in the simple case of joint trajectory tracking (i.e. $h(q, t) = \frac{1}{2}\|q - q_r(t)\|^2$), these two control methods respectively lead to

$$\Gamma = -k\Delta(\mu D(q - q_r) + \dot{q}) \qquad (5.3.47)$$

5.3 APPLICATION TO PROMINENT CONTROL SCHEMES

for GVC, and

$$\Gamma = -k\Delta(\mu D(q - q_r) + (\dot{q} - \dot{q}_r)) \qquad (5.3.48)$$

for NVC, which are simple linear decoupled proportional-derivative loops.

5.3.2.3 Analysis of classical resolved motion rate control

Let $q_r(t)$ be the ideal trajectory associated with the task e, i.e.

$$e(q_r(t), t) = 0. \qquad (5.3.49)$$

Differentiating this equation yields

$$\dot{e}(q_r(t), t) = \frac{\partial e}{\partial t}(q_r(t), t) + J_T(q_r(t), t)\dot{q}_r(t) = 0 \qquad (5.3.50)$$

which allows us to define the ideal velocity along the ideal trajectory

$$\dot{q}_r(t) = -J_T^{-1}(q_r(t), t)\frac{\partial e}{\partial t}(q_r(t), t). \qquad (5.3.51)$$

Remark. This approach is also commonly used for solving redundancy problems by taking the following general solution of (5.3.50), when J_T is in fact the $(m \times n)$ matrix J_1:

$$\dot{q}_r(t) = -J_1^g \frac{\partial e}{\partial t} + (I - J_1^g J_1)z \qquad (5.3.52)$$

where J_1^g is a generalized inverse of J_1 (see Chapter 4), and z is the gradient of a secondary cost to be minimized (all functions are evaluated at $q = q_r(t)$).

From this point of view, $\dot{q}_r(t)$ is the true desired velocity and we should take, in accordance with the original spirit of the velocity control approach,

$$\dot{q}_d = \dot{q}_r \qquad (5.3.53)$$

leading to the control

$$\Gamma = -k\Delta(\dot{q} - \dot{q}_r(t)). \qquad (5.3.54)$$

However, two important problems occur at this stage.

1. Generally, $q_r(t)$ is not known beforehand; assuming that the control is able to maintain $q(t)$ close to $q_r(t)$, the exact ideal expression (5.3.51) is in fact approximated by

$$\dot{q}_r(t) \approx \hat{\dot{q}}_r(t) = -J_T^{-1}(q(t), t)\frac{\partial e}{\partial t}(q(t), t). \qquad (5.3.55)$$

2. As mentioned previously, some position feedback is needed to stabilize the system and maintain a small position error. This leads to attempting to

implement

$$\dot{q}_d = \dot{q}_r + \mu D(q - q_r) \tag{5.3.56}$$

instead of (5.3.53), so as to ensure the convergence of $q - q_r$ to zero when $\dot{q}_d = \dot{q}$.

However, as $q_r(t)$ is unknown, we again need an approximation for $(q - q_r)$. Assuming that $(q - q_r)$ is small, let us perform a first-order expansion of $e(q_r(t), t)$:

$$e(q_r(t), t) = 0 = e(q(t), t) + J_T(q(t), t)(q(t) - q_r(t)) + \ldots \tag{5.3.57}$$

Thus

$$(q - q_r)(t) \approx \widehat{(q - q_r)}(t) = -J_T^{-1}(q(t), t) e(q(t), t). \tag{5.3.58}$$

Using (5.3.55) and (5.3.58) in (5.3.56), we finally obtain

$$\dot{q}_d = \dot{q}_r + \mu D \widehat{(q - q_r)} = -J_T^{-1}\left(\mu D e + \frac{\partial e}{\partial t}\right) \tag{5.3.59}$$

where all functions are evaluated at $(q(t), t)$.

We thus fall back on the NVC scheme of section 5.3.2.2.2.

Let us end this presentation with a classical example. In order to avoid the problem of inverse kinematics, it is common in classical robot controllers to consider a velocity control approach with a desired velocity initially specified in Cartesian space. Let x_r be a six-dimensional position/orientation vector to be reached by the robot, and x the actual one. The desired velocity in x space is chosen as

$$\dot{x}_d = -\mu D(x - x_r) \tag{5.3.60}$$

which is transformed into a desired velocity in the joint space by

$$\dot{q}_d = \left(\frac{\partial x}{\partial q}\right)^{-1} \dot{x}_d. \tag{5.3.61}$$

This solution is directly obtained by choosing in (5.3.59)

$$e = x - x_r. \tag{5.3.62}$$

5.3.2.4 Conclusion on velocity control

These classical control schemes belong to the general class of controls (5.2.63). They are simple but not specifically designed to handle high-performance systems or direct-drive robots. However, they may be useful for controlling classical industrial robots, or when high performance is not required unconditionally.

5.3.3 Adaptive control

5.3.3.1 Introduction

In this section we shall merely give a short overview of adaptive control techniques applied to rigid robots, without going deeply into adaptation methods, which form the subject of much specialized literature. However, let us recall that two main classes of adaptive control techniques are commonly reported.

1. The first is based on the adaptation of feedback gains (position and velocity in the case of robots); the adaptation method is deduced from a stability analysis of the closed-loop system, which is based either on the second Lyapunov method, or on Popov's hyperstability theory. This approach is sometimes understood as a particular form of the well-known model referenced adaptive control (MRAC) technique. This kind of method does not attempt to increase explicitly our knowledge of the system to be controlled; for this reason it may reasonably be considered as a robust control technique aimed at determining non-linear control gains which ensure system stability despite modelling errors. This type of analysis is very similar to that presented in Chapter 6, and we shall not proceed further here.

2. The second, which may truly be called 'adaptive', is based on the idea of learning: knowledge of the system is improved with time through identification of the parameters of a model of the system. Of the methods belonging to this class of adaptive control, the 'indirect' approach is known to lead to the more general stability results, and we shall consider only this method in the following.

Before describing the general structure of adaptive methods, let us try to analyse the reasons that have led to the relative popularity of these approaches. Historically, non-linear decoupling was the first idea to appear for realizing accurate control; however, the drawbacks due to modelling errors, as explained in section 5.1, rapidly arose. From this came the idea of on-line correction of the models used, by taking advantage of the information held in the observation of input and output variables. The idea of applying this principle of creating a kind of 'model feedback' was reinforced by the existence of important work and results on adaptive control theory and applications. However, most of those results were obtained for linear systems, and extension to robotics was not quite straightfoward.

Various adaptive control schemes have thus been proposed, and we shall now try to combine them within a common structure.

5.3.3.2 Adaptive control synthesis

5.3.3.2.1 Background. A method of expressing uncertainty in the robot model consists in parametrizing it by a vector θ, the components of which are *a priori* unknown. Starting from the robot model (5.1.11), this consists in choosing the following representation:

$$\Gamma = \hat{M}(\theta; q)\ddot{q} + \hat{N}(\theta; q, \dot{q}, t). \quad (5.3.63)$$

We must note immediately that equation (5.3.63) is only for *control* purposes: it may not necessarily be complex or realistic. However, it must respect the basic structure of a robot, that is to say a set of disturbed and coupled double integrators.

The physical meaning of θ depends on the chosen parametrization, i.e. the analytic way in which the variable θ appears inside $\hat{M}$ and $\hat{N}$. Classically, a requirement needed by adaptive control theory is that the equality

$$\hat{M}(\theta; q(t))\ddot{q}(t) + \hat{N}(\theta; q(t), \dot{q}(t), t) = M(q(t))\ddot{q} + N(q(t), \dot{q}(t), t) \quad (5.3.64)$$

should be possible (for any actual trajectory $\{q(t)\}$) and compatible with slow variations or rare abrupt changes in θ. Later we shall give some examples of parametrization, some of which will not satisfy this condition.

Let us now pose the *control* problem: obviously, if θ^*, the solution of (5.3.64), *exists* and is *known*, a natural control choice would be the linear decoupling technique (5.3.10), with the θ-dependent dynamic model:

$$\Gamma(\theta^*) = -k\hat{M}(\theta^*; q)J_T^{-1}G\left(\mu De + J_T\dot{q} + \frac{\partial e}{\partial t}\right)$$
$$- \hat{M}(\theta^*, q)J_T^{-1}\hat{f} + \hat{N}(\theta^*; q, \dot{q}, t). \quad (5.3.65)$$

Now, as θ^* is unknown, the adaptive version of this control scheme is obtained by replacing θ^* in (5.3.65) by a *running estimate* $\hat{\theta}(t)$. The related control is denoted by $\Gamma(\hat{\theta}(t))$. In fact, this method ultimately consists in trying to control an 'adaptive model' of the robot through a non-linear decoupling technique. It again belongs to the general structure (5.2.63).

5.3.3.2.2 Identification. The adaptive control scheme will be completed when the estimation algorithm for $\hat{\theta}(t)$ has been specified. This estimation would be based both on model (5.3.58) and on actual measurements, by trying to find the best 'fit' between them. This is a classical identification problem, for which various resolution methods exist, depending on the framework (deterministic or stochastic) and on real-time constraints (off-line versus on-line).

Although it is not our purpose to analyse identification methods, we may combine most of the existing techniques by posing the problem as that of

5.3 APPLICATION TO PROMINENT CONTROL SCHEMES

recursive estimation of a state variable $\theta^*(t)$ through the following two equations.

1. *State*:
$$\dot{\theta}^*(t) = \eta(t) \tag{5.3.66}$$

where $\eta(t)$ is a centred white noise, which is generally a way of allowing $\theta(t)$ to vary with time, without this model being necessarily a realistic representation of the evolution of $\theta(t)$.

2. *Measurement*:
$$\Gamma(\hat{\theta}(t)) = \hat{M}(\theta^*(t); q(t))\ddot{q}(t) + \hat{N}(\theta^*(t); q(t), \dot{q}(t), t) + v(t) \tag{5.3.67}$$

where $v(t)$ is another centred white noise which represents modelling errors in (5.3.63) and measurement noise. $\Gamma(\hat{\theta}(t))$ indicates that the control used is the adaptive one computed from the *estimate* $\hat{\theta}(t)$. Note that here a knowledge of $\ddot{q}$ is needed.

(a) *Linear case*. If the chosen parametrization is such that model (5.3.63) is linear with respect to θ^*, the measurement equation (5.3.67) is linear in θ^*, and may thus be written

$$y(t) = \phi^T(t)\theta^*(t) + v(t). \tag{5.3.68}$$

A general identification algorithm thus has the form

$$\begin{aligned}\dot{\hat{\theta}}(t) &= P(t)\phi(t)(y(t) - \phi^T(t)\hat{\theta}(t)) \\ &= P(t)\phi(t)[\Gamma(\hat{\theta}(t)) - \hat{M}(\hat{\theta}(t); q(t))\ddot{q} - \hat{N}(\hat{\theta}(t), q(t), \dot{q}(t), t)]\end{aligned} \tag{5.3.69}$$

where $P(t)$ is the *gain* of the estimation algorithm. Various existing methods of recursive identification differ essentially in the choice of $P(t)$, which in all cases however has to be a positive matrix. Two classes of methods are particularly well known.

(i) The optimal method, in the sense of the minimization of $\mathbf{E}[(\theta^*(t) - \hat{\theta}(t))^T(\theta^*(t) - \hat{\theta}(t))]$, is given by the Kalman filter, in which the Kalman gain

$$K(t) = P(t)\phi(t) \tag{5.3.70}$$

is the solution of the equations

$$K(t) = C(t)\phi(t)R^{-1}(t) \tag{5.3.71}$$

$$\dot{C}(t) = -K(t)R(t)K^T(t) + Q(t) \tag{5.3.72}$$

where R and Q are respectively the covariance matrices of η and v, assumed to be independent. This method leads to algorithms belonging to the large family of 'recursive least squares'.

(ii) The choice of $P(t)$ as

$$P(t) = \rho(t)I \qquad (5.3.73)$$

corresponds to a class of methods known as 'stochastic approximation', or 'gradient' algorithms.

(b) *Non-linear case*. If now the parametrization is such that the measurement equation (5.3.67) is non-linear, a classical method of solving the problem is to linearize this equation at time t around a given value $\theta(t)$. This approach is best justified when θ is close to the ideal value, $\theta^*(t)$. An obvious choice for the linearization point is $\theta(t) = \hat{\theta}(t)$. Linearization of equation (5.3.67) around $\hat{\theta}(t)$ then leads to the first-order approximation

$$\Gamma(\hat{\theta}) \approx \hat{M}(\hat{\theta}; q)\ddot{q} + \hat{N}(\hat{\theta}; q, \dot{q}, t)$$

$$+ \frac{\partial}{\partial \theta}(\hat{M}(\theta; q)\ddot{q} + \hat{N}(\theta; q, \dot{q}, t))\bigg|_{\theta = \hat{\theta}} (\theta^* - \hat{\theta}) \qquad (5.3.74)$$

where all variables are time-dependent. It is easy to verify that (5.3.67) and (5.3.74) coincide exactly when model (5.3.64) is linear in θ^*.

Now equation (5.3.68) holds again, by setting

$$\phi^T = \frac{\partial}{\partial \theta}(\hat{M}(\theta; q)\ddot{q} + \hat{N}(\theta; q, \dot{q}, t))\bigg|_{\theta = \hat{\theta}} \qquad (5.3.75)$$

and

$$y = \Gamma(\hat{\theta}) - \hat{M}(\hat{\theta}; q)\ddot{q} - \hat{N}(\hat{\theta}; q, \dot{q}, t)$$

$$+ \frac{\partial}{\partial \theta}(\hat{M}(\theta; q)\ddot{q} + \hat{N}(\theta; q, \dot{q}, t))\bigg|_{\theta = \hat{\theta}} \hat{\theta} \qquad (5.3.76)$$

where all variables are again time-dependent. We thus come back to the linear case, and algorithms of type (5.3.69). One of the well-known techniques is 'extended Kalman filter', which combines linearization and Kalman recursive filtering.

It must be emphasized that the *explicit* expression of the applied control never appears in these algorithms. They can therefore be applied to adaptive as well as to non-adaptive control schemes (this means that the identification results may or may not be used on-line in the control scheme). However, in the special case of the adaptive linear decoupling control obtained by replacing $\theta^*(t)$ by $\hat{\theta}(t)$ in (5.3.65), it is easy to see that

$$\Gamma(\hat{\theta}) - \hat{M}(\hat{\theta}; q)\ddot{q} - \hat{N}(\hat{\theta}; q, \dot{q}, t) = -\hat{M}(\hat{\theta}; q)J_T^{-1}(kG(\mu De + \dot{e}) + \ddot{e}) \qquad (5.3.77)$$

where all variables depend on t.

5.3 APPLICATION TO PROMINENT CONTROL SCHEMES

The consequence is that, if $\dot{e}(t)$ and $\ddot{e}(t)$ are also measured, the identification algorithm (5.3.69) may be implemented in the form

$$\dot{\hat{\theta}} = -P\phi\hat{M}(\hat{\theta};q)J_T^{-1}(kG(\mu De + \dot{e}) + \ddot{e}). \tag{5.3.78}$$

In this form, it is clearly seen that the estimate $\hat{\theta}(t)$ stops evolving as soon as the control effectively realizes decoupling in task space, i.e. when

$$kG(\mu De + \dot{e}) + \ddot{e} = 0. \tag{5.3.79}$$

Finally, however, it should be recalled that, in general, it is difficult to obtain direct measurements of $\dot{e}$ and $\ddot{e}$. It should also be emphasized that in practice all the identification algorithms are implemented in discrete-time.

5.3.3.3 Parametrization

We shall give three examples of parametrization:

1. *Pa 1.* The simplest (and the crudest!) parametrization of model (5.3.63) is obtained by setting

$$\hat{M}(\theta;q) = \hat{M}(\theta_1); \qquad \hat{N}(\theta;q,\dot{q},t) = \hat{N}(\theta_2) \tag{5.3.80}$$

and

$$\theta^T = (\theta_1 \quad \theta_2)^T. \tag{5.3.81}$$

This consists in making the assumption that M and N are quasi-invariant functions, which depend only on an unknown vector of parameters, θ, to be estimated. For example, θ_1 may be composed of the $n(n+1)/2$ distinct entries of the symmetric matrix $\hat{M}$ and θ_2 of the n distinct components of vector $\hat{N}$. The dimension of θ is then $[n(n+1)/2 + n]$, i.e. 27 in the frequent case where $n = 6$. A smaller dimension $(2n)$ is obtained when choosing $\hat{M}$ diagonal *a priori*. Further, if M is almost constant, it is sometimes of benefit to choose for $\hat{M}$ a given constant symmetric positive definite matrix and to reduce the identification problem to the estimation of θ_2. The advantage of this method appears in the control, since ensuring the positivity of $\hat{M}$ is essential to the robustness of the control scheme (cf. Chapter 6). Finally, the choice $\{\hat{N} = 0; \theta = \theta_1\}$ means that the robot is approximated by an invariant second-order linear system.

2. *Pa 2.* When the velocity is high, the inluence of Coriolis and centrifugal forces becomes important, and an approximation of N is then

$$N(q,\dot{q},t) \approx \begin{pmatrix} \dot{q}^T V_1(q)\dot{q} \\ \vdots \\ \dot{q}^T V_n(q)\dot{q} \end{pmatrix}. \tag{5.3.82}$$

A possible parametrization is then

$$\hat{M}(\theta; q) = \hat{M}(\theta_1) \tag{5.3.83}$$

and

$$\hat{N}(\theta; q, \dot{q}, t) = \begin{pmatrix} \dot{q}^T B_1(\theta_2) \dot{q} \\ \vdots \\ \vdots \\ \dot{q}^T B_n(\theta_2) \dot{q} \end{pmatrix} \tag{5.3.84}$$

where the structure of matrices B_i is analogous to that of matrices V_i.

This approach in fact consists of isolating the terms which depend only on q, and considering them as unknown parameters to be estimated. This is also a linearization of robot equations for small range motions. Finally, note that we may add to $\hat{N}$ in (5.3.84) a last parameter vector, θ_3, in order to model gravity forces or other disturbances responsible for static position errors which depend only on q. Thus: $\theta^T = (\theta_1 \quad \theta_2 \quad \theta_3)^T$.

3. *Pa* 3. The most realistic parametrizations, which are also the most compatible with the classical assumptions of adaptive control theory, are obtained from the analytic expression of a complete dynamic model (5.1.11). It may be shown that this kind of model is linear with regard to some physical parameters such as inertia, mass, or geometric ones. However, these parameters do not generally appear in independent forms, and the final vector θ may in fact contain only a set of combinations of physical parameters. It is also possible to include in θ other unknown physical coefficients, such as viscous friction.

The associated control schemes may truly be called 'adaptive non-linear decoupling' controls, although linear decoupling always remains approximate in practice, even with on-line identified parameters. From the point of view of adaptive control, these methods are of interest because the system is truly linear with respect to the parameters θ, and these parameters are theoretically invariant. This ensures compatibility with basic assumptions of adaptive control theory (see section 5.3.3.4). On the other hand, an obvious drawback to this approach may be its computing cost. However, it is possible to limit the number of parameters to be identified on-line, by performing a good off-line identification. For example, an efficient adaptive linear decoupling control would use a fairly good robot model which would include explicitly the contribution of the payload, the only parameter remaining to be estimated on-line being its mass. Nevertheless, the overall computing cost remains high, compared to that of simpler control methods.

5.3 APPLICATION TO PROMINENT CONTROL SCHEMES

5.3.3.4 A brief stability analysis

In the same way that many adaptive control schemes may be combined into a unique approach, a global stability analysis may be proposed, following the same lines as global stability proofs found in adaptive control theory. The basic assumption required for this kind of analysis is the existence of θ^*, an unknown vector of 'true' parameters, for which model (5.3.63) coincides with the true equation of the robot. This assumption may appear to be strong; in fact, it is possible to weaken it slightly without modifying the main stability results. This means that the equation

$$\Gamma = \hat{M}(\theta^*; q)\ddot{q} + \hat{N}(\theta^*; q, \dot{q}, t) \qquad (5.3.85)$$

exactly represents the robot's behaviour. For the purpose of analysis, we must also assume that the chosen parametrization is linear with respect to θ. A consequence is that equation (5.3.85) is linear in θ^*, and may thus be written

$$y(t) = \phi^T(t)\theta^*. \qquad (5.3.86)$$

Let $\hat{\theta}(t)$ be the estimated value of θ^* at time t, provided by an identification algorithm of type (5.3.69). For simplicity only, suppose that its gain $P(t)$ has the simplest form:

$$P(t) = \rho(t)I, \qquad (\rho(t) > \rho_{\min} > 0). \qquad (5.3.87)$$

Let us define the minimization error as

$$\tilde{\theta}(t) = \theta^* - \hat{\theta}(t). \qquad (5.3.88)$$

Then, using (5.3.69) and (5.3.87)

$$\dot{\tilde{\theta}} = -\rho(t)\phi(t)[y(t) - \phi^T(t)\hat{\theta}(t)] \qquad (5.3.89)$$

and, by (5.3.86)

$$\dot{\tilde{\theta}} = -\rho(t)\phi(t)\phi^T(t)\tilde{\theta}(t). \qquad (5.3.90)$$

Premultiplying (5.3.90) by $\tilde{\theta}^T(t)$ yields

$$\frac{1}{2}\frac{d}{dt}\|\tilde{\theta}(t)\|^2 = -\rho(t)\|\phi^T(t)\tilde{\theta}(t)\|^2 \leq 0. \qquad (5.3.91)$$

This expression means that the norm of the estimation error decreases, and asymptotically tends towards a constant value, not necessarily equal to zero. An important point to be underlined is that this does *not* imply that $\hat{\theta}(t)$ converges towards θ^*. As $\rho(t) > 0$, we then have:

$$\lim_{t \to +\infty} \phi^T(t)\tilde{\theta}(t) = 0. \qquad (5.3.92)$$

This equation characterizes the asymptotic performances of the identification algorithm. This result is the best to be expected when it is not possible to prove that $\phi(t)$ satisfies 'persistent excitation' conditions. On the other hand

$$\phi^T \tilde{\theta} = \Gamma(\hat{\theta}) - \hat{M}(\hat{\theta}; q)\ddot{q} - \hat{N}(\hat{\theta}; q, \dot{q}, t) \tag{5.3.93}$$

and thus, by (5.3.77)

$$\phi^T \tilde{\theta} = -\hat{M}(\hat{\theta}; q) J_T^{-1} \varepsilon \tag{5.3.94}$$

where

$$\varepsilon = \ddot{e} + kG(\mu De + \dot{e}). \tag{5.3.95}$$

Thus, from (5.3.92), it is clear that, if:

(1) $\hat{M}(\hat{\theta}; q)$ is bounded below and positive for any t;

(2) $J_T(q, t)$ is non-singular, and $\| J_T(q, t) \|$ is uniformly bounded, then

$$\lim_{t \to +\infty} \varepsilon(t) = 0. \tag{5.3.96}$$

Assuming that k, μ, G, D are chosen so as to ensure exponential stability of the homogeneous equation associated with (5.3.90), we finally obtain

$$\lim_{t \to +\infty} e(t) = 0. \tag{5.3.97}$$

This last expression shows that, under the previous assumptions, the adaptive control asymptotically ensures perfect task realization.

Remarks.
1. Ensuring the positivity of $\hat{M}(\hat{\theta}; q)$ is of major importance. This may be realized either by freezing $\hat{\theta}(t)$ as soon as $\hat{M}$ tends to become singular (i.e. when its determinant is close to zero) or, as mentioned previously, by making the *a priori* assumption that $\hat{M}$ is a known constant matrix.

2. We see the admissibility requirements appearing again in condition (2).

5.3.3.5 Discussion

Despite being fairly simple the previous analysis has demonstrated the important fact that the convergence of $\hat{\theta}(t)$ towards θ^* is *not necessary* to ensure good asymptotic behaviour of the control. In fact, it is known that this convergence depends on classical conditions of identifiability (related to the choice of parametrization) and persistent excitation (related to robot motion), which are not necessarily satisfied in practice.

Another point concerns the assumptions of invariance of θ^* and of correct modelling; in fact, it is now possible to show that the asymptotic stability

5.3 APPLICATION TO PROMINENT CONTROL SCHEMES

property of the system remains true when bounded additive disturbances occur, and when θ^* varies 'slowly'. However, these results are not sufficient to conclude that stability is ensured when the basic hypothesis (5.3.85) is strongly violated. This unfortunately happens when parametrizations *Pa 1* or *Pa 2* are used. Roughly speaking, this means that, if the rate of variation of the parameters is comparable with the rate of variation of the system state vector, the ability of tracking $\theta^*(t)$ cannot be ensured, and all the previous results become invalidated. This is why only parametrization *Pa 3* is capable of leading to good identification results.

On the other hand, we must emphasize the *asymptotic* nature of the previous stability results. In fact, we have not obtained any information about the *transient* behaviour of the control method. For example, nothing ensures that the robot remains inside the task admissibility domains during the transient stage. It also appears that good behaviour of adaptive control schemes depends heavily on the performance of the identification algorithms (particularly on the decreasing rate of $\phi^T \tilde{\theta}$). The best way to meet these conditions and also to avoid undesirable burst transient phenomena frequently encountered when using adaptive controls still consists in ensuring a fast convergence of $\hat{\theta}(t)$ towards θ^*. A practical consequence is that it may be better to activate an identification stage only when measurements provide highly informative and useful data (i.e. during fast and large robot motions, or during a strong acceleration/braking phase), and to use identification results only *after* parameter convergence. In fact, it seems that a good use of 'adaptive' control lies in a refined *cooperation* between identification and control, which does not imply using every output (estimated parameters) of an on-line recursive identification algorithm in the control computation. This problem goes beyond the scope of this book since it is still not fully understood and is a current central topic in adaptive control studies.

Let us conclude this section and this chapter with a return to robustness considerations. It has been frequently observed in practice that some adaptive control schemes based on *Pa 1* or *Pa 2* parametrizations might present correct behaviour (from the point of view of task realization), even with poor parameter estimation. This point will be examined in the next chapter, and we may at present merely emphasize the fact that bad estimation (or modelling) of $M(q)$ and $N(q, \dot{q}, t)$ *does not necessarily* lead to bad control performance! In fact, it will be shown further that the robustness of control depends directly on the size of the feedback gains. For example, overestimation of $\hat{M}(\hat{\theta}; q)$ (i.e. increasing its eigenvalues) is equivalent to increasing the position and velocity gains in the overall control, which is a way of improving control robustness. This is why in several cases the unexpectedly good behaviour of adaptive control results less from the accuracy of the estimation algorithms than from the intrinsic robustness of the general control scheme (5.2.63).

5.4 BIBLIOGRAPHIC NOTE

Concerning components of robotic systems, let us note that robot actuator properties are considered in general in André et al. (1983) while direct-drive robots are analyzed in Asada and Kanade (1983a), Asada et al. (1983b). More generally, the problems related to the approaches to robot control described in this chapter have been extensively studied in the literature, although the broad-based analysis given here appears to be original. The general approach proposed here has its origin in Samson (1987a, b), also cited in Boissonnat et al. (1988).

As robustness and non-linear aspects will be presented in Chapter 6, and all kinds of sensor-based control (force, proximity) in Chapter 7, we concentrate here on classical control techniques. General books on robot control exist: from Paul (1982), Vukobratovic and Potkonjak (1982a) and Vukobratovic and Stokic (1982b) to Craig (1986) and Asada and Slotine (1986). The class of computed torque methods has been studied for some time (Markiewitz 1973; Khalil 1978; Khalil et al. 1979; Khatib 1980). An experimental evaluation may be found in An et al. (1987a), and a description of this approach in the 'operational' space is presented in Khatib (1985a). The paper (Luh et al. 1980b) is of special interest, because it was the first approach to combine computed torque methods with the use of an orientation error expression.

On the other hand, much literature on robot control is devoted to adaptive methods; often, the idea is to directly apply the existing approaches in adaptive control theory (Landau 1979; Landau and Dugard 1986; Larminat 1984) to robotic systems (Dubowsky and Desforges 1979) and an interesting parametrization was proposed in Horowitz and Tomizuka (1980). Many examples of applications of adaptive control are reported in the literature: use of self-tuning methods in Koivo and Paul (1980), Lelic and Wellstead (1987), Liu and Lin (1987), or of related techniques in Takegaki and Arimoto (1981), Lim and Eslami (1987). In Middleton and Goodwin (1988), a method for avoiding the use of an acceleration measurement is presented. Basic references on the subject are also Egardt and Samson (1982), Ljung and Sonderstrom (1983), Ljung (1987), Praly (1983), and, for robotics only, Mufti (1985), Craig (1987).

More recently, structural properties of mechanical systems have also been exploited by Slotine and Li (1987) to design and analyse other algorithms. At the present time, it is interesting to note that the trend is towards using adaptive control techniques based on explicit model identification in the control of *flexible* arms.

Also recently, the problems of linearization and decoupling of rigid robots was analysed from the point of view of non-linear control theory, including

5.4 BIBLIOGRAPHIC NOTE

the case of non-fully driven robots (D'andrea and Levine 1986; D'andrea-Novel (1987).

For general information, note that sessions devoted to robot control exist in the major conferences on robotics or on automatic control; see for example the IEEE Conferences on Robotics and Automation, or on Automatic Control, the IFAC Symposia, the Conferences on Decision and Control (CDC), the International Symposium on Robotics Research, and so on.

6
STABILITY ANALYSIS

6.1 PRELIMINARY RESULTS

6.1.1 Introduction

The object of this chapter is to describe a powerful analysis tool, the application of which to the robot field will provide us with robustness conditions for all the elements belonging to the control class defined by the general control scheme (5.2.69). Intuitively, 'robustness' denotes, as briefly evoked in Chapter 5, stability results which remain true in the presence of modelling errors or of certain classes of disturbance; we shall also find that such results will be based on adequate choices of feedback gains in the system.

This chapter is organized as follows: the first part is devoted to posing the problem and analysing the simpler case of invariant linear systems. The second part presents a general stability theorem, valid for a class of non-linear systems. Finally, this analysis is applied to the case of robots.

6.1.2 Statement of the problem

Let $C_{\rho, T}$ be a compact connected set in $\mathbf{R}^n \times [0, T]$, such that

(1) $(q(0), 0) \in C_{\rho, T}$;

(2) $e(q, t)$ is ρ-admissible in $C_{\rho, T}$ during $[0, T]$.

We are interested in studying the solutions of the general differential equation

$$\ddot{e} + k(q, \dot{q}, t) A_1(q, t) A_2(q, t) (\mu B(q, t) De + \dot{e} + v(q, t)) + s(q, \dot{q}, t) = 0 \tag{6.1.1}$$

where

$$e = e(q, t); \qquad J_T(q, t) = \frac{\partial e}{\partial q}(q, t), \tag{6.1.2}$$

$$\dot{e} = J_T(q, t)\dot{q} + \frac{\partial e}{\partial t}(q, t), \tag{6.1.3}$$

$$\ddot{e} = J_T(q, t)\ddot{q} + f(q, \dot{q}, t), \tag{6.1.4}$$

6.1 PRELIMINARY RESULTS

with

$$f(q, \dot{q}, t) = \begin{pmatrix} \dot{q}^T W_1(q, t) \dot{q} \\ \vdots \\ \dot{q}^T W_n(q, t) \dot{q} \end{pmatrix} + 2 \frac{\partial^2 e}{\partial q \partial t}(q, t) \dot{q} + \frac{\partial^2 e}{\partial t^2}(q, t), \quad (6.1.5)$$

$W_i(q, t)$ being the partial derivative of $J_T^T(q, t)$ with respect to q (cf. section 5.1.2.5.2).

The following assumptions are associated with (6.1.1).

(H1) $\mu > 0$ and D is an $n \times n$ symmetric positive matrix.

(H2) On an open set of $\mathbf{R}^n \times \mathbf{R}$ containing $C_{\rho, T}$:
 (a) the $(n \times n)$ matrix functions $A_1(q, t)$, $A_2(q, t)$ and $B(q, t)$ are C^1;
 (b) $A_1(q, t)$ and $B(q, t)$ are positive;
 (c) $A_2(q, t)$ is symmetric and positive;
 (d) the vector function $v(q, t)$ is C^1.

(H3) On an open set of $\mathbf{R}^n \times \mathbf{R} \times \mathbf{R}^n$ including $C_{\rho, T} \times \mathbf{R}^n$ $((q, t) \in C_{\rho, T}; \dot{q} \in \mathbf{R}^n)$:
 (a) the $(n \times 1)$ vector function $s(q, \dot{q}, t)$ is C^1;
 (b) the scalar function $k(q, \dot{q}, t)$ is positive and C^1.

It is easy to see that (6.1.1) contains in particular the expression representing the differential behaviour of the model in task space (5.1.22) subjected to the general control structure (5.2.69).

The differential equation (6.1.1) may also be written in the classical form

$$\dot{X} = E(X, t), \quad (6.1.6)$$

where

$$X = \begin{pmatrix} q \\ \dot{q} \end{pmatrix}; \quad E = \begin{pmatrix} E_1 \\ E_2 \end{pmatrix} \quad (6.1.7)$$

and,

$$E_1 = \dot{q} \quad (6.1.8)$$

$$E_2 = -J_T^{-1}(q, t)[f(q, \dot{q}, t) + k(q, \dot{q}, t)A_1(q, t)A_2(q, t)(\mu B(q, t)De(q, t)$$

$$+ J_T(q, t)\dot{q} + \frac{\partial e}{\partial t}(q, t) + v(q, t))] + s(q, \dot{q}, t). \quad (6.1.9)$$

If they exist, the solutions of this equation are the C^2 trajectories $\{q(t)\}$ which satisfy (6.1.1) for $t \geq 0$. Clearly, from the limited range of assumptions H2–H3, we shall only study the solutions starting from a point $q_0 \in C_{\rho, T}$. Under the sufficient condition that the function $e(q, t)$ is C^3 on its domain, (which includes $C_{\rho, T}$), the H2–H3 assumptions of differentiability of the other functions, *and* the admissibility properties of $e(q, t)$ imply that $E(X, t)$ is C^1 on an open set including $C_{\rho, T} \times \mathbf{R}^n$; $((q, t) \in C_{\rho, T}; \dot{q} \in \mathbf{R}^n)$.

Let us now apply Theorem 0.6 to establish the following result.

Proposition P6.1 *Under H2–H3, and assuming that $e(q, t)$ is C^3 and ρ-admissible on $C_{\rho, T}$ during $[0, T]$, then the solution $\{q(t)\}$ of (6.1.6) locally exists and is unique when $(q(0), 0) \in C_{\rho, T}$.*

Remark. Local existence and uniqueness of solutions can be proved with slightly weaker assumptions than those used in Proposition P6.1. For example, it is not strictly necessary for $e(q, t)$ to be C^3 on $C_{\rho, T}$ and for $s(q, \dot{q}, t)$ and $v(q, t)$ to be C^1. The choice of strong and simple assumptions is a way of avoiding cumbersome technical details which are not essential to the analysis.

The analysis developed in this chapter will show how the solutions extend to time T, such that

$$\text{for any } t \in [0, T]: \|e(t)\| = \|e(q(t), t)\| < \rho. \qquad (6.1.10)$$

This will be done by imposing further conditions on the scales of the gain function $k(q, \dot{q}, t)$, the scalar μ, and the initial conditions $\{e(0); \dot{e}(0)\}$.

Moreover, sufficient conditions for ensuring asymptotic smallness of $\|e(t)\|$ will be established.

6.1.3 Case of an invariant linear equation

Before dealing with the most general expression of (6.1.1), we study the particular case where it reduces to an invariant linear equation. This first step shows that the analysis of this simple case already contains the main features of the analysis of the non-linear case, of course with fewer difficulties. A supplementary advantage of this intermediate step is to demonstrate the basic properties which are needed for establishing the main results of the control analysis.

Assume that:

(1) $e(q, t) = q$;

(2) A_1, A_2 and B are constant matrices;

(3) k and μ are constant scalars;

(4) v and s are any bounded disturbances, which depend only on time.

Then (6.1.1)–(6.1.5) reduce to the second-order linear equation

$$\ddot{e} + kA_1 A_2(\mu BDe + \dot{e} + v(t)) + s(t) = 0 \qquad (6.1.11)$$

6.1 PRELIMINARY RESULTS

or, equivalently, in state-space form,

$$\dot{X} = AX + B \tag{6.1.12}$$

where

$$X = \begin{pmatrix} e \\ \dot{e} \end{pmatrix}, \tag{6.1.13}$$

$$A = \begin{pmatrix} 0 & I \\ -k\mu A_1 A_2 BD & -k A_1 A_2 \end{pmatrix}, \tag{6.1.14}$$

$$B = \begin{pmatrix} 0 \\ -k A_1 A_2 v - s \end{pmatrix}. \tag{6.1.15}$$

The first result, known from linear system theory, is that, for any initial condition $X(0)$, the solution $X(t), t \geq 0$ exists and is unique.

The characteristic polynomial associated with (6.1.12) is given by

$$P(\lambda) = \det(A - \lambda I). \tag{6.1.16}$$

Knowing that

$$\det\begin{pmatrix} A & D \\ C & B \end{pmatrix} = \det A \, \det(B - CA^{-1}D), \tag{6.1.17}$$

we find

$$P(\lambda) = \det(\lambda^2 I + k\lambda A_1 A_2 + k\mu A_1 A_2 BD) \tag{6.1.18}$$

We are interested in studying the *stability* of (6.1.12). (Either *asymptotic* or *exponential* stability, since the two concepts coincide in the case of linear invariant systems.) More precisely, if the system is exponentially stable, then

(1) $\|X(t)\|$ exponentially converges to zero for all $X(0)$, if $v = s = 0$;

(2) $\|X(t)\|$ is asymptotically uniformly bounded if $v(t)$ and $s(t)$ are bounded.

A necessary and sufficient condition for stability is that *the 2n roots of* $P(\lambda)$ *have negative real parts*.

Let us now study how the properties of the matrices appearing in (6.1.11) and the size of k influence the stability of the system.

An initial result is the following.

Lemma 6.1 *Let M_1 and M_2 be two real-valued matrices with M_1 non-singular. Let $\{\mu_i; i = 1, \ldots, n\}$ be the eigenvalues of M_1, and $\{v_i; i = 1, \ldots, n\}$ the eigenvalues of M_2. Then, if k is large enough, the set of solutions $\{\lambda_i; i = 1, \ldots, 2n\}$ of the equation*

$$\det(\lambda^2 I + k\lambda M_1 + k M_1 M_2) = 0 \tag{6.1.19}$$

split into two disjoint subsets $\{\lambda_i; i \in I_1\}$ and $\{\lambda_i; i \in I_2\}$ such that:

(1) *for all* $i \in I_1$ *there exists a* j, $1 \leq j \leq n$ *and* $\lambda_i \to -v_j$ *when* $k \to +\infty$;

(2) *for all* $i \in I_2$ *there exists a* j, $1 \leq j \leq n$ *and* $\lambda_{i+n}/k \to -\mu_j$ *when* $k \to +\infty$.

The proof is given in Appendix A6.1.

Before applying this result to (6.1.18), we need the following lemma.

Lemma 6.2 *If A_1 and A_2 are two positive real-valued $n \times n$ matrices and A_2 is symmetric, then the eigenvalues of $A_1 A_2$ have positive real parts.*

The proof is given in Appendix A6.1.

We may now relate the matrices M_1, M_2 of (6.1.19) to the matrices of (6.1.18) through

$$M_1 = A_1 A_2 \qquad (6.1.20)$$

$$M_2 = \mu BD. \qquad (6.1.21)$$

By assumptions H1, H2 of section 6.1.2, and Lemma 6.2, we thus deduce that the eigenvalues of M_1 and M_2 have positive real parts. Using Lemma 6.1, and recalling that the stability condition is that the $2n$ roots of $P(\lambda)$ have negative real parts, we then immediately have the following stability result.

Lemma 6.3 *There exists a minimal value $k_0 \geq 0$ such that, if $k > k_0$, then system (6.1.11) is exponentially stable.*

Of course, cases exist in which exponential stability of system (6.1.11) may be obtained with the single condition of gain positivity, (i.e. $k_0 = 0$).

Corollary 6.3.1 *Sufficient conditions for the minimal gain k_0 to be zero (i.e. for (6.1.11) to be exponentially stable for any gain $k > 0$) are:*

(1) $A_1 A_2$ *and BD are symmetric positive definite matrices; or*

(2) $A_1 A_2 = \alpha I$ ($\alpha > 0$) *and the eigenvalues of μBD are real and positive; or*

(3) $\mu BD = \alpha I$ ($\alpha > 0$) *and the eigenvalues of $A_1 A_2$ are real and positive.*

The proof is given in Appendix A6.1.

However, it is also easy to find *unstable* systems belonging to the class (6.1.11) with k positive and none of these three conditions being satisfied. Therefore, the minimum value k_0, beyond which the system is stable, is not always zero.

By Lemma 6.1, it is sufficient that eigenvalues of matrices $-M_1 = -A_1 A_2$ and $-M_2 = -\mu BD$ have negative real parts (i.e. matrices $-M_1$ and $-M_2$ are 'stable') to ensure applicability of Lemma 6.3. We may then wonder why

6.1 PRELIMINARY RESULTS

stronger positivity conditions have been imposed on matrices A_1, A_2, B, D. The main reason will become clear in the next section; but it can already be observed that the extension of the present stability results to the non-linear case needs to impose on matrices $-M_1(q(t), t)$ and $-M_2(q(t), t)$ stronger conditions than simple stability at any time t.

This fact may be easily demonstrated by recalling a stability condition of the system

$$\dot{X} = -F(X, t)X. \qquad (6.1.22)$$

Indeed, to ensure stability of (6.1.22), it is known that it is not sufficient that, for any $t \geq 0$ and $X \in \mathbf{R}^n$, $-F(X, t)$ be stable. However, if $F(X, t)$ is known to be strictly positive and such that for all $X \in \mathbf{R}^n$ and $t \geq 0$,

$$\lambda_{\min} \tfrac{1}{2}(F(X, t) + F^T(X, t)) > \delta > 0 \qquad (6.1.23)$$

then

$$X^T \dot{X} = \frac{1}{2} \frac{d}{dt} \|X\|^2 = -X^T F(X, t) X \leq -\delta \|X\|^2 \qquad (6.1.24)$$

which implies that $\|X(t)\|$ exponentially converges to zero.

It should be emphasized that this result is true for any rate of variation of $F(X, t)$. In fact, this result foreshadows the important role which will be played by *matrix positivity* properties in establishing stability results for non-linear systems.

Another point is related to the specificity of application to robot control: any dynamic robot equation exhibits a matrix with such a property: the kinetic energy matrix $M(q)$. Moreover (cf. Chapter 5), when the task consists in minimizing a cost function $h(q, t)$, an admissibility requirement is that $J_T(q, t) = (\partial^2 h / \partial q^2)(q, t)$ is positive in the neighbourhood of the minimum. Since the property of matrix positivity appears naturally in the statement of the robot control problem, it is logical to exploit it in the analysis.

Remark. Singular perturbation theory is another way of establishing the stability result (6.1.3). It is possible to rewrite the system (6.1.11) in the form

$$\dot{e} = z \qquad (6.1.25)$$

$$\eta \dot{z} = -\mu A_1 A_2 BDe - A_1 A_2 z - A_1 A_2 v(t) - \eta s(t) \qquad (6.1.26)$$

with

$$\eta = \frac{1}{k}. \qquad (6.1.27)$$

It may be shown that, when $\eta \to 0$, this system splits into a transient 'fast' part, quickly fading, and a 'slow' part representing the asymptotic behaviour of the system.

We again find that the extinction time of the fast part depends on the very stable eigenvalues of the system, modules of which tend to $+\infty$ when $\eta \to 0$ (by Lemma 6.1, the product of these eigenvalues by η tends towards the opposite of the eigenvalues of $A_1 A_2$).

A way to isolate the 'slow' part consists of setting $\eta = 0$ in (6.1.26), which yields

$$\dot{e} \approx -\mu BDe + v(t). \tag{6.1.28}$$

This shows that, when η is small, the system (6.1.11) becomes asymptotically equivalent to a first-order one. The roots of its characteristic polynomial are of the opposite sign to the eigenvalues of μBD. Equation (6.1.28) shows further that it is possible to make $e(t)$ asymptotically as small as desired by increasing the size of μ.

The proof of all these results is included in the general stability theorem established in the next section.

6.2 A GENERAL STABILITY THEOREM

We now return to the general differential equation (6.1.1) for which we shall give a stability result that generalizes those obtained in the linear case. Although the statements and proofs are more complex, the philosophy of the results themselves is similar in the two cases: roughly, we shall see that, provided that initial errors are small enough, there exists a function $k(.)$ such that for any gain function greater than $k(.)$ the solution $e(t)$ of (6.1.11) remains 'small'. In fact, 'smallness' of $e(t)$ will have to be related to the admissibility radius ρ, such that $e(t)$ is certain to lie *inside* the sphere B_ρ. This section will deal only with the statement of the general results; application to robot control will be described later.

6.2.1 Preliminaries

Recall that the differential equation is

$$\ddot{e} + k(q, \dot{q}, t) A_1(q, t) A_2(q, t) [\mu B(q, t) De + \dot{e} + v(q, t)] + s(q, \dot{q}, t) = 0 \tag{6.1.1}$$

with assumptions H1–H3 given in section 6.1.

6.2.1.1 Notation

In order to obtain a concise statement of the main theorem, we give below the notation which will be used.

$$C_{A_1} \geq \sup_{q, t \in C_{\rho, T}} (\|\tfrac{1}{2}(A_1(q, t) + A_1^T(q, t))^{-1}\| \, \|A_1(q, t)\|) \tag{6.2.1}$$

$$a_2^m(q, t) = \lambda_{\min}(A_2(q, t)) \quad \text{(smallest eigenvalue of } A_2(q, t)) \tag{6.2.2}$$

6.2 A GENERAL STABILITY THEOREM

$$a_2^M(q, t) = \lambda_{\max}(A_2(q, t)) \quad \text{(greatest eigenvalue of } A_2(q, t)) \quad (6.2.3)$$

$$C_{A_2} \geq \sup_{q, t \in C_{\rho, T}} \left(\frac{a_2^M(q, t)}{a_2^m(q, t)}\right) \quad (6.2.4)$$

$$0 < \sigma_2^m \leq \inf_{q, t \in C_{\rho, T}} (a_2^m(q, t)) \quad (6.2.5)$$

$$\sigma_2^M \geq \sup_{q, t \in C_{\rho, T}} ((a_2^M(q, t)) \quad (6.2.6)$$

$$b^m(q, t) = \lambda_{\min}[\tfrac{1}{2}(B(q, t) + B^T(q, t))] \quad (6.2.7)$$

$$b^M(q, t) = \lambda_{\max}[\tfrac{1}{2}(B(q, t) + B^T(q, t))] \quad (6.2.8)$$

$$0 < \beta^m \leq \inf_{q, t \in C_{\rho, T}} (b^m(q, t)) \quad (6.2.9)$$

$$b_{\max} \geq \sup_{q, t \in C_{\rho, T}} (\|B(q, t)\|) \quad (6.2.10)$$

$$d^m = \lambda_{\min}(D) \quad (6.2.11)$$

$$d^M = \lambda_{\max}(D) \quad (6.2.12)$$

$$v \geq \sup_{q, t \in C_{\rho, T}} (\|v(q, t)\|). \quad (6.2.13)$$

6.2.1.2 Definition of the gain function

In the statement of the theorem, we shall use the positive scalar function $k_{\alpha, \mu, \phi, \delta}(q, \dot{q}, t)$ which is defined in the following way:

$$k_{\alpha, \mu, \phi, \delta}(q, \dot{q}, t) = \sup[k_{\alpha, \mu}^1(q, \dot{q}, t); k_{\alpha, \mu, \phi, \delta}^2(q, \dot{q}, t)] \quad (6.2.14)$$

with

$$k_{\alpha, \mu}^1(q, \dot{q}, t) = \|A_3(q, t)^{-1}\| \left\{ \mu d^M \left(\frac{a_2^M(q, t)}{a_2^m(q, t)}\right)^{1/2} \left[\frac{\beta^m}{\alpha} \left(\frac{\sigma_2^m}{a_2^m(q, t)}\right)^{1/2}\right. \right.$$
$$+ (n + 1)\|B(q, t)\| + n \frac{\alpha}{\beta^m} \left(\frac{a_2^m(q, t)}{\sigma_2^m}\right)^{1/2} \|B(q, t)\|^2 \right]$$
$$\left. + n \frac{\alpha}{\beta^m} \left(\frac{a_2^m(q, t)}{\sigma_2^m}\right)^{1/2} \|\dot{B}(q, \dot{q}, t)\| + \frac{1}{2} \frac{\|\dot{A}_2(q, \dot{q}, t)\|}{a_2^m(q, t)} \right\} \quad (6.2.15)$$

and

$$k_{\alpha, \mu, \phi, \delta}^2(q, \dot{q}, t) = \left(C_{A_1} C_{A_2}(\sigma_2^M)^{1/2} \frac{v}{\phi} + 1\right) \|A_3^{-1}(q, t)\|$$
$$\times \left\{\mu d^M \left(\frac{a_2^M(q, t)}{a_2^m(q, t)}\right)^{1/2} \left(\|B(q, t)\| + \frac{\beta^m}{\alpha}\right) + \delta\right\}$$
$$+ \frac{(a_2^M(q, t))^{1/2}}{\phi} \|A_3^{-1}(q, t)\| \|s(q, \dot{q}, t)\| \quad (6.2.16)$$

where $\alpha > 1$; $\phi > 0$; $\delta > 0$ are free parameters, and

$$A_3(q, t) = A_2^{1/2}(q, t)(\tfrac{1}{2}(A_1(q, t) + A_1^T(q, t))A_2^{1/2}(q, t), \quad (6.2.17)$$

$A_2^{1/2}(q, t)$ being the symmetric positive square root matrix of $A_2(q, t)$; we denote also

$$\dot{B}(q, \dot{q}, t) = \frac{d}{dt} B(q, t). \quad (6.2.18)$$

$$\dot{A}_2(q, \dot{q}, t) = \frac{d}{dt} A_2(q, t). \quad (6.2.19)$$

We finally define the positive scalar μ_ρ:

$$\mu_\rho = \frac{1}{\rho} \left(\frac{d^M}{d^m}\right)^{1/2} \frac{1}{\left(1 - \frac{1}{\alpha}\right)\beta^m d^m} \left[\left(\frac{\sigma_2^M}{\sigma_2^m}\right)^{1/2} C_{A_1} C_{A_2} v + \frac{\phi}{(\sigma_2^m)^{1/2}}\right]. \quad (6.2.20)$$

6.2.2 Stability Theorem

We are now ready to state the following theorem.

Theorem 6.1 *If*:

$$\mu > \mu_\rho; \quad (6.2.21)$$

$$k(q, \dot{q}, t) > k_{\mu, \alpha, \phi, \delta}(q, \dot{q}, t); \quad (6.2.22)$$

the initial conditions $q(0)$ and $\dot{q}(0)$ are such that $(q(0), 0) \in C_{\rho, T}$, and

$$\|e(0)\| + \frac{\|\dot{e}(0)\|}{\mu b_{\max} d^M} < \left(\frac{d^m}{d^M}\right)^{3/2} \left(1 - \frac{1}{\alpha}\right) \frac{\beta^m}{b_{\max}} \left(\frac{\sigma_2^M}{\sigma_2^m}\right)^{1/2} \rho \quad (\Rightarrow \|e(0)\| < \rho); \quad (6.2.23)$$

then:

1. *(existence).* The solution $q(t)$ of the differential equation (6.1.1) exists and is unique on $[0, T]$;
2. *(stability).* for any $t \in [0, T]$: $\|e(t)\| < \rho$, $\|\dot{e}(t)\|$ and $\|\dot{q}(t)\|$ are uniformly bounded on $[0, T]$ with respect to initial conditions;
3. *(asymptotic behaviour).* On $[0, T]$, $e(t)$ satisfies the equation:

$$\dot{e}(t) = -L(t)e(t) + \eta(t) \quad (6.2.24)$$

where

(a) $L(t)$ is the state matrix of a uniformly exponentially stable system; more precisely, by writing $\|X\|_D^2 = X^T DX$, then

$$\dot{X}(t) = -L(t)X(t) \Rightarrow \|X(t)\|_D < \|X(0)\|_D \exp\left(-\mu d^m \left(1 - \frac{1}{\alpha}\right)\beta^m t\right) \quad (6.2.25)$$

6.2 A GENERAL STABILITY THEOREM

Moreover

$$L(t) \to \mu B(q(t), t) D \text{ uniformly on } [0, T] \text{ when } \alpha \to +\infty; \quad (6.2.26)$$

(b) $\|\eta(t)\| < \sup\left[\left(\frac{\sigma_2^M}{\sigma_2^m}\right)^{1/2} \|\dot{e}(0) + \mu B(0) D e(0)\| \exp(-\delta t); \right.$

$$\left. \left(\frac{\sigma_2^M}{\sigma_2^m}\right)^{1/2} C_{A_1} C_{A_2} v + \frac{\phi}{(\sigma_2^m)^{1/2}} \right]. \quad (6.2.27)$$

The proof is given in Appendix A6.2.

This theorem may be extended to the time interval $[0, +\infty[$ as follows.

Corollary 6.1.1 *If*:

the function $e(q, t)$ is ρ-admissible on $C_\rho = \sup_{0 \le T < +\infty} C_{\rho, T}$ during $[0, +\infty[$;

$$\left\| \left(\frac{\partial e}{\partial q}(q, t)\right)^{-1} \right\| \text{ and } \left\|\frac{\partial e}{\partial t}(q, t)\right\| \text{ are bounded on } C_\rho;$$

bounds $C_{A_1}, C_{A_2}, \sigma_2^m, \sigma_2^M, \beta^m, b_{\max}$ and v exist for $(q, t) \in C_\rho$;

then:

1. *The results of the theorem are valid on $[0, +\infty[$;*

2. *If the function $\|s(q, \dot{q}, t)\|$ is uniformly bounded with respect to q and t when $(q, t) \in C_\rho$, then $k_{\mu, \alpha, \phi, \delta}(q, \dot{q}, t)$ is uniformly bounded with respect to q and t when $(q, t) \in C_\rho$;*

3. *If, moreover, $k(q, \dot{q}, t)$ is uniformly bounded with respect to q and t when $(q, t) \in C_\rho$, then $k(q(t), \dot{q}(t), t)$ and $\|\ddot{e}(t)\|$ are uniformly bounded on $[0, +\infty[$ with respect to the initial conditions.*

6.2.3 Discussion

This theorem allows us to analyse the effects of every component appearing in the differential equation (6.1.1) on the stability and the asymptotic behaviour of the error $e(t)$. Details of such an analysis will not be relevant in the most general case, and we shall give them later only when studying robotic applications. For the present, a global interpretation of this result is that Theorem 6.1 generalizes the stability result of Lemma 6.3. In fact, it shows that if the initial errors $\|e(0)\|$ and $\|\dot{e}(0)\|$ are small enough, (their size being commensurable with the admissibility radius ρ of $e(q, t)$), and if $\mu > \mu_\rho$, where μ is proportional to $1/\rho$, then there exists a *minimal non-linear gain function* such that, if $k(q, \dot{q}, t)$ is greater than this minimal gain, then $e(t)$ stays inside the sphere $B(0, \rho)$.

The proof of the theorem is adapted from singular perturbation techniques. This appears through (6.2.24), which allows us to specify the asymptotic behaviour of the solution $e(t)$. By (6.2.24) and (6.2.25), if the gain $k(q, \dot{q}, t)$ is 'large enough', then $e(t)$ approximately follows the equation of evolution

$$\dot{e}(t) \approx -\mu B(q(t), t) De(t) + \eta(t). \tag{6.2.28}$$

In this equation, $\eta(t)$ may be understood as a disturbance preventing $e(t)$ from converging to zero.

By choosing a small value for the free parameter ϕ, (6.2.27) shows that $v(t)$ controls the asymptotic behaviour of $\eta(t)$, with the asymptotic upper bound

$$\|\eta(t)\| < \left(\frac{\sigma_2^M}{\sigma_2^m}\right)^{1/2} C_{A_1} C_{A_2} v. \tag{6.2.29}$$

Furthermore, a very important consequence of (6.2.28) is that (when $k(q, \dot{q}, t)$ is large enough), the asymptotic *dynamic behaviour* of $e(t)$ mainly depends on $-\mu B(q(t), t)D$, and is almost independent of $A_1(q, t)$ and $A_2(q, t)$. (6.2.28) can also be used to show that $e(t)$ may be made asymptotically as small as desired, whatever the bounds of the disturbances $v(q(t), t)$ and $s(q(t), \dot{q}(t), t)$, by increasing the parameter μ. Intuitively, and in a manner similar to the linear case, a large gain k cancels the effects of the disturbance s, while a large parameter μ cancels the effects of v. ('Large' here is used in the sense that the *relative* values $k/k_{\mu, \alpha, \phi, \delta}$ and μ/μ_ρ are large.)

6.2.4 Two complementary results

For robotic applications, it is useful to add two complementary lemmas to the previous theorem: the first shows that it is not necessary for the gain $k(q, \dot{q}, t)$ to be *always* greater than $k_{\mu, \alpha, \phi, \delta}(q, \dot{q}, t)$ in order to ensure that $e(t)$ lies inside the sphere $B(0, \rho)$; the second establishes that a *constant* gain k, when large enough, may be sufficient to guarantee stability of (6.1.1) in a domain the size of which increases with the size of k.

Lemma 6.4 *If condition* (6.2.22) *is replaced by the weaker one*:

$$k(q, \dot{q}, t) > k_{\mu, \alpha, \phi, \delta}(q, \dot{q}, t)$$

whenever

$$\|e(q, t)\| + \frac{\|\dot{e}(q, \dot{q}, t)\|}{\mu b_{\max} d^M} \geq \left(\frac{d^m}{d^M}\right)^{3/2} \left(1 - \frac{1}{\alpha'}\right) \frac{\beta^m}{b_{\max}} \left(\frac{\sigma_2^M}{\sigma_2^m}\right)^{1/2} \rho,$$

$$(\alpha > \alpha' > 1) \tag{6.2.30}$$

$k(q, \dot{q}, t) > 0$ *elsewhere*;
then (1)–(3) *of Theorem* 6.1 *remain true*.

The proof is given in Appendix A6.1.

In other words as long as $\|e(t)\|$ and $\|\dot{e}(t)\|$ are small, the gain k may have any positive value. The uniform boundedness of $\|\dot{e}(t)\|$ and that of $\|e(t)\|$ with respect to ρ are ensured with the sole condition that k again becomes greater than $k_{\mu, \alpha, \phi, \delta}$ when expression (6.2.30) is true. It will be seen further that, in robotics applications, this result increases the freedom of choice of the gain $k(q, \dot{q}, t)$. It shows also that large control gains are not necessary in all circumstances.

Lemma 6.5 *There exists a positive scalar $k_{\mu, \rho, T}$ such that, if condition (6.2.22) is replaced by*

$$k(q, \dot{q}, t) > k_{\mu, \rho, T}, \qquad (6.2.31)$$

then the results of Theorem 6.1 remain valid.

In particular, any *constant* gain greater than $k_{\mu, \rho, T}$ ensures the boundedness of $\|e(t)\|$ by ρ.

This result is simply a consequence of the fact that, for all the solutions $\{q(t)\}$, $t \in [0, T]$ which satisfy the initial constraint (6.2.23), and which are obtained with any gain satisfying (6.2.22), the function $t \to k_{\mu, \alpha, \phi, \delta}(q(t), \dot{q}(t), t)$ is uniformly bounded by a positive real scalar $k_{\mu, \rho, T}$. The dependence with respect to ρ and T results from the fact that all points $(q(t), t)$ belongs to $C_{\rho, T}$.

It should finally be emphasized that the condition (6.2.22) about initial values of $\|e(0)\|$ and $\|\dot{e}(0)\|$ shows that the stability domain increases with ρ. As $C_{\rho, T}$ increases with ρ also, the bound $k_{\mu, \rho, T}$ increases with ρ. A consequence is that the greater the constant gain k, the larger the stability domain ensured by the analysis. The previous study performed for the linear invariant case shows also that there may exist a minimal gain value k_0 which ensures an infinitely large stability domain, when the task is ∞-admissible.

6.3 PRINCIPLES OF APPLICATION TO ROBOT CONTROL

6.3.1 Evolution equation of the tracking error $e(q, t)$

A first step in applying the previous analysis to the case of robots is to derive a differential equation of general form (6.1.1). This may be done by applying the general control scheme (5.2.63) to the robot model in task space (5.1.22). We recall that (5.2.63) has the form

$$\Gamma = -k\hat{M} J_q^{-1} G(\mu D \hat{e} + J_q \hat{\dot{q}} + J_t) + \hat{I} \qquad (6.3.1)$$

where the functions depend on the time and on the measured (or estimated) values, $\hat{q}$ and $\hat{\dot{q}}$, of q and $\dot{q}$. Let us express the measurement errors as C^1 time-dependent additive functions:

$$\hat{q}(t) = q(t) + b_q(t), \tag{6.3.2}$$

$$\hat{\dot{q}}(t) = \dot{q}(t) + b_{\dot{q}}(t), \tag{6.3.3}$$

$$\hat{e}(t) = e(t) + b_e(t). \tag{6.3.4}$$

We may then set

$$k'(q, \dot{q}, t) = k(\hat{q}, \hat{\dot{q}}, t), \tag{6.3.5}$$

$$\hat{M}'(q, t) = \hat{M}(\hat{q}, t), \tag{6.3.6}$$

$$J_q'(q, t) = J_q(\hat{q}, t), \tag{6.3.7}$$

$$J_t'(q, t) = J_t(\hat{q}, t) + \mu D b_e(t) + J_q(\hat{q}, t) b_{\dot{q}}(t), \tag{6.3.8}$$

$$\hat{l}'(q, \dot{q}, t) = \hat{l}(\hat{q}, \hat{\dot{q}}, t), \tag{6.3.9}$$

and thus write the control expression in the form:

$$\Gamma = -k'(q, \dot{q}, t)\hat{M}'(q, t)J_q'^{-1}(q, t)G[\mu De(q, t) + J_q'(q, t)\dot{q} + J_t'(q, t)] + l'(q, \dot{q}, t), \tag{6.3.10}$$

where all functions now depend on the time and the actual state.

Using this control expression in the robot model in task space, the form of which is (see relation 5.1.22)

$$\ddot{e} + J_T M^{-1} N - f = J_T M^{-1} \Gamma, \tag{6.3.11}$$

leads to the following expression for the tracking error:

$$\ddot{e} + k' J_T M^{-1} \hat{M}' J_q'^{-1} G[\mu De + J_q' \dot{q} + J_t'] + J_T M^{-1}(N - \hat{l}) - f = 0. \tag{6.3.12}$$

Recalling that

$$\dot{e}(q, \dot{q}, t) = J_T(q, t)\dot{q} + \frac{\partial e}{\partial t}(q, t), \tag{6.3.13}$$

we finally obtain

$$\ddot{e} + k' J_T M^{-1} \hat{M}' J_q'^{-1} G J_q' J_T^{-1} \left[\mu J_T J_q'^{-1} De + \dot{e} + J_T J_q'^{-1} J_t' - \frac{\partial e}{\partial t} \right] + J_T M^{-1}(N - \hat{l}) - f = 0. \tag{6.3.14}$$

6.3 PRINCIPLES OF APPLICATION TO ROBOT CONTROL

It is easy to see that this equation has the form (6.1.1), with the following correspondences:

$$A_1 A_2 = J_T M^{-1} \hat{M}' J_q'^{-1} G J_q' J_T^{-1}, \qquad (6.3.15)$$

$$B = J_T J_q'^{-1}, \qquad (6.3.16)$$

$$v = J_T J_q'^{-1} J_t' - \frac{\partial e}{\partial t}, \qquad (6.3.17)$$

$$s = J_T M^{-1}(N - \hat{l}) - f. \qquad (6.3.18)$$

6.3.2 Application of the Stability Theorem

Knowing expressions (6.3.15)–(6.3.18) which specify the particular situation of robot control, we have to verify the validity of assumptions H1–H3, which are needed to apply Theorem 6.1.

6.3.2.1 Validity of Assumptions (H1)–(H3)

Assumption H1 is obviously satisfied by choosing μ positive and D symmetric positive.

Assumptions H2(d) and H3(a) impose only differentiability properties for J_q', J_t', and $\hat{l}$; hence they do not represent a strong constraint. Assumption H3(b) means that the gain function $k'(q, \dot{q}, t)$ has to be positive and C^1 on $C_{\rho,T}$. This is easily satisfied. Finally notice that, owing to the explicit expression of $J_q'^{-1}$ in (6.3.15)–(6.3.17), we need the following supplementary condition:

H4: $J_q'(q, t)$ is invertible on $C_{\rho, T}$.

The most sensitive and constraining assumptions in H2 are items (b) and (c) which impose positivity conditions on the A_1, A_2 and B matrix functions.

1. *Positivity of $A_1(q, t)$ and $A_2(q, t)$*. The relation (6.3.15) gives the form of the product $A_1 A_2$. Some freedom thus remains in the choice of the matrices A_1 and A_2 which have to be positive. A_2 is also required to be symmetric.

We shall give some examples of possible choices, based on different constraints imposed on $\hat{M}(q, t)$, $J_q'(q, t)$ and G. To simplify we shall neglect the measurement error b_q on joint positions so that

$$\hat{M}'(q, t) = \hat{M}(q, t); \quad J_q'(q, t) = J_q(q, t). \qquad (6.3.19)$$

Then

$$A_1 A_2 = J_T M^{-1} \hat{M} J_q^{-1} G J_q J_T^{-1}. \qquad (6.3.20)$$

Example 6.1 Consider the following user's choices of $\hat{M}$ and J_q:

$$\hat{M}(q, t) = M(q), \qquad (6.3.21)$$

$$J_q(q, t) = J_T(q, t). \qquad (6.3.22)$$

This means that the inertia (or kinetic energy) matrix and the task Jacobian matrix are assumed to be perfectly known and explicitly used in the control scheme.

In this ideal situation, (6.3.15) becomes

$$A_1(q, t) A_2(q, t) = G. \tag{6.3.23}$$

$$A_1 = G; \qquad A_2 = I. \tag{6.3.24}$$

It can be seen that any positive matrix G can be chosen in the control law.

Example 6.2 Assuming that $\hat{M}$ and J_q are chosen such that

$$\hat{M}(q, t) = M(q) \tag{6.3.25}$$

$$J_q(q, t) = I. \tag{6.3.26}$$

Then, choosing again $G > 0$ allows us to set

$$A_1 = J_T G J_T^T \tag{6.3.27}$$

$$A_2 = J_T^{-T} J_T^{-1} = (J_T J_T^T)^{-1}. \tag{6.3.28}$$

Recall that J_T^{-1} exists on $C_{\rho, T}$, from the admissibility assumptions.

Example 6.3 If $\hat{M}$ and J_q are chosen such that

$$M^{-1}(q)\hat{M}(q, t) > 0 \quad \text{on } C_{\rho, T}, \tag{6.3.29}$$

$$J_q(q, t) = I. \tag{6.3.30}$$

Then by choosing G symmetric and positive, we may set

$$A_1 = J_T M^{-1} \hat{M} J_T^T, \tag{6.3.31}$$

$$A_2 = J_T^{-T} G J_T^{-1}. \tag{6.3.32}$$

Here, the condition on $\hat{M}$ is weaker than the ideal equality with M; for example $\hat{M} = I$ is a possible choice. However, this relaxing of the constraint on $\hat{M}$ has to be balanced by the additional constraint of the symmetry of G.

Example 6.4 If $\hat{M}$ and G are chosen such that

$$M^{-1}(q)\hat{M}(q, t) > 0 \tag{6.3.33}$$

$$G = I, \tag{6.3.34}$$

we may set

$$A_1 = J_T M^{-1} \hat{M} J_T^T \tag{6.3.35}$$

$$A_2 = (J_T J_T^T)^{-1}. \tag{6.3.36}$$

6.3 PRINCIPLES OF APPLICATION TO ROBOT CONTROL

Remark. A condition equivalent to (6.3.33) is: $\hat{M}M^T > 0$, because we may write
$$\hat{M}M^T = M(M^{-1}\hat{M})M^T.$$

Example 6.5 Another possible and important case is
$$\hat{M}(q, t) \text{ symmetric positive,} \qquad (6.3.37)$$
$$G = I. \qquad (6.3.38)$$

Then we may set
$$A_1 = J_T M^{-1} J_T^T \qquad (6.3.39)$$
$$A_2 = J_T^{-T} \hat{M} J_T^{-1}. \qquad (6.3.40)$$

This example is quite interesting: it shows that the stability analysis of section 6.2 may be applied as soon as the approximation $\hat{M}$ of the kinetic energy matrix is chosen symmetric and positive on $C_{\rho, T}$.

In particular, this shows that to apply Theorem 6.1, Corollary 6.1.1, and Lemmas 6.4 and 6.5, we do not need the modelling error $\| M(q) - \hat{M}(q, t) \|$ to be small.

Condition (6.3.37), which is hardly restrictive, is satisfied by all common control schemes presented in Chapter 5, such as computed torque and velocity control methods. In the case of adaptive control schemes, it coincides with the condition on $\hat{M}(\hat{\theta}, t)$ given in section 5.3.3.4, which is one of those ensuring the convergence of the error to zero.

2. *Positivity of $B(q, t)$*. By (6.3.16), the positivity condition on $B(q, t)$ is equivalent to
$$J_T(q, t) J_q'^{-1}(q, t) > 0, \qquad \text{for any } (q, t) \in C_{\rho, T}. \qquad (6.3.41)$$

If, as previously, we assume that the measurement noise on q is small, this condition becomes
$$J_T(q, t) J_q^{-1}(q, t) > 0, \qquad \text{for any } (q, t) \in C_{\rho, T}; \qquad (6.3.42)$$

or equivalently
$$J_q^T(q, t) J_T(q, t) > 0, \qquad \text{for any } (q, t) \in C_{\rho, T}. \qquad (6.3.42')$$

since $J_q^T J_T = J_q^T (J_T J_q^{-1}) J_q$.

In the case of J_q chosen as in Examples 6.2 and 6.3, we see that this condition implies that the choice $J_q = I$ can be made only when $J_T(q, t)$ is positive on $C_{\rho, T}$.

Condition (6.3.42) is of extreme importance: in a sense it characterizes the *minimal information* which the user must have about the system and the task in order to design an efficient control scheme. Indeed, when this condition is satisfied, it is sufficient (from Theorem 6.1) to choose a gain $k(q, \dot{q}, t)$ (constant or non-linear) and a scalar μ, both large enough, to ensure the stability of the system and the smallness of $e(t)$. This condition, the satisfaction of which may not be obvious in some cases, is essential to ensure robustness of the control. Roughly J_q may be used in the control as if it were the true task Jacobian J_T, on condition that $J_q^T J_T > 0$.

Note that the information required to satisfy condition (6.3.42) concerns *only* the task Jacobian $J_T(q, t)$, and does not involve the inertia characteristics of the robot, which appear elsewhere (for example in conditions on A_1 and A_2). As the task does not depend *a priori* on the inertia parameters of the robot, this means that, theoretically, it is not necessary to know anything about them to design an efficient control. If, moreover, the task does not depend on the robot *kinematics* (for example, when tracking a trajectory in joint space, i.e. $J_T = I$), then it is possible to control the robot without knowing anything about it! This is the reason why in some cases very simple control schemes (for example constant decoupled P.D. loops) may lead to acceptable results, especially in the absence of noise and when torque limitations are not reached. Note that this is the case for many industrial robot controllers.

We shall return later to the practical incidence of condition (6.3.42) in applications.

6.3.2.2 Summary

We have shown that the stability results provided by Theorem 6.1 might be applied to any kind of control of the general form (6.3.10), on condition that:

(1) $\hat{M}(q, t)$ be chosen symmetric and positive;

(2) $J_q(q, t)$ be such that $\quad J_T(q, t) J_q^{-1}(q, t) > 0$

The matrix G should be chosen positive, and equal to identity when $\hat{M}(q, t)$ is *not* a good approximation to the true $M(q)$. Provided that these conditions are fulfilled, the stability properties and the accuracy of the control schemes mainly depend on the choices made for the gain function $k(q, \dot{q}, t)$ and for the scalar μ. We already know that, in the ideal decoupling method, and when there is no measurement noise, it is sufficient for k and μ to be *positive* in order to ensure that $e(t)$ converges to zero. For other control schemes, Theorem 6.1 proves the existence of *lower bounds* on $k(q, \dot{q}, t)$ and μ, above which the boundedness of $\|e(q, t)\|$ is ensured. The theorem also shows how an adequate increase in $k(q, \dot{q}, t)$ and μ allows us to make $\|e(t)\|$ asymptotically as small as desired whatever the modelling errors. Equations (6.2.14) and (6.2.20) give an indication of how the other terms in the expression of Γ influence the 'sizes' of μ and $k(q, \dot{q}, t)$. Further inspection would show for example that, the further we move from the ideal decoupling control, the greater μ and $k(q, \dot{q}, t)$ may have to be in order to maintain $\|e(t)\|$ small. In this sense, it may be understood that the use of large gains is a way of improving robustness with regard to modelling errors.

It is again to be emphasized that 'large' only refers to the ratio of the actual control gains and smaller gains which ensure the boundedness of $\|e(t)\|$ by ρ. Moreover the lower gains used in the very general theorem 6.1 were specifically tailored for the singular perturbation technique used in the proof of the theorem. We do not claim that these gains are absolute minima. In fact it is possible to exhibit smaller gains by restricting the set of control schemes to a smaller subset. The case of perfect decoupling controls, for which the lower bounds for the gains k and μ are easily shown to be equal to zero, is a good illustration of this. It is also possible in some cases to take advantage of additional structural properties of mechanical systems (in relation to energy dissipation, for example) and such cases have been reported in the literature (see bibliographic note). In practice, the choice of the gains will depend on several elements that are specific to each application; for example, task characteristics, motion velocity, transmission ratios, expected accuracy, amounts of dry friction and of gravity torques which are non-mechanically compensated.

Corollary 6.3.1 extends the results of Theorem 6.1 to the time interval $[0, +\infty[$. The main fact to be noted is the importance of $\|s(q, \dot{q}, t)\|$ being bounded in q and t in order to obtain boundedness of $e(q, t)$ with a bounded gain $k(q, \dot{q}, t)$. Equation (6.3.18) shows that, in order to be in the ideal situation where $s(q, \dot{q}, t)$ vanishes, it is necessary to have

$$\hat{l} = N - MJ_T^{-1}f. \qquad (6.3.43)$$

In practice, except when the task consists of tracking a trajectory in joint space, it will be difficult to satisfy this equality because of the complexity of $f(q, \dot{q}, t)$ (cf. (5.1.20)). However, the uniform boundedness of $\|J_T\|$, $\|M^{-1}\|$, $\|f\|$ and $\|\hat{l} - N\|$ is sufficient to ensure the uniform boundedness of s with respect to q and t. This allows us in particular to introduce into $\hat{l}(q, \dot{q}, t)$ an integral feedback term (for which the control then becomes a proportional-integral-derivative-like control) on condition of its boundedness. Simple saturation mechanisms can be used for this purpose.

6.3.3 The Gain $k(q, \dot{q}, t)$

We may distinguish two situations, depending on whether this gain is constant or varying non-linearly.

6.3.3.1 Constant gain

The stability analysis demonstrates the importance of the choice of $k(q, \dot{q}, t)$ in ensuring control performances. Roughly it indicates that, the larger this gain is, the less sensitive is the resulting behaviour of the system to the choice of other terms. This fact follows from (6.2.24) and (6.2.26). More precisely, if:

(1) k is large ($\Rightarrow \alpha$ is large);

(2) $J_q \approx J_T (\Rightarrow B \approx I)$;

(3) $J_t \approx \dfrac{\partial e}{\partial t}$;

(4) $b_q, b_{\dot{q}}, b_e$ are small ($\Rightarrow \eta(t) \approx 0$ asymptotically),

then we obtain an asymptotic first-order linear invariant equation for the error

$$\dot{e} + \mu D e = 0. \tag{6.3.44}$$

Thus the error no longer depends on the characteristics of the robot, or the choices made for $\hat{M}$ and $\hat{l}$. The closed-loop controlled system behaves asymptotically like a first order linear system, the dynamics of which can be tuned through the choice of μ and D.

If we now combine this result with that of Lemma 6.5 which shows that stability properties can be preserved by using a constant gain, it is tempting to conclude that *it is sufficient to choose a large constant gain k to ensure good behaviour of the control and accurate task completion in all cases.*

From a theoretical point of view this assertion is not false: large gains do indeed confer some robustness properties on the control, as established in the analysis of section 6.2. However, supplementary elements have to be taken into account, since large constant gains also present drawbacks in certain cases, such as sensitivity to noise, or high energy in the control torques. These aspects will be studied in more detail later. Nevertheless, the choice of a large constant gain remains the simplest possible choice, and, for this reason is encountered frequently in practice.

Let us provide some indication about the tuning of parameters μ and k, when they are chosen to be constant. In the case of perfect linearization, and as seen in section 5.3.1, the evolution equation of the error is given by

$$\ddot{e} + kG(\mu De + \dot{e}) = 0. \tag{6.3.45}$$

and a natural choice, which preserves decoupling, is $G = D = I$. Then (6.3.45) reduces to a set of n one-dimensional equations with the same dynamics. In each of these PD loops, k is the velocity feedback gain and μ is the ratio between position and velocity gains. Critical damping (i.e. double eigenvalues in each equation) is obtained when $k = 4\mu$, which is the lowest value that avoids oscillations. With larger values of k, the system is overdamped.

The scalar μ fixes the convergence rate of $e(t)$ towards zero. Recall that $e(t)$ is homogeneous to a position error. Intuitively, the dynamics of $e(t)$ have to be faster than the inner dynamics of the ideal trajectory $q_r(t)$ (which do not

6.3 PRINCIPLES OF APPLICATION TO ROBOT CONTROL

depend on the chosen control scheme). For example, a large motion of the end-effector of a robot may take about 1s. Then a reasonable value for μ may be 10, in order to set the response time of $e(t)$ to 0.1 s.

In the general case, when perfect (or almost perfect) linearization is not implemented, it becomes very difficult to give rules for the lower limits of k. Usually this gain will be tuned experimentally or heuristically by considering the most unfavourable case (i.e. the configuration requiring the largest value). We may simply note that, if Coriolis, centrifugal, friction, and gravity effects are negligible with regard to inertia effects, we have, roughly,

$$\ddot{q} \approx M^{-1}(q)\Gamma. \qquad (6.3.46)$$

This shows that the global gain of the system is 'divided' by the kinetic energy matrix; it is thus minimal when $M(q)$ is maximal. By choosing for example,

$$\hat{M}(q) = D = I \qquad (6.3.47)$$

it appears that sufficient damping requires

$$k \geq 4\mu\lambda_{\max}(M) \qquad (6.3.48)$$

where $\lambda_{\max}(M)$ is an upper bound of the eigenvalues of M.

6.3.3.2 Non-linear gain

Theorem 6.1 allows us to use a non-linear gain function $k(q, \dot{q}, t)$. More precisely it says that in order to ensure $\|e(t)\| < \rho$, where ρ is a positive number smaller than the admissibility radius $\rho_{\max}$, it is sufficient to use a non-linear gain greater than the minimum $k_{\alpha, \mu, \phi, \delta}(q, \dot{q}, t)$ given by (6.2.14).

Given $\hat{M}(q, t)$, $\hat{l}(q, \dot{q}, t)$, $J_q(q, t)$, $J_t(q, t)$, and knowing approximately the robot characteristics ($M(q)$ and $l(q, \dot{q}, t)$) and the task function $e(q, t)$, it is generally fairly easy to find a function which is always greater than $k_{\alpha, \mu, \phi, \delta}(q, \dot{q}, t)$. However (6.2.14) has more of a *qualitative* than a quantitative value, because the conditions of the theorem are only sufficient. For this reason, use of (6.2.14) may lead to the derivation of a gain function larger than necessary, i.e. very 'conservative'.

This is why it appears to be reasonable to try to determine *a priori* a non-linear gain function, starting from practical data, and then to verify that it is coherent with the results of the analysis given in section 6.2.

We shall attempt to provide some simple basic ideas for finding such a gain function:

1. As long as $\|e(t)\|$ is smaller than the accepted error, we are free to decrease the gain. On the other hand, when $\|e(t)\|$ tends to increase beyond the accepted limit it means that modelling errors or disturbances begin to have a discernible negative effect. A larger gain is then needed to reduce

this effect. The size of $\|e(t)\|$ and gain variations may thus be related in a rather simple way. Note that increasing k with $\|e(t)\|$ is consistent with reducing the effects of static errors due to gravity and dry friction.

2. When Coriolis and centrifugal torques are not exactly modelled in the compensating term $\hat{N}(q, \dot{q}, t)$ appearing in $\hat{l}(q, \dot{q}, t)$, their influence increases with the robot velocity $\dot{q}(t)$ (cf. (2.1.8)). Increasing the gain simultaneously with $\|\dot{q}(t)\|$ tends to reduce the related disturbances.

Points (1) and (2) suggest, among other possibilities, the choice of a composite gain function depending on both $\|e(t)\|$ and $\|\dot{q}(t)\|$. A possible candidate is, for example:

$$k(q(t), \dot{q}(t), t) = k_0 (1 + \|e(t)\|_{D_1})^{n_1} (1 + \|\dot{q}(t)\|_{D_2})^{n_2} \qquad (6.3.49)$$

where n_1 and n_2 are positive, and $\|\cdot\|_{D_i}$ is the norm associated with a given positive diagonal matrix D_i.

From Lemma 6.5, we already know that a minimal value of k exists above which the stability of the system is theoretically ensured. In addition, Lemma 6.4 shows that it is *not necessary* to have $k(q, \dot{q}, t)$ *systematically* large, and thus to set k_0 itself large. Take for example the simple case of joint trajectory tracking. Then:

$$e(t) = q - q_r(t), \qquad (6.3.50)$$

$$J_T = J_q = I, \qquad (6.3.51)$$

$$f = 0. \qquad (6.3.52)$$

Let us make the simple choices

$$\hat{M}(q, t) = I, \qquad (6.3.53)$$

$$\hat{l}(q, \dot{q}, t) = 0, \qquad (6.3.54)$$

$$G = I. \qquad (6.3.55)$$

Then we get

$$A_1 A_2 = M^{-1}, \qquad (6.3.56)$$

$$B = I, \qquad (6.3.57)$$

$$v = 0, \qquad (6.3.58)$$

$$s = M^{-1} N. \qquad (6.3.59)$$

It is easy to see that a function greater than the gain (6.2.14) can be obtained in the form

$$\alpha \|M\| (\beta + \|M^{-1} N\|), \quad \alpha, \beta > 0 \qquad (6.3.60)$$

which is itself upper-bounded by

$$k(.) = \|M\| (a_1 + a_2 \|N\|), \quad a_1, a_2 > 0. \qquad (6.3.61)$$

6.3 PRINCIPLES OF APPLICATION TO ROBOT CONTROL

As M is the kinetic energy matrix, each of its components increases with $\|q\|^2$ at the most, and the norm of M may be bounded by a function of the form

$$a_3 + a_4\|q\| + a_5\|q\|^2. \tag{6.3.62}$$

With $q_r(t)$ chosen to be bounded and because of Lemma 6.4, this may be replaced by

$$a_6 + a_7\|e\| + a_8\|e\|^2. \tag{6.3.63}$$

Note that, if all joints are rotational, $\|M\|$ may be simply bounded by a positive constant.

In the same manner, it is possible to derive upper bounding functions for:

1. *Coriolis and centrifugal terms.* Returning to the definition of this term (cf. (2.1.8)), and recalling that matrices $V_i(q)$ cannot increase faster than $\|q\|$, we obtain the following upper bound:

$$\|\dot{q}\|(a_9 + a_{10}\|q\|). \tag{6.3.64}$$

which can itself be upper-bounded by a function of the form:

$$\|\dot{q}\|(a_{11} + a_{12}\|e(t)\|). \tag{6.3.65}$$

2. *Gravity and friction terms.* Gravity forces cannot increase faster in norm than $\|q\|$, and, typically, friction forces do not increase faster than $\|\dot{q}\|$. Thus, an upper bound of those terms is

$$a_{13} + a_{14}\|q\| + a_{15}\|\dot{q}\| \tag{6.3.66}$$

and we then may choose as an upper-bounding function

$$a_{16} + a_{17}\|e\| + a_{18}\|\dot{q}\|. \tag{6.3.67}$$

Using (6.3.63), (6.3.65), (6.3.66) and (6.3.67) in (6.3.61) shows that, in this case, the gain function given by (6.3.49) satisfies the conditions of Lemma 6.4, with

$$k_0 > 0, \tag{6.3.68}$$

$$n_1 \geq 3, \tag{6.3.69}$$

$$n_2 \geq 1. \tag{6.3.70}$$

and matrices D_1 and D_2 chosen large enough. Note that the minimum value k_0 of the non-linear gain k may be chosen as small as desired.

Remark. If the robot has only rotational joints, condition (6.3.69) reduces to

$$n_1 \geq 1. \tag{6.3.71}$$

6.3.3.3 Concluding remarks

Use of a non-linear gain $k(q, \dot{q}, t)$ provides an additional degree of freedom in the design of the control scheme. As an alternative to the use of a constant gain, it may allow the overall performance of the system to be improved. The reason is that a constant gain must be tuned with respect to the most unfavourable case, while a non-linear gain can be designed so as to vary with the robot configuration, and assume large values only when it is really needed. The drawbacks of large gains will be discussed later. Finally, note that non-linear gains, such that those given by (6.3.49), do not greatly penalize the computing cost. This fact has to be taken into account when designing a robot controller, in comparison with the amount of computation needed to form a refined model of the robot.

6.4 EFFECTS OF MEASUREMENT ERRORS AND DISTURBANCES

We shall analyse the effects of measurement errors from two complementary points of view: the first consists of using some of the results of the previous stability theorem, and the second is based on a stochastic analysis.

6.4.1 Analysis based on the stability theorem

6.4.1.1 Noise at the measured joint position

The most important effect of the error b_q is related to the fundamental condition (6.3.1):

$$J_T(q, t) J_q^{-1}(\hat{q}(t), t) > 0 \tag{6.4.1}$$

where

$$\hat{q}(t) = q(t) + b_q(t). \tag{6.4.2}$$

Suppose that it is possible to choose $J_q(q, t) = J_T(q, t)$. When $b_q = 0$, the condition of positivity (6.4.1) is trivially satisfied. When $b_q \neq 0$, let us perform a first-order expansion of $J_q(\hat{q}, t)$; then

$$J_T(q, t) J_q^{-1}(\hat{q}(t), t) = (I + \Delta_i(t))^{-1} \tag{6.4.3}$$

where

$$\Delta_i(t) = \left(\sum_{i=1}^{n} b_{q,i}(t) \frac{\partial^2 e}{\partial q^2} (q(t), t) \right) J_T^{-1}(q(t), t), \tag{6.4.4}$$

$b_{q,i}$ being the ith component of b_q.

This equation shows that, if b_q is large enough, or if $\| J_T^{-1} \|$ tends to become large (for example near a Jacobian singularity), then $\Delta_i(t)$ may have an order of magnitude sufficient to destroy the positivity of $(I + \Delta_i)$, which has dramatic consequences for the system stability.

6.4.1.2 Measurement errors in joint velocity ($b_{\dot{q}}$) and task vector (b_e)

These two errors act within the disturbance vector:

$$v(t) = J_T J_q^{-1}(\hat{q}(t), t) + \mu D b_e(t) + J_q(\hat{q}(t), t) b_{\dot{q}}(t)) - \frac{\partial e}{\partial t}(q(t), t). \quad (6.4.5)$$

Choose the favourable situation where $\{b_q = 0;\ J_q = J_T;\ J_t = \partial e/\partial t;\ D = I\}$. Then

$$v(t) = \mu_e(t) + J_T(q(t), t) b_{\dot{q}}(t). \quad (6.4.6)$$

First examine the effect of b_e when assuming $b_{\dot{q}} = 0$. Then

$$v = \sup_{t \in [0, T]} \|v(t)\| = \mu b_e^{\max} \quad (6.4.7)$$

where

$$b_e^{\max} = \sup_{t \in [0, T]} \|b_e(t)\|. \quad (6.4.8)$$

Equation (6.2.20), which determines the minimal value, μ_ρ, of μ for which the Stability Theorem may be applied, here takes the form

$$\mu_\rho = \frac{1}{\rho} \frac{1}{1 - (1/\alpha)} \left(\left(\frac{\sigma_2^M}{\sigma_2^m} \right)^{1/2} C_{A_1} C_{A_2} v + \frac{\phi}{(\sigma_2^m)^{1/2}} \right), \quad \alpha > 1;\quad k > 0. \quad (6.4.9)$$

Then, using (6.4.7),

$$\mu_\rho = \frac{1}{\rho} \frac{1}{1 - (1/\alpha)} \left(\left(\mu \frac{\sigma_2^M}{\sigma_2^m} \right)^{1/2} C_{A_1} C_{A_2} b_e^{\max} + \frac{\phi}{(\sigma_2^m)^{1/2}} \right). \quad (6.4.10)$$

By choosing $\phi \ll 1$, the basic condition $\mu > \mu_\rho$ leads to

$$b_e^{\max} < \rho \frac{1}{1 - (1/\alpha)} \left(\frac{\sigma_2^m}{\sigma_2^M} \right)^{1/2} \frac{1}{C_{A_1} C_{A_2}}. \quad (6.4.11)$$

This inequality shows in particular that the maximum size allowed for $b_e(t)$ is *proportional* to the task-admissibility radius ρ. If the task is ∞-admissible, we may deduce that it is always possible to ensure the stability of the system, for any bounded size of $b_e(t)$. On the other hand, with low admissibility (for example when approaching a singularity), it is recommended that $b_e(t)$ be made as small as possible.

We may now analyse similarly the effect of the joint velocity error by setting $b_e = 0$ in (6.4.6). Then

$$v \leq \sup \|J_T\| b_{\dot{q}}^{\max} \quad (6.4.12)$$

and reapplication of condition (6.2.21) of the theorem yields

$$\mu > \frac{1}{\rho} \left(\frac{\sigma_2^M}{\sigma_2^m} \right)^{1/2} C_{A_1} C_{A_2} \sup \| J_T \| b_{\dot{q}}^{\max}. \tag{6.4.13}$$

This expression shows that it is theoretically always possible to bound the effects of the velocity error $b_{\dot{q}}$ and to ensure stability by choosing μ *large enough*. The minimum size of μ is also *inversely proportional* to the task admissibility radius ρ. Finally, if both the size of b_e and the value of μ are compatible with realization of condition (6.2.21), and when $k(q, \dot{q}, t)$ is chosen large enough, then, from (6.2.24)–(6.2.27), the tracking error $e(t)$ approximately and asymptotically has the following evolution equation:

$$\dot{e} = -\mu J_T J_q^{-1} De + \eta, \tag{6.4.14}$$

with

$$\| \eta(t) \| < \left(\frac{\sigma_2^M}{\sigma_2^m} \right)^{1/2} C_{A_1} C_{A_2} v + \frac{\phi}{(\sigma_2^m)^{1/2}}. \tag{6.4.15}$$

Roughly, (6.4.14) shows (by setting $\dot{e} = 0$) that the size of $\| e(t) \|$ is proportional to that of $\| \eta(t) \|$. By choosing $\phi \ll 1$ (which may oblige us to have a large gain $k(q, \dot{q}, t)$ if $\| s(q, \dot{q}, t) \|$ is not small), (6.4.15) shows that the size of η itself is proportional to the size of the disturbance vector $v(t)$. Since the sizes of $e(t)$ and $v(t)$ are asymptotically proportional, sufficient conditions for ensuring good control accuracy (i.e. $\| e(t) \|$ small asymptotically) are that $k(q, \dot{q}, t)$ should be large enough and that $v(t)$ should be asymptotically small.

From (6.4.5), the latter condition holds if the errors $b_e, b_q, b_{\dot{q}}$ are small, and if

$$J_T J_q^{-1} J_t - \frac{\partial e}{\partial t} \approx 0 \tag{6.4.16}$$

which shows that the ideal choice for J_t would be

$$J_t = J_q J_T^{-1} \frac{\partial e}{\partial t}. \tag{6.4.17}$$

Remarks

1. In fully decoupled control, we find again the ideal choice: $J_t = \partial e/\partial t$. For example, when the task consists of tracking a joint reference trajectory $q_r(t)$ with $e = q - q_r$, condition (6.4.16) is satisfied with the simple choice $\{ J_q = I; J_t = \partial e/\partial t = -\dot{q}_r(t) \}$.
2. When $(\partial e/\partial t)(q_r(t), t) \to 0$ as $t \to +\infty$, it is possible to choose $J_t(t) = 0$. Then, when the errors are small, we have $v(t) \approx -(\partial e/\partial t)(q(t), t)$, and the value of μ has to be tuned as a function of ρ and of the maximum size of $\| \partial e/\partial t \|$.

6.4 MEASUREMENT ERRORS AND DISTURBANCES

6.4.2 Stochastic analysis

6.4.2.1 Introduction

A common feeling is that the use of large gains leads to large values of the control torques. This assertion is false as a generalization; indeed the analysis shows that, if the measurements of $e(t)$, $q(t)$, and $\dot{q}(t)$ are error-free and if $e(0) = \dot{e}(0) = 0$, then the larger the gain k, the more the control looks like the ideal open-loop control:

$$\Gamma_r(t) = M(q_r(t))\ddot{q}_r(t) + l(q_r(t), \dot{q}_r(t), t) \tag{6.4.18}$$

which is the minimum-energy control ensuring perfect task execution (i.e. $e(t) = 0$; $t \in [0, T]$). This property remains true even when the robot model $(\hat{M}, \hat{l})$ is poorly approximated, and when $J_q \neq J_T$ and $J_t \neq \partial e/\partial t$. Again the sufficiency of the conditions stated in Theorem 6.1 (mainly $\hat{M} > 0$ and $J_T J_q^{-1} > 0$) must be emphasized.

This property of large gains is the basic reason for the robust behaviour of the control schemes that we have analysed; however, when significant measurement errors exist large gains may indeed induce large control torques, as shown in the following.

6.4.2.2 Energy consequences of measurement noise

In this subsection we shall only attempt to indicate some basic practical principles, so we shall merely consider the one-dimensional case

$$\Gamma = m\ddot{q}. \tag{6.4.19}$$

Let us also assume that the task consists of tracking the reference trajectory $q_r(t)$. The following control is chosen:

$$\Gamma = -k(\mu\hat{e} + \dot{\hat{e}}) + m\ddot{q}_r \tag{6.4.20}$$

where

$$\hat{e} = \hat{q} - q_r = q - q_r + b_q = e + b_e \tag{6.4.21}$$

(so $b_e = b_q$), and

$$\dot{\hat{e}} = \dot{\hat{q}} - \dot{q}_r = \dot{e} + b_{\dot{q}}. \tag{6.4.22}$$

Equations (6.4.19)–(6.4.22) yield the following evolution equation:

$$\ddot{e} + \frac{k}{m}(\mu e + \dot{e}) + \frac{k}{m}(\mu b_e + b_{\dot{q}}) = 0. \tag{6.4.23}$$

Its representation in state form is

$$\dot{X} = AX + W \tag{6.4.24}$$

with

$$X = \begin{pmatrix} e \\ \dot{e} \end{pmatrix}; \quad A = \begin{pmatrix} 0 & 1 \\ -\mu\dfrac{k}{m} & -\dfrac{k}{m} \end{pmatrix}; \quad W = \begin{pmatrix} 0 \\ -\mu\dfrac{k}{m} \end{pmatrix} b_e + \begin{pmatrix} 0 \\ \dfrac{k}{m} \end{pmatrix} b_{\dot{q}}.$$

(6.4.25)

Assume that the measurement errors b_e and $b_{\dot{q}}$ may be modelled as zero-mean stationary independent white noise, with variances

$$\sigma_e = E(b_e^2); \quad \sigma_{\dot{q}} = E(b_{\dot{q}}^2). \tag{6.4.26}$$

Let Σ be the asymptotic covariance matrix of X:

$$\Sigma = \lim_{t \to +\infty} E(X(t) X^T(t)) \tag{6.4.27}$$

which is the solution of

$$A\Sigma + \Sigma A^T + R = 0 \tag{6.4.28}$$

where

$$R = E(W(t) W^T(t)) = \begin{pmatrix} 0 & 0 \\ 0 & \dfrac{k^2}{m^2}(\mu^2 \sigma_e + \sigma_{\dot{q}}) \end{pmatrix}. \tag{6.4.29}$$

From (6.4.28) and (6.4.29), we finally get

$$\Sigma = \begin{pmatrix} E(e^2) & 0 \\ 0 & E(\dot{e}^2) \end{pmatrix} = \begin{pmatrix} 1/2\left(\mu\sigma_e + \dfrac{\sigma_{\dot{q}}}{\mu}\right) & 0 \\ 0 & 1/2\left(\mu\sigma_e + \dfrac{\sigma_{\dot{q}}}{\mu}\right)\dfrac{\mu k}{m} \end{pmatrix}. \tag{6.4.30}$$

This expression shows that:

1. The energy, $E(e^2)$, of the position error does *not* depend on the size of the gain k. It may also be noticed that increasing the value of μ tends to reduce the effects of the velocity measurement noise, (which is generally more significant than the position measurement noise). These two points are in accordance with the deterministic analysis.

2. The energy, $E(\dot{e}^2)$, of the velocity error, is *proportional* to the size of the gain k. This fact is related to the increase of the bandwidth of the controlled system when k is increased. This phenomenon enhances the high-frequency vibrations. If, as initially assumed, the robot were perfectly rigid, the amplitude of these vibrations would depend only on the size of the measurement noise, and would result in increasing the long-term mechanical fatigue of the structure. However, in practice, the consequences are more dramatic, because, as the systems are not perfectly rigid, there is a risk of exciting their resonance frequencies. This 'side effect' of large gains

was not exhibited in the deterministic analysis because it does not interfere, in the ideal case of rigid robots, with the problem of maintaining the error small.

Let us now examine the effects of measurement noise on the energy of the *control* itself. From (6.4.20), we define $\tilde{\Gamma}$ as:

$$\tilde{\Gamma} = \Gamma - E(\Gamma) = -k(\mu e + \dot{e}) - k(\mu b_e + b_{\dot{q}}) \qquad (6.4.31)$$

where $E(\Gamma) = m\ddot{q}_r$ is the mathematical expectation of the control (and also its ideal value when the errors are zero). Note that the energy of this ideal control depends only on the choice of the reference trajectory. If $b_e = b_{\dot{q}} = 0$, then $e(t)$ and $\dot{e}(t)$ exponentially converge to zero, and therefore $\lim_{t \to +\infty} \tilde{\Gamma}(t) = 0$. $\tilde{\Gamma}$ may thus be understood as the *noisy* part of the control, which arises from the measurement noise. Let us compute the energy of $\tilde{\Gamma}$, from (6.4.30) and (6.4.31):

$$E(\tilde{\Gamma}^2) = \left[\frac{k^3}{2m} + k^2\left(\frac{\mu}{2} + 1\right)\right](\mu^2 \sigma_e + \sigma_{\dot{q}}). \qquad (6.4.32)$$

This expression shows that:

3. The energy of the noisy part of the control is approximately proportional to the *cube* of the gain k. In other words, the energy consumed by the inefficient part of the control increases very quickly with the gain k. If the noise is important, if k is chosen very large, or when $\ddot{q}_r \to 0$, then this inefficient part may become stronger than the useful one. With electric drives, the resulting large variations of the controlled current are filtered by the mechanical load, and the possible consequences for the structure are the same as in case (2); in addition, overheating problems may occur in conventional D.C. motors with strong current demands.

6.5 EFFECTS OF DISCRETIZATION AND MEASUREMENT DELAYS

6.5.1 Introduction

As stated previously, three basic assumptions underlie the control analysis:

equations (5.1.11)–(5.1.14) fully and exactly represent the dynamics of the system;
the control vector is computed in continuous time;
all the measurements $e(t)$, $q(t)$, and $\dot{q}(t)$ are instantaneously available.

In practice, none of these assumptions is fully satisfied: the first, for example, implies perfect rigidity of the system, which is never true, owing to structural flexibility or elasticity in the gears. The second is violated by all robot controllers of digital type, and the third neglects the response times of the sensors and of the communication channels. All these facts tend to limit the field of application of the theoretical results given in this chapter. However, if the situations very far from the above assumptions (highly flexible structures, or low sampling rate, for example) are ignored, the main limitation lies in the need for upper bounding the gains, especially k. For example, as presented in the last section, this avoids excitation of resonance frequencies of a non-rigid system; we shall see below that a study of discretization effects leads to the same conclusion.

6.5.2 Limits due to discretization

Consider the simple system obtained with perfect decoupling:

$$\ddot{e} = u_e, \tag{6.5.1}$$

where u_e is the auxiliary control related to the ideal motor torque by

$$\Gamma = M(q) J_T^{-1}(q, t)(u_e - f(q, \dot{q}, t)) + l(q, \dot{q}, t). \tag{6.5.2}$$

The continuous control is then chosen as

$$u_e = -k(\mu e + \dot{e}) \tag{6.5.3}$$

and e exponentially converges to zero for any $k > 0$ and $\mu > 0$. Owing to the ideal decoupling, we may consider independently each component of e, which is equivalent for the analysis to stating that e is one-dimensional. With this assumption let us discretize the control with a sampling period h, and a zero-order hold; we obtain

$$u_e(t) = -k(\mu e(t_n) + \dot{e}(t_n)); \qquad t_n < t \leq t_{n+1}, \tag{6.5.4}$$

where u_e and l are now one-dimensional and

$$t_n = nh; \qquad n \text{ integer.} \tag{6.5.5}$$

The related discrete-time state system is

$$\begin{pmatrix} e \\ \dot{e} \end{pmatrix}(t_{n+1}) = A \begin{pmatrix} e \\ \dot{e} \end{pmatrix}(t_n) \tag{6.5.6}$$

with

$$A = \begin{bmatrix} 1 - \mu k \dfrac{h^2}{2} & h\left(1 - k\dfrac{h}{2}\right) \\ -\mu k h & 1 - kh \end{bmatrix}. \tag{6.5.7}$$

6.5 EFFECTS OF DISCRETIZATION AND MEASUREMENT DELAYS

This system is stable if the eigenvalues of A are inside the unit circle. Using the Jury criterion (cf. Chapter 0) yields the two conditions

$$0 < k < \frac{2}{h}, \tag{6.5.8}$$

$$0 < \mu < \frac{2}{h}. \tag{6.5.9}$$

The values of k and μ are thus upper bounded by *twice* the sampling frequency, and we find again the absence of upper limitations in the continuous case ($h = 0$). These expressions show that one advantage in practice of using high sampling rates lies in the freedom of using larger gains without destabilizing the system.

6.5.3 Effects of measurement delays

Just to give an idea of these effects, we shall consider a first-order system with a position feedback. The study of a more complex second-order system would lead to similar results, though by means of tedious computations. Thus consider the equation

$$\dot{e} = u_e \tag{6.5.10}$$

where, in a first step, the control is a discrete position feedback with no delay in the measurement of $e(t)$:

$$u_e(t) = -ke(t_n); \qquad t_n < t \leq t_{n+1}. \tag{6.5.11}$$

The discretized closed-loop system then is

$$e(t_{n+1}) = (1 - kh)e(t_n), \tag{6.5.12}$$

which is stable if and only if

$$0 < k < \frac{2}{h}. \tag{6.5.13}$$

We find the same upper bound as in the case of a second-order system. Suppose now that the measurement of the position error, e, is available only after a delay of $\delta \leq h$ or that the computation of the control lasts δ. The control variable becomes

$$u_e(t) = -ke(t_n); \qquad t_n + \delta < t \leq t_{n+1} + \delta. \tag{6.5.14}$$

Thus

$$e(t_{n+1} + \delta) - e(t_n + \delta) = -khe(t_n), \tag{6.5.15}$$

and

$$e(t_{n+1} + \delta) = e(t_{n+1}) - \delta k e(t_n) \tag{6.5.16}$$

Therefore

$$e(t_n + \delta) = e(t_n) - \delta k e(t_{n-1}). \tag{6.5.17}$$

From (6.5.15) and (6.5.17) the equation of the discretized closed-loop system is

$$\begin{pmatrix} e(t_{n+1}) \\ e(t_n) \end{pmatrix} = B \begin{pmatrix} e(t_n) \\ e(t_{n-1}) \end{pmatrix} \tag{6.5.18}$$

with

$$B = \begin{pmatrix} 1 - k(h - \delta) & -k\delta \\ 1 & 0 \end{pmatrix}. \tag{6.5.19}$$

Reapplication of the Jury criterion leads to the following stability conditions of system (6.5.18)–(6.5.19):

$$0 < k < \frac{1}{\frac{1}{2}h - \delta} \quad \text{if} \quad h \geq 4\delta, \tag{6.5.20}$$

$$0 < k < \frac{1}{\delta} \quad \text{if} \quad \delta \leq h \leq 4\delta. \tag{6.5.21}$$

Let us examine the relation between the sampling frequency $f_s = 1/h$ and the maximum possible value k_{max} of the gain k. In Fig. 6.1 $k_{max}(f_s)$ is plotted on the basis of (6.5.20) and (6.5.21). From the perspective of this figure it appears that, in the presence of a measurement delay, it is beneficial to increase the sampling frequency up to a given value (here $1/4\delta$); above this, k_{max} remains constant and equal to $1/\delta$.

6.6 USE OF THE POSITIVITY CONDITION OF $J_T J_q^{-1}$

The analysis of the control scheme (6.3.10) shows the critical influence of the positivity of the matrix $B(q, t) = J_T J_q^{-1}$ on the closed-loop stability of the

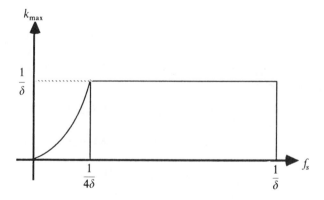

Fig. 6.1 Variation of the maximum possible gain

6.6 USE OF THE POSITIVITY CONDITION OF $J_T J_q^{-1}$

system. In practice, the attempt to satisfy this condition has some consequences for the choice of the task function $e(q, t)$ and the control. Some typical situations are analyzed below.

6.6.1 Case 1: J_T is known

6.6.1.1 General case

When the task is such that the knowledge of an *explicit* expression of J_T is available, a first reasonable choice for J_q is

$$J_q = J_T \qquad (6.5.22)$$

and then $B(q, t)$ is the identity matrix. This choice is the one made in decoupling control schemes and Newton-type controls.

Example 6.6 Consider a tracking problem in a space of operational coordinates, where $x_r(t)$ is the trajectory to be followed by the position coordinates $x(q)$. A possible choice for the task function is

$$e(q, t) = x(q) - x_r(t). \qquad (6.5.23)$$

Then

$$J_T = \frac{\partial x}{\partial q} \qquad (6.5.24)$$

and

$$\frac{\partial e}{\partial t} = -\dot{x}_r(t). \qquad (6.5.25)$$

The function $x(q)$, being known, we may choose

$$J_q = \frac{\partial x}{\partial q}. \qquad (6.5.26)$$

Setting $J_t = -\dot{x}_r(t)$ and $G = I$ in the general control scheme (6.3.10), we obtain the control vector

$$\Gamma = -k\hat{M}\left(\left(\frac{\partial x}{\partial q}\right)^{-1}(\mu D(x - x_r) - \dot{x}_r) + \dot{q}\right) + l_c. \qquad (6.5.27)$$

Remark. In some cases it is sufficient to choose for J_q an approximation $\partial \hat{x}/\partial q$ to J_T; indeed, a sufficient condition for the positivity of $B(q, t)$ is that $\|J_T - J_q\|$ is small, and such that $\|J_T - J_q\| \|J_q^{-1}\| < 1$ on $C_{\rho, T}$.

Example 6.7 As seen in Chapter 3, a more general task function for the tracking problem, when $x_r(t)$ does not necessarily belong to the range $R(x(q))$, is the gradient of the following cost function:

$$h(q, t) = \tfrac{1}{2}(x(q) - x_r(t))^T W(x(q) - x_r(t)); \qquad (6.5.28)$$

where W is symmetric positive definite.
Thus

$$e(q, t) = \frac{\partial x^T}{\partial q} W(x(q) - x_r(t)) \qquad (6.5.29)$$

and

$$J_T(q, t) = \frac{\partial x^T}{\partial q}(q) W \frac{\partial x}{\partial q}(q) + \sum_{i=1}^{n} \frac{\partial^2 x_i}{\partial q^2}(q)[W(x(q) - x_r(t))]_i \qquad (6.5.30)$$

(the subscript i denoting the ith component of a vector);

$$\frac{\partial e}{\partial t}(q, t) = -\frac{\partial x^T}{\partial q}(q) W \dot{x}_r. \qquad (6.5.31)$$

In this example, the computation of J_T involves the computation of the second derivatives of $x(q)$ and is therefore more complex than in the previous case. However, if $x_r \in R(x)$, and if $x(q(t))$ remains close to $x_r(t)$, then

$$J_T \approx \frac{\partial x^T}{\partial q} W \frac{\partial x}{\partial q}. \qquad (6.5.32)$$

This means that the right hand side of (6.5.32) is a good approximation to $J_T(q, t)$ in the neighbourhood of the ideal trajectory $q_r(t)$. We may thus choose

$$J_q = \frac{\partial x^T}{\partial q} W \frac{\partial x}{\partial q}. \qquad (6.5.33)$$

Setting $J_t = -\partial x^T/\partial q\, W\dot{x}_r$, $G = I$, and $W = D$, the control vector is

$$\Gamma = -k\hat{M}\left(\left(\frac{\partial x^T}{\partial q} D \frac{\partial x}{\partial q}\right)^{-1}\left(\mu D e - \frac{\partial x^T}{\partial q} D \dot{x}_r\right) + \dot{q}\right) + l_c. \qquad (6.5.34)$$

Using (6.5.29) in (6.5.34) again gives the control (6.5.27), which is not surprising since the tracking goals are similar in the two cases when $x_r \in R(x)$. However, if this last condition is not satisfied, the choice (6.5.33) may be far from the ideal one, as $x(q_r(t)) - x_r(t)$ may be non-zero. Full computation of J_T would then be needed in a decoupling control scheme. We shall see later that gradient-type control may lead to a simpler solution.

6.6.1.2 Avoiding the inversion of J_T

Recall that the general control expression (6.3.10) uses J_q^{-1}. When J_T is known, it is possible to choose

$$J_q = J_T^{-T}. \qquad (6.5.35)$$

6.6 USE OF THE POSITIVITY CONDITION OF $J_T J_q^{-1}$

Then, $B(q, t) = J_T J_T^T$, is positive definite when the task is admissible. If we use the choice (6.5.35) in Example 6.6, the control vector (6.5.27) is replaced by

$$\Gamma = -k\hat{M}(\mu J_T^T D(x - x_r) + \dot{q}) + l_c. \quad (6.5.36)$$

It should be emphasized that the concomitant of this simplification is that the decoupling of the system is not realized. In fact, as seen below, the control scheme (6.5.36) is of gradient type, when the task is expressed as the minimization of a cost function.

6.6.1.3 Case of location trajectory tracking

Let us consider the particular case where the last frame F_6 has to track a reference frame F_r. An associated task function is thus

$$e(q, t) = \begin{pmatrix} x(q) - x_r(t) \\ P_r(q, t) \end{pmatrix} \quad (6.5.37)$$

where, as usual, x is the position of the origin of the frame with respect to F_0, and P_r is a parametrization of the attitude error between F_6 and F_r. It was shown in Chapter 1 (section 1.2.5.4) that properties of positivity can be associated with certain parametrizations. For example when the parametrization PL derived from the matrix representation or when the parametrization Q derived from the quaternion representation is used, we can write

$$J_T = \begin{pmatrix} I_3 & 0 \\ 0 & J_p \end{pmatrix} J \quad (6.5.38)$$

where J is the basic Jacobian matrix associated with the end-effector of frame F_6, and J_p is a matrix that is positive when the error angle θ between F_6 and F_r is less than:

$\frac{1}{2}\pi$ with the parametrization PL;
π with the parametrization Q.

Moreover J_p is either the identity matrix or $\frac{1}{2}Id$ along the reference trajectory. The positivity condition of $J_T J_q^{-1}$ is thus satisfied by choosing in both cases

$$J_q = J, \quad (6.3.39)$$

and there is no need to compute J_p.

6.6.2 Case 2: J_T is positive definite

This situation occurs, for example, when the aim of the task is to minimize a cost function $h(q, t)$, and the gradient of h is used as a task function. Then

$$J_T(q, t) = \frac{\partial^2 h}{\partial q^2}(q, t) \quad (6.5.40)$$

which is positive in the neighbourhood of the ideal trajectory $q_r(t)$ when the task is admissible. In that case the freedom of choice for J_q is large, and the simplest choice is

$$J_q = I. \tag{6.5.41}$$

This choice is that of gradient-type control schemes.

Example 6.8 We consider the same cost and task functions as in Example 6.7. When choosing $J_q = I$, $J_t = 0$, and $G = D = I$, the control is

$$\Gamma = -k\hat{M}\left(\mu \frac{\partial x^T}{\partial q} W(x - x_r) + \dot{q}\right) + l_c. \tag{6.5.42}$$

We find again control (6.5.36), with the difference that it may be applied even when $x_r(t)$ does not belong to $R(x(q))$. The resulting control is simpler in this case than in Example 6.7, where the choice $J_q = J_T$ was made.

Remark
Since this control is of gradient type, it may be used even when the task becomes non-admissible near a Jacobian singularity, for example when $x_r(t)$ reaches the boundary of $x(q)$. However, in such situations the resulting behaviour of the system is no longer predictable next to the singularity. We have seen in Chapter 3 that it is then possible to make the task more regular by replacing the cost function (6.5.28) by

$$h'(q, t) = h(q, t) + \tfrac{1}{2}\lambda \|q - y(t)\|^2; \quad \lambda > 0, \tag{6.5.43}$$

$y(t)$ being, for example, a filtered vector of $q(t)$. The task function is then

$$e = \frac{\partial x^T}{\partial q} W(x - x_r) + \lambda(q - y) \tag{6.5.44}$$

and the control (6.5.42) becomes

$$\Gamma = -k\hat{M}\left(\mu\left(\frac{\partial x^T}{\partial q} W(x - x_r) + \lambda(q - y)\right) + \dot{q}\right) + l_c. \tag{6.5.45}$$

Without being much more complex, this new control fully fits the scope of the analysis.

6.6.3 Redundant Case

It is not necessary that the task function $e(q, t)$ be the gradient of a potential function in order to make J_T positive in the neighbourhood of the ideal trajectory. In Chapter 4, we defined the task function:

$$e = B^T e_1 + \alpha\left(I - \left(\frac{\partial e_1^T}{\partial q}\right)\left(A \frac{\partial e_1^T}{\partial q}\right)^{-1} A\right) \frac{\partial h^T}{\partial q} \tag{6.5.46}$$

where $\alpha > 0$, e_1 is the main task function and h the secondary cost to be minimized under the constraint $e_1 = 0$. We showed in Lemma 4.4 that certain choices of A and B had the property of making $J_T > 0$ in the neighbourhood of $q_r(t)$ (on condition of choosing α small enough). One possibility is

$$A = B = \frac{\partial e_1}{\partial q}; \qquad (6.5.47)$$

another one is

$$A = \frac{\partial W}{\partial q}; \quad B^T = \frac{\partial W^+}{\partial q}. \qquad (6.5.48)$$

where W is a known matrix function such that $R(W^T) = R(\partial e_1^T/\partial q)$ and $(\partial e_1/\partial q) W^T > 0$. The advantage of these choices is to avoid the complex computation of J_T since J_q can then be taken to be equal to the identity matrix.

Example 6.9 Choose, as a main task function

$$e_1(q, t) = x(q) - x_r(t); \quad \dim x(q) < n. \qquad (6.5.49)$$

The secondary cost function to be minimized is $h(q, t)$, and we choose A and B as in (6.5.48) with $W = \partial x/\partial q$. Then

$$e = \frac{\partial x^+}{\partial q}(x - x_r) + \alpha\left(I - \frac{\partial x^+}{\partial q}\frac{\partial x}{\partial q}\right)\frac{\partial h^T}{\partial q}. \qquad (6.5.50)$$

By setting $J_q = I$, $J_t = -\left(\frac{\partial x}{\partial q}\right)^+ \dot{x}_r$, $G = D = I$, we obtain the following control

$$\Gamma = -k\hat{M}\left(\frac{\partial x^+}{\partial q}(\mu(x - x_r) - \dot{x}_r) + \alpha\mu\left(I - \frac{\partial x^+}{\partial q}\frac{\partial x}{\partial q}\right)\frac{\partial h^T}{\partial q} + \dot{q}\right) + l_c \qquad (6.5.51)$$

which reduces to the control (6.5.27) of Example 6.6 with $D = I$ when the system is not redundant. By choosing A and B as in (6.5.47), and setting $J_t = 0$, we would similarly find the control (6.5.36). When the main task is not redundant, the choice (6.5.47) thus leads to gradient-type control, while the choice (6.5.48) gives a Newton-type control scheme.

6.7 BIBLIOGRAPHIC NOTE

All the original part of the work described in this chapter is due to Samson, and has been partly reported in the following references: Samson (1983a–c; 1987a, b). Singular perturbation theory is a general tool for studying multiple time-scale systems and robustness; this approach has been much analysed in the automatic control literature since the 1960s, see, for example, Young et al. (1977), Young (1978), and Ioannov and Kokotovic (1983). Basic mathematical tools may be found in Birkhoff and Rota (1989), Bogoliubov and Mitropolski (1961), Desoer and Vidyasagar (1975), and Solomon and Davison (1983).

In the field of robot control, robustness issues have also been approached using variable structure systems (sliding modes): see Slotine and Sastry (1983), Slotine (1985), Utkin (1978), Young et al. (1977), Young (1978), and Khatib et al. (1987). Robustness is also considered by Spong et al. (1984, 85), Spong and Vidyasagar (1987), Bremer and Truckenbrodt (1984), and Ha and Gilbert (1985). Taking advantage of some of the structural properties of mechanical systems in control design is considered in Takegaki and Arimoto (1981), Slotine and Li (1987), Paden and Panja (1988). Non-linear control studies by Freund (1982), Isidori (1985, 86), D'Andrea-Novel and Levine (1986), D'Andrea-Novel (1987) or Marino (1986), for example, may also be cited as promising techniques for the analysis and design of new control schemes.

APPENDIX A6.1 PROOFS OF LEMMAS OF CHAPTER 6

A6.1.1 Proof of Lemma 6.1

The left-hand member of (6.1.19) is a $2n$-degree polynomial in λ, so has $2n$ roots. To prove the first assertion of the lemma, let us assume that $|\lambda|$ is bounded by an arbitrary positive number r_1. Under this condition, $\lambda I + k M_1$ is regular for a large enough k. This can be seen from

$$\lambda I + k M_1 = k M_1 \left(\frac{\lambda}{k} M_1^{-1} + I \right). \qquad (A6.1.1)$$

The sum in the right hand side of the equation is invertible when

$$\frac{|\lambda|}{k} \| M_1^{-1} \| < 1 \qquad (A6.1.2)$$

and a sufficient condition is:

$$k > \frac{r_1}{\| M_1 \|}. \qquad (A6.1.3)$$

Assuming now that this condition is satisfied, (6.1.19) can be rewritten as

$$\lambda^2 + k\lambda M_1 + k M_1 M_2 = (\lambda I + k M_1)[(\lambda I + M_2) - (\lambda I + k M_1)^{-1} \lambda M_2] \qquad (A6.1.4)$$

so that the solutions of (6.1.19) are the solutions of

$$\det [(\lambda I + M_2) - (\lambda I + k M_1)^{-1} \lambda M_2] = 0. \qquad (A6.1.5)$$

APPENDIX A6.1 PROOFS OF LEMMAS OF CHAPTER 6

The matrix inversion formula then gives

$$(\lambda I + kM_1)^{-1} = \frac{1}{\det(\lambda I + kM_1)} \text{Cof}(\lambda I + kM_1) \quad (A6.1.6)$$

where $\text{Cof}(\lambda I + kM_1)$ is the matrix made up of the principal minors of $(\lambda I + kM_1)$. Every entry of this matrix is a polynomial $P_{ij}(k, \lambda)$ with degree at most $n - 1$ with respect to k. $\det(\lambda I + kM_1)$ is also a polynomial $-Q(k, \lambda)$ with degree n with respect to k. An entry of the matrix

$$[(\lambda I + M_2) - (\lambda I + kM_1)^{-1} \lambda M_2] \quad (A6.1.7)$$

can be written as

$$\delta_{ij}\lambda + [M_2]_{ij} + \frac{P_{ij}(k, \lambda)}{Q(k, \lambda)} \lambda [M_2]_{ij}, \quad (A6.1.8)$$

where δ_{ij} is the Kronecker symbol. From this expression we can rewrite (A6.1.5) as

$$\sum_{\sigma \in S_n} \prod_{i=1}^{i=n} \left[(\delta_{i\sigma(i)}\lambda + [M_2]_{i\sigma(i)}) + \frac{P_{ij}(k, \lambda)}{Q(k, \lambda)} \lambda [M_2]_{ij} \right] = 0 \quad (A6.1.9)$$

$$= \sum_{\sigma \in S_n} \sum_{I \in P([1,n])} \prod_{i \in I} (\delta_{i\sigma(i)}\lambda + [M_2]_{i\sigma(i)}) \prod_{i \in I^c} \frac{P_{i\sigma(i)}(k, \lambda)}{Q(k, \lambda)} \lambda [M_2]_{i\sigma(i)} \quad (A6.1.10)$$

$$= \det(\lambda I + M_2) + \sum_{\sigma \in S_n} \sum_{I \in P([1,n]), I \neq \emptyset} \prod_{i \in I} \frac{P_{i\sigma(i)}(k, \lambda)}{Q(k, \lambda)} \lambda [M_2]_{i\sigma(i)}$$

$$\times \prod_{i \in I^c} (\delta_{i\sigma(i)}\lambda + [M_2]_{i\sigma(i)}) \quad (A6.1.11)$$

where $P([1, n])$ is the set of all subsets that can be extracted from the integer interval $[1, n]$, S_n is the set of permutations of the same interval, and I^c is the complementary of I. Since I is not empty, the second term of the sum in (A6.1.11) has a degree, with respect to k, at most equal to -1. As a consequence, the polynomial with variable λ in (A6.1.11) converges to $\det(\lambda I + M_2)$ when k tends to $+\infty$. Moreover the convergence is uniform with respect to λ on every disk with radius r_1, so that

$$\det((\lambda I + M_2) - (\lambda I + kM_1)^{-1} \lambda M_2) \quad (A6.1.12)$$

is a sequence, indexed by k, of holomorphic functions converging uniformly to $\det(\lambda I + M_2)$ on every disk $D_1 = \{\lambda; |\lambda| < r_1\}$, when k tends to $+\infty$.

Let us now recall a classical theorem from holomorphic functions theory.

Theorem (Rouché)
Let $f(z)$ and $g(z)$ be two functions holomorphic on an open subset Ω of $\mathbf{C}$ and let D be a compact subset of Ω such that $|f(z)| > |g(z)|$ on the boundary of D; then $f(z) + g(z)$ and $g(z)$ have the same number of zeros contained in D.

Let $f(z) = \det(zI + M_2)$ and $f_k(z) = \det((zI + M_2) - (zI + kM_1)^{-1}zM_2)$. f and f_k are holomorphic on an arbitrary large disk with radius r provided that $k > r/\|M_1\|$. Let D be a compact disk with radius $r_1 < r$ containing all the zeros of $f(z)$ in its interior so that $f(z) \neq 0$ on its boundary Γ. Since Γ is compact, $m = \inf\{|f(z)|; z \in \Gamma\} \neq 0$.

As $f_k(z)$ converges uniformly to $f(z)$ on Γ there exists a positive real number K such that

$$k > K \Rightarrow |f_k(z) - f(z)| < m < |f(z)|$$

on Γ. From Rouché's theorem, $f(z)$ and $f_k(z) = f_k(z) - f(z) + f(z)$ have the same number of zeros contained in D. This shows that when k is large enough, n of the $2n$ roots of $\det(\lambda^2 + k\lambda M_1 + kM_1M_2)$ belong to D and the n others are outside D. Since D can be taken arbitrarily large, these last roots go to infinity with k.

Let us now consider a zero z_0 of $f(z)$. This function being a polynomial, its zeros are isolated points. Let $B_{z_0,\varepsilon}$ be a disk centred in z_0 with radius ε containing only one zero of $f(z)$. For a large enough k, $f_k(z)$ has a number of zeros in $B_{z_0,\varepsilon}$ equal to the multiplicity of z_0 as zero of $f(z)$. Since $B_{z_0,\varepsilon}$ can be made as small as desired, the zeros of $f_k(z)$ contained in $B_{z_0,\varepsilon}$ converge to z_0. This ends the proof of the first part of the lemma.

The proof of the second assertion is very similar. Let $\delta = \lambda/k$. Then there exists an r_2 such that if $\delta > r_2$ and k is large enough, then $\delta I + M_2/k$ is non-singular so that the solutions of (6.1.19) are those of

$$\det\left[(\delta I + M_1) - \frac{1}{k}\delta M_2\left(\delta I + \frac{1}{k}M_2\right)^{-1}\right] = 0. \quad (A6.1.13)$$

Once the uniform convergence of this determinant to $\det(\delta I + (1/k)M_2)$ on every compact set outside the disk $\{z \leq r_2\}$ is proved, Rouché's theorem applies with $\Omega = \{\delta/r_2 < \delta < r_3\}$ and D a compact subset of Ω containing all the zeros of $\det(\delta I + (1/k)M_2) = 0$. The second assertion of the Lemma follows immediately.

A6.1.2 Proof of Lemma 6.2

Let λ be an eigenvalue of $A = A_1A_2$ and x an associated eigenvector. Then we get:

$$x^*A_2Ax = \lambda x^*A_2x$$
$$= x^*A_2A_1A_2x \quad (A6.1.14)$$

where x^* is the conjugate of the row vector x^T. Computing λ we have:

$$\lambda = \frac{x^*A_2A_1A_2x}{x^*A_2x} \quad (A6.1.15)$$

Now, owing to the properties of A_1 and A_2, $x^* A_2 x$ is a positive real number and $x^* A_2 A_1 A_2 x$ is a complex number with positive real part, thus λ has a positive real part.

A6.1.3 Proof of Corollary 6.3.1

Case 1. $M_1 = A_1 A_2$ and $M_2 = \mu BD$ are symmetric positive definite. Let λ be any eigenvalue of $\mathbf{A}$ and x an associated eigenvector:

$$\mathbf{A}x = \lambda x \qquad (A6.1.16)$$

where $\mathbf{A}$ is given by (6.1.14); then

$$(\lambda^2 I + k\lambda M_1 + k M_1 M_2)x = 0. \qquad (A6.1.17)$$

Left-multiplying this equation by $x^* M_1^{-1}$ gives

$$\lambda^2 \frac{x^* M_1^{-1} x}{\|x\|^2} + k\lambda + k \frac{x^* M_2 x}{\|x\|^2} = 0. \qquad (A6.1.18)$$

When k is positive, all the coefficients of this equation in λ are positive; as a consequence the real part of λ is negative.

Cases (2) and (3). The eigenvalues of $M_2 = \mu BD$ are real and positive, and $M_1 = A_1 A_2 = \alpha I$ ($\alpha > 0$). Let δ_i be an eigenvalue of M_2, and x an associated eigenvector. It is easy to see that x also satisfies (A6.1.17) if λ is such that

$$\lambda^2 + k\alpha\lambda + k\alpha\delta_i = 0. \qquad (A6.1.19)$$

As the coefficients of (A6.1.19) are positive, the real parts of its two solutions are negative. By taking all the n eigenvalues δ_i of M_2, we again find the $2n$ roots of $P(\lambda)$.

The proof of Case (3) is similar.

A6.1.4 Proof of Lemma 6.4

Let $q(t)$ be any solution satisfying condition (6.2.23) of Theorem 6.1. The associated error is $e(t) = e(q(t), t)$. To prove Lemma 6.4, it is sufficient to show that, on any time interval $[t_0, t_1[$ ($t_0 \geq 0$; $t_1 \leq T$) where the inequality

$$\|e(t)\| + \frac{\|\dot{e}(t)\|}{\mu b_{max} d^M} < \left(\frac{d^m}{d^M}\right)^{3/2} \left(1 - \frac{1}{\alpha'}\right) \frac{\beta^m}{b_{max}} \left(\frac{\sigma_2^M}{\sigma_2^m}\right)^{1/2} \rho \qquad (A6.1.20)$$

is not satisfied, the theorem may be applied.

Indeed, elsewhere, this inequality leads to the uniform boundedness of $\|\dot{e}(t)\|$ and to that of $\|e(t)\|$ by ρ. Owing to continuity properties of $e(t)$ and $\dot{e}(t)$, (A6.1.20) becomes an *equality* at time t_0. The initial condition (6.2.23) of the theorem is thus satisfied at t_0. Moreover, owing to the condition of Lemma 6.4, k is greater than $k_{\mu,\alpha,\phi,\delta}$ on $[t_0\, t_1[$. Taking t_0 as a new initial time shows that Theorem 6.1 may be applied on $[t_0\, t_1[$.

APPENDIX A6.2 PROOF OF THEOREM 6.1

The notation used is given in section 6.2.1.1.

Given assumptions H2–H3 of section 6.1.2, and the assumptions of the theorem, we study the solutions of the equation

$$\ddot{e} + k(q, \dot{q}, t)A_1(q, t)A_2(q, t)[\mu B(q, t)De + \dot{e} + v(q, t)] + s(q, \dot{q}, t) = 0. \tag{A6.2.1}$$

The function $e(q, t)$ is assumed to be ρ-admissible on the connected set $C_{\rho,T}$ during $[0, T]$. Therefore, there is a diffeomorphism (of class C^3 because we have specified e to be C^3) between an open set containing $C_{\rho,T}$ and an open set containing $B(0, \rho) \times [0, T]$. Owing to this diffeomorphism, we may consider (A6.2.1) as an equation in the variable e instead of the variable q, as long as e stays inside the sphere $B(0, \rho)$.

For any initial conditions which satisfy the condition (6.2.23), local existence and uniqueness of the solution $e(t)$ result from regularity properties imposed on the functions appearing in (A6.2.1). In order to show that the solution $e(t)$ may be extended on $[0, T]$, it is sufficient to show that, on any interval $[0, \tau[$ $(0 < \tau \leq T)$ on which $e(t)$ exists, $\|e(t)\|$ is uniformly bounded by ρ, and $\|\dot{e}(t)\|$ is uniformly bounded by a constant which does not depend on τ.

Consider in a first step the following class of matrix differential equations:

$$\dot{Q} = -k(q(t), \dot{q}(t), t)A_1(q(t), t)A_2(q(t), t)Q$$
$$+ \mu(Q + B(q(t), t))D(Q + B(q(t), t)) - \dot{B}(q(t), \dot{q}(t), t) \tag{A6.2.2}$$

with the initial condition

$$Q(0) = 0. \tag{A6.2.3}$$

Since $q(t)$ and $\dot{q}(t)$ exist on $[0, \tau[$, we know that the solution $Q(t)$ exists and is unique on an interval $[0, \tau_1[$ $(\tau_1 \leq \tau)$.

Let q_i be the ith column of Q:

$$q_i = Qr_i \tag{A6.2.4}$$

where r_i is the vector with all components equal to zero except for the ith one which is unity.

Left-multiplying (A6.2.2) by $q_i^T A_2$ and right-multiplying by r_i gives

$$\frac{1}{2}\frac{d}{dt}(q_i^T A_2 q_i) = -k q_i^T A_2 A_1 A_2 q_i + z_i \tag{A6.2.5}$$

APPENDIX A6.2 PROOF OF THEOREM 6.1

with
$$z_i = \mu q_i^T A_2(Q + B)D(Q + B)r_i - q_i^T A_2 \dot{B} r_i - \tfrac{1}{2} q_i^T \dot{A}_2 q_i. \quad \text{(A6.2.6)}$$

Set
$$y_i = U q_i \quad \text{(A6.2.7)}$$

where U is the symmetric positive square root matrix of A_2:
$$A_2 = U^2 \quad (U = A_2^{1/2}). \quad \text{(A6.2.8)}$$

Equation (A6.2.5) can be rewritten in the form
$$\tfrac{1}{2} \frac{d}{dt} \|y_i\|^2 = -k y_i^T U A_1 U y_i + z_i \quad \text{(A6.2.9)}$$

with
$$z_i = \mu y_i^T U(Q+B)D(Q+B)r_i - y_i^T U \dot{B} r_i - \tfrac{1}{2} y_i^T U^{-1} \dot{A}_2 U^{-1} y_i. \quad \text{(A6.2.10)}$$

Using the notation of section 6.2.1, and applying classical norm inequalities to (A6.2.10) we get

$$\begin{aligned} \|z_i\| \leq {} & \mu d^M \left(\frac{a_2^M}{a_2^m}\right)^{1/2} \|Q\| \|y_i\|^2 + \mu d^M (a_2^M)^{1/2} \|Q\| \|B\| \|y_i\| \\ & + \mu d^M \left(\frac{a_2^M}{a_2^m}\right)^{1/2} \|B\| \|y_i\|^2 + \mu d^M (a_2^M)^{1/2} \|B^2\| \|y_i\| \\ & + (a_2^M)^{1/2} \|\dot{B}\| \|y_i\| + \tfrac{1}{2}\left(\frac{\|\dot{A}_2\|}{a_2^m}\right) \|y_i\|^2. \end{aligned} \quad \text{(A6.2.11)}$$

Let us prove by contradiction that Q is uniformly bounded: let us assume that there exists a first time t_1 $(0 < t_1 \leq \tau_1)$ and a column q_i of Q such that

$$\|y_i(t_1)\| = (\sigma_2^m)^{1/2} \beta^m (\alpha n)^{-1} \quad \text{(A6.2.12)}$$

and that for any j $(1 \leq j \leq n)$, and for any $t \in [0, t_1[$,

$$\|y_j(t)\| < (\sigma_2^m)^{1/2} \beta^m (\alpha n)^{-1}. \quad \text{(A6.2.13)}$$

Since
$$\|Q\| \leq \sum_{j=1}^n \|q_j\| \quad \text{and} \quad \|q_j\| \leq \|U^{-1}\| \|y_j\| \quad \text{(A6.2.14)}$$

we have necessarily at time t_1

$$\begin{aligned} \|Q(t_1)\| &\leq \|U^{-1}(t_1)\| (\sigma_2^m)^{1/2} \beta^m \alpha^{-1} \\ &\leq \left(\frac{\sigma_2^m}{a_2^m(t_1)}\right)^{1/2} \beta^m \alpha^{-1}. \end{aligned} \quad \text{(A6.2.15)}$$

By using (A6.2.15) in (A6.2.11), we get

$$\|z_i(t_1)\| \le \left\{ \mu d^M \left(\frac{a_2^M}{a_2^m}\right)^{1/2} \left[\left(\frac{\beta^m}{\alpha}\right)\left(\frac{\sigma_2^m}{a_2^m}\right)^{1/2} + (n+1)\|B\| \right.\right.$$

$$\left.+ n\left(\frac{\alpha}{\beta^m}\right)\left(\frac{a_2^m}{\sigma_2^m}\right)^{1/2}\|B\|^2\right] + n\left(\frac{\alpha}{\beta^m}\right)\left(\frac{a_2^M}{\sigma_2^m}\right)^{1/2}\|\dot B\|$$

$$\left.+ \frac{1}{2}\frac{\|\dot A_2\|}{a_2^m}\right\}_{t=t_1} \|y_i(t_1)\|^2. \qquad (A6.2.16)$$

On the other hand,

$$ky_i^T U A_1 U y_i = ky_i^T U(\tfrac{1}{2}(A_1 + A_1^T)) U y_i > k_{\alpha,\mu,\phi,\delta}$$

$$\times \frac{\|y_i\|^2}{\|U^{-1}(\tfrac{1}{2}(A_1 + A_1^T))^{-1} U^{-1}\|}. \qquad (A6.2.17)$$

Taking into account the expression of $k_{\alpha,\mu}^1$ given by (6.2.15), and knowing that $k_{\alpha,\mu,\phi,\delta} \ge k_{\alpha,\mu}^1$, we deduce, from (A6.2.16) and (A6.2.17),

$$(-ky_i^T U A_1 U y_i + z_i)|_{t=t_1} < 0 \qquad (A6.2.18)$$

and from (A6.2.9),

$$\frac{d}{dt}\|y_i\|_{t=t_1}^2 < 0. \qquad (A6.2.19)$$

As y_i is continuously differentiable on $[0, \tau_1[$, the inequality (A6.2.19) contradicts the existence of t_1 since $\|y_i\|$ can only reach, for the first time, the constant $(\sigma_2^m)^{1/2} \beta^m (\alpha n)^{-1}$ with a positive or zero derivative. Since t_1 does not exist, we have proved that,

for any $t \in [0, \tau_1[$, and for any $i \in \{1, \ldots, n\}$:

$$\|y_i(t)\| < (\sigma_2^m)^{1/2} \beta^m (\alpha n)^{-1}. \qquad (A6.2.20)$$

Now let $[0, \tau_2[, (\tau_2 \le \tau_1)$ denote the largest time interval contained in $[0, \tau_1[$ for which $\|e(t)\| < \rho$. From (A6.2.7) and (A6.2.20), we have

for any $t \in [0, \tau_2[$, $\|Q(t)\| < \frac{\beta^m}{\alpha}. \qquad (A6.2.21)$

Note that this upper bound does not depend on τ_2.

In the second part of the proof we shall use the matrix function $L(t)$ defined in the following way:

$$L(t) = \mu(Q(t) + B(t)D; \quad L(0) = \mu B(0)D. \qquad (A6.2.22)$$

APPENDIX A6.2 PROOF OF THEOREM 6.1

So far, $L(t)$ is defined on $[0, \tau_2[$ and is uniformly bounded on this interval. Indeed, from (A6.2.21) and (A6.2.22),

$$\|L(t)\| < \mu d^M \left(b_{\max} + \frac{\beta^m}{\alpha} \right). \tag{A6.2.23}$$

Moreover, from (A6.2.21) and (A6.2.22)

$$x^T D L(t) x = \mu x^T D (Q(t) + B(t)) D x \geq \mu \|x\|^2 \left[b^m(t) - \frac{\beta^m}{\alpha} \right] (d^m)^2 \tag{A6.2.24}$$

with

$$b^m(t) - \frac{\beta^m}{\alpha} \geq \left(1 - \frac{1}{\alpha} \right) \beta^m > 0. \tag{A6.2.25}$$

It follows from (A6.2.24) and (A6.2.25) that the first-order differential system

$$\dot{x} = -L(t)x \tag{A6.2.26}$$

is exponentially stable. Indeed, combining (A6.2.24), (A6.2.25) and (A6.2.26), we have

$$\frac{1}{2}\frac{d}{dt}(x^T D x) \leq -\mu \left(1 - \frac{1}{\alpha} \right) \beta^m (d^m)^2 \|x\|^2. \tag{A6.2.27}$$

This inequality obviously implies the exponential convergence of $x(t)$ to zero. Relation (6.2.25) of the theorem is derived from (A6.2.27).

From (A6.2.2) and (A6.2.22), it is easy to show that $L(t)$ also satisfies

$$\dot{L} = -kA_1 A_2 L + L^2 + k\mu A_1 A_2 BD. \tag{A6.2.28}$$

Let us define the auxiliary variable η in the following way:

$$\eta(t) = \dot{e}(t) + L(t)e(t); \quad \eta(0) = \dot{e}(0) + \mu B(0) D e(0). \tag{A6.2.29}$$

This equation is (6.2.24) of the theorem. Since $e(t)$, $\dot{e}(t)$, and $L(t)$ exist on $[0, \tau_2[$, we already know that $\eta(t)$ exists on this time interval.

From (A6.2.1), (A6.2.28) and (A6.2.29), it is not difficult to show that $\eta(t)$ also satisfies the relation

$$\dot{\eta}(t) = (-k(t)A_1(t)A_2(t) + L(t))\eta(t) - k(t)A_1(t)A_2(t)v(t) - s(t). \tag{A6.2.30}$$

Left-multiplying (A6.2.30) by $\eta^T(t) A_2(t)$ yields

$$\frac{1}{2}\frac{d}{dt}\|z\|^2 = -kz^T U A_1 U z + w \tag{A6.2.31}$$

where

$$z = U\eta \quad (U = A_2^{1/2}) \tag{A6.2.32}$$

and

$$w = -kz^T U A_1 A_2 v + z^T U L U^{-1} - z^T U s. \tag{A6.2.33}$$

Using (A6.2.22) to obtain an upper bound of $\|L\|$, we deduce from (A6.2.33):

$$\|w\| \leq \mu d^M \left(\frac{a_2^M}{a_2^m}\right)^{1/2} \left(\|B\| + \frac{\beta^m}{\alpha}\right) \|z\|^2 + k(a_2^M)^{3/2} v \|A_1\| \|z\|$$
$$+ (a_2^M)^{1/2} \|s\| \|z\| \tag{A6.2.34}$$

where v denotes the upper bound of $v(t)$ on $[0, T]$. Therefore, if $\|z\| \geq (C_{A_1} C_{A_2} (\sigma_2^M)^{1/2} + \phi)$, then

$$\|w\| \leq \left[\mu d^M \left(\frac{a_2^M}{a_2^m}\right)^{1/2} \left(\|B\| + \frac{\beta^m}{\alpha}\right) + \frac{k(a_2^M)^{3/2} v \|A_1\|}{(C_{A_1} C_{A_2} (\sigma_2^M)^{1/2} v + \phi)}\right.$$
$$\left. + \frac{(a_2^M)^{1/2} \|s\|}{(C_{A_1} C_{A_2} (\sigma_2^M)^{1/2} v + \phi)}\right] \|z\|^2. \tag{A6.2.35}$$

On the other hand,

$$k z^T U A_1 U z \geq k \frac{\|z\|^2}{\|U^{-1} (\frac{1}{2}(A_1 + A_1^T))^{-1} U^{-1}\|}. \tag{A6.2.36}$$

From (A6.2.31), (A6.2.35) and (A6.2.36), we deduce that, if $\|z\| \geq (C_{A_1} C_{A_2} (\sigma_2^M)^{1/2} v + \phi)$, then

$$\frac{1}{2} \frac{d}{dt} \|z\|^2 \leq -k \left[\frac{1}{\|U^{-1}(\frac{1}{2}(A_1 + A_1^T))^{-1} U^{-1}\|}\right.$$
$$\left. - (a_2^M)^{3/2} v \frac{\|A_1\|}{(C_{A_1} C_{A_2} (\sigma_2^M)^{1/2} v + \phi)}\right] \|z\|^2$$
$$+ \left[\mu d^M \left(\frac{a_2^M}{a_2^m}\right)^{1/2} \left(\|B\| + \frac{\beta^m}{\alpha}\right)\right.$$
$$\left. + \frac{(a_2^M)^{1/2} \|s\|}{(C_{A_1} C_{A_2} (\sigma_2^M)^{1/2} v + \phi)}\right] \|z\|^2. \tag{A6.2.37}$$

By noticing that

$$\frac{1}{\|U^{-1}(\frac{1}{2}(A_1 + A_1^T))^{-1} U^{-1}\|} - (a_2^M)^{3/2} v \|A_1\| \frac{1}{(C_{A_1} C_{A_2} (\sigma_2^M)^{1/2} v + \phi)}$$
$$\geq \frac{v}{(C_{A_1} C_{A_2} (\sigma_2^M)^{1/2} v + \phi)} \times \frac{1}{\|U^{-1}(\frac{1}{2}(A_1 + A_1^T))^{-1} U^{-1}\|} \tag{A6.2.38}$$

and using the fact that $k \geq k^2_{\alpha, \mu, \phi, \delta}$ (given by (6.2.16)), we deduce from (A6.2.37):

$$\frac{1}{2} \frac{d}{dt} \|z\|^2 < -\delta \|z\|^2 \quad \text{when} \quad \|z\| \geq (C_{A_1} C_{A_2} (\sigma_2^M)^{1/2} v + \phi). \tag{A6.2.39}$$

APPENDIX A6.2 PROOF OF THEOREM 6.1

This shows that, on $[0, \tau_2[$,

$$\|z\| \leq \max(\|z(0)\|e^{-\delta t}; \quad (C_{A_1}C_{A_2}(\sigma_2^M)^{1/2}v + \phi)). \quad (A6.2.40)$$

Moreover, since $\eta = U^{-1}z$, we have

$$\|\eta\| \leq \|U^{-1}\|\|z\| \leq \|z\|(\sigma_2^m)^{-1/2} \quad \text{on} \quad [0, \tau_2[. \quad (A6.2.41)$$

From (A6.2.29), we also have

$$\|z(0)\| \leq (\sigma_2^M)^{1/2}\|\dot{e}(0) + \mu B(0)De(0)\|. \quad (A6.2.42)$$

Therefore, from (A6.2.40)–(A6.2.42),

$$\|\eta\| \leq \max\left[\left(\frac{\sigma_2^M}{\sigma_2^m}\right)^{1/2}\|\dot{e}(0) + \mu B(0)De(0)\|e^{-\delta t};\right.$$

$$\left.\left(\frac{\sigma_2^M}{\sigma_2^m}\right)^{1/2}C_{A_1}C_{A_2}v + \phi(\sigma_2^m)^{-1/2}\right]. \quad (A6.2.43)$$

for $t \in [0, \tau_2[$. This shows that $\|\eta\|$ is uniformly bounded on $[0, \tau_2[$. Relation (6.2.27) of the theorem is simply (A6.2.43).

To complete the proof, return to (A6.2.29), which may be written

$$\dot{e}(t) = -L(t)e(t) + \eta(t). \quad (A6.2.44)$$

From (A6.2.22) and (A6.2.44)

$$\frac{1}{2}\frac{d}{dt}\|r\|^2 = -\mu r^T D^{1/2}(Q + B)D^{1/2}r + r^T D^{1/2}\eta \quad (A6.2.45)$$

where

$$r = D^{1/2}e. \quad (A6.2.46)$$

We have

$$\|r^T D^{1/2}\eta\| \leq (d^M)^{1/2}N\|r\| \quad (A6.2.47)$$

where $N = \sup_{t \in [0, \tau_2[}\|\eta(t)\|$.

Moreover, by using (A6.2.21), and for $t \in [0, \tau_2[$,

$$\mu r^T D^{1/2}(Q + B)D^{1/2}r \geq \mu d^m\left(1 - \frac{1}{\alpha}\right)\beta^m\|r\|^2. \quad (A6.2.48)$$

Relations (A6.2.45), (A6.2.47) and (A6.2.48), yield

$$\frac{1}{2}\frac{d}{dt}\|r\|^2 \leq 0 \quad \text{when} \quad \|r\| \geq (d^M)^{1/2}\frac{N}{\mu d^m(1 - (1/\alpha))\beta^m}. \quad (A6.2.49)$$

It follows that, on $[0, \tau_2[$,

$$\|r\| \leq \max\left\{\|r(0)\|; (d^M)^{1/2}\frac{N}{\mu d^m(1 - (1/\alpha))\beta^m}\right\}. \quad (A6.2.50)$$

From (A6.2.46), we also have

$$\|r(0)\| \leq (d^M)^{1/2} \|e(0)\| \quad (A6.2.51)$$

and

$$\|e\| \leq \|r\|(d^m)^{-1/2}. \quad (A6.2.52)$$

Therefore, on $[0, \tau_2[$

$$\|e\| \leq \left(\frac{d^M}{d^m}\right)^{1/2} \max\left\{\|e(0)\|; \frac{N}{\mu d^m(1-(1/\alpha))\beta^m}\right\}. \quad (A6.2.53)$$

From (A6.2.43), we have also

$$N \leq \max\left\{\left(\frac{\sigma_2^M}{\sigma_2^m}\right)^{1/2}(\|\dot{e}(0)\| + \mu b_{\max}d^M\|e(0)\|;\right.$$

$$\left.\left(\frac{\sigma_2^M}{\sigma_2^m}\right)^{1/2}C_{A_1}C_{A_2}v + \phi(\sigma_2^m)^{-1/2}\right\}. \quad (A6.2.54)$$

From (A6.2.53) and (A6.2.54), and using conditions (6.2.21) ($\mu > \mu_\rho$) and (6.2.23) which give the maximum authorized size for the initial errors $e(0)$ and $\dot{e}(0)$, we get

$$\text{for any } t \in [0, \tau_2[: \quad \|e(t)\| < \rho. \quad (A6.2.55)$$

Since $e(t)$ is continuous on $[0, \tau[$, (A6.2.55) shows that $\tau_1 = \tau_2$. Otherwise we would have, from the definition of τ_2, $\|e(\tau_2)\| = \rho$.

Since the boundedness of Q on $[0, \tau_1[$ is uniform and does not depend on τ_1, we finally have $\tau_1 = \tau$.

We have shown that, if $e(t)$ exists on a time interval $[0, \tau[$, then $\|e(t)\|$ is uniformly bounded by ρ on this interval. Uniform boundedness of $\|\dot{e}(t)\|$ on the same interval is simply a consequence of (A6.2.29) and of the uniform boundedness of $\|e(t)\|$, $\|\eta(t)\|$, and $\|L(t)\|$. Therefore, $e(t)$ exists and is unique on the whole interval $[0, T]$ (in fact on $[0, T[$, and, by continuity, on $[0, T]$), and all the results obtained through the proof on $[0, \tau_2[$ remain true on $[0, T]$.

The statements of the theorem are contained in these results.

7
SENSOR-BASED TASKS

INTRODUCTION

This chapter focuses on the utilization of exteroceptive sensors' outputs in the design of dedicated task functions. Since the sensors' outputs provide some measurements of the interactions between the robot and the environment, the control schemes associated with such task functions can be viewed as working in closed-loop with the robot's environment. Sensors which exclusively lend themselves to symbolic information processing are not considered here because they do not provide the type of measurements needed to form a task vector. Also, since the control loop has to be continuously fed with up-to-date information, only high data rate sensors can be used. This excludes, for example, from the analysis, high-level visual functions (sorting, recognition, inspection), which are activated from time to time, while it includes force sensing, proximity or local range sensing. Another consequence is that, often, only local measurements of the interactions between the robot and its environment are used. Despite these limitations, the utilization at the control level of task functions derived from exteroceptive sensors' outputs is an effective way of broadening the working capabilities of a robot.

We may distinguish between two large classes of sensors. The first one regroups non-contact sensors which have a very large domain of application. For example, local (end-effector) sensing may allow the controller to compensate for terminal errors arising from elasticities in a large light structure for robotics applications in space. It is also often easier in many industrial applications to compensate for relative positioning errors by moving the robot rather than the workpiece. Furthermore, precise clamping of a part at a reference position is sometimes difficult and deformation or uncertainties at the workpiece itself may lead to the need for final relative adjustments. Among the most common industrial examples, the mounting of a windshield on an automobile and the tracking of a seam in arc welding may be cited. Another important problem that requires local sensing, mainly in non-manufacturing applications, is on-line obstacle avoidance. This problem is encountered when the environment is not well known, imperfectly modelled or liable to unexpected changes, as for example in remote manipulation (or telerobotics) or in hostile worlds (space, undersea or nuclear environments). All examples above require non-contact sensing, such as range finding or proximity measurement.

A second class of local sensors regroups *contact* sensors, including force sensors in particular. The main applications of these systems are assembly processes and contouring problems. Fine control of grasping hands also demands the use of force and tactile sensors.

From a practical point of view, the flexibility and accuracy of a robot control system depend on its ability to take such sensory information into account. A systematic way of introducing this information into the controller consists of two parts.

1. Consider sensors as devices which allow the values taken by dedicated task functions (called sensor-based task functions) to be measured on-line. All analysis tools developed in the previous chapters then apply since we remain in the task-function framework.
2. As far as possible, unify the techniques of obtaining the sensor-based task functions themselves independently of the type of sensors used. This involves a particular modelling effort. We shall see, for example, that proximity and force information may be treated in the same manner.

Such are the objectives of this chapter which is organized as follows. After introducing the important concept of interaction screw which can be associated with a large class of sensors, we provide the reader with a practical method of building proximity-based tasks from a point of view consistent with the analysis of contact phenomena between solids. Then we give some comments on the control aspects and consider some examples of specific robotic tasks that use proximity sensors. The last section is devoted to the analysis of force-based control, and the Appendix presents some technical points concerning optical proximity sensors.

7.1 MODELLING OF THE INTERACTIONS WITH THE ENVIRONMENT

7.1.1 Variations of a general sensory signal

7.1.1.1 Introduction

To take part in the design of a task function, a sensory signal must possess a certain number of properties related to those of any task function $e(q, t)$. Obviously, the functional associated with the sensory signal should depend on the variables q and t. Another property is the twice differentiability of this functional. We will therefore restrict the analysis to sensors which can be modelled by a C^2 functional denoted as $s(q, t)$. For example, this excludes binary sensors, which are preferably used for simple detection or alarm purposes, but includes tactile sensors when used as a set of local deformation sensors.

7.1 MODELLING OF THE INTERACTIONS

As emphasized throughout the previous chapters the Jacobian matrix of a task function is a fundamental characteristic of the task on which the control design critically depends. The aim of this section is to describe a general form of this Jacobian matrix when sensory signals are used.

7.1.1.2 The screw product associated with the variations of a sensory signal

We shall use notation taken from Chapters 1 and 2, i.e.

(1) F_i; $i = 0, \ldots, n$: the frame related to link i, defined by its origin, O_i, and the basis of vectors $\{x_i, y_i, z_i\}$; F_0 is the fixed reference frame in the three dimensional space;

(2) F_A: a frame arbitrarily related to a point A, defined by its origin, A, and the basis of vectors $\{x_A, y_A, z_A\}$; when A belongs to a rigid body, F_A may also represent a frame linked to this body;

(3) V_{A/F_B}: velocity vector of point A with respect to the frame F_B;

(4) ω_{F_A/F_B}: angular velocity vector of frame F_A with respect to frame F_B.

A screw is characterized by its field evaluated at a certain point and its vector, for example, $H = (H(O), \omega)$. The velocity screw of a rigid body (frame F_K) with respect to another one (frame F_L) is denoted T_{KL}.

Remark. Throughout the chapter, we shall use the following notation: $[M]_F$ will denote the expression of M (vector or matrix of coordinates) in the frame F. To avoid burdening the notation when all the terms entering an equation have to be expressed in a single frame F, we shall simply write $|_F$ at the end of the equation. When no indication is given, all terms are, if needed, implicitly assumed to be expressed in the same single frame.

Let us thus consider the situation shown in Fig. 7.1. (S) is a sensor rigidly fixed on the ith link, with the associated frame F_S. The local environment, i.e. the part of the environment to which (S) is sensitive is represented by an object (T) (like 'target'), with the associated frame F_T. Let the one-dimensional function s be the useful part of the signal provided by (S) (i.e. excluding noise and other disturbances). We shall further assume in the following that:

H1: *Given (S) and (T), the signal s is a function of the relative position and orientation of (S) with respect to (T).*

If the sensor is motionless with respect to the target, then the signal remains constant. Furthermore, since the position/attitude of F_S with respect to the fixed reference frame F_0 is a function of the robot joint coordinates q, and

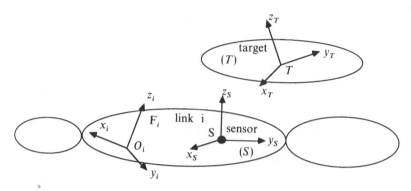

Fig. 7.1 Sensor, target, and frames

since we can use the variable t to parametrize the position/attitude of the target with respect to F_0, it appears that the relative position of F_S and F_T, which is an element $\bar{r}$ of SE_3 (cf. Chapter 1), is itself a function of q and t, and so is the signal s. The assumption H1 is therefore compatible with our initial requirement according to which the sensory signals must be functions of q and t. In practice this assumption is reasonable for a large class of proximity sensors, range finders or force sensors. We may thus write

$$s(\bar{r}(q,t)) = s(F_S, F_T). \tag{7.1.1}$$

As a C^2 function with values in $\mathbf{R}$ and domain in SE_3, s admits a derivative represented by an element of se_3, or a screw (cf. section 1.3.2.6), and we may write the differential as

$$\dot{s} = \frac{\partial s}{\partial \bar{r}} \bullet \frac{d\bar{r}}{dt} = H \bullet T_{ST} \tag{7.1.2}$$

where $\bullet$ denotes the screw product and $H = \partial s/\partial \bar{r}$ is a screw which characterizes the variation of the signal s. We will denote the vector of this screw as u and write $H = (H(.), u)$ where $H(.)$ is the corresponding vector field.

The velocity screw $d\bar{r}/dt = T_{ST}$ is also a function of q and t and it is possible to write:

$$\frac{d\bar{r}}{dt} = T_{ST} = \frac{\partial \bar{r}}{\partial q} \dot{q} + \frac{\partial \bar{r}}{\partial t}. \tag{7.1.3}$$

We shall see that H will play an important role both in the design of a sensor-based task and in the analysis of the properties of this task. For a given situation (S, T), H depends only on the characteristics of the sensor itself and of the target (geometry, reflectance, etc.). It should be emphasized that it is *independent* of the robot itself.

7.1.1.3 Derivation of $\frac{\partial s}{\partial q}(q, t)$ and $\frac{\partial s}{\partial t}(q, t)$

From Chapter 1 (eqn 1.3.20), we have

$$T_{ST} = T_{S0} - T_{T0} \tag{7.1.4}$$

If the sensor (S) is carried by the link L_i of the robot, we have

$$T_{S0} = T_{i0} \tag{7.1.5}$$

Then, (7.1.2) becomes

$$\dot{s} = H \bullet T_{i0} - H \bullet T_{T0}. \tag{7.1.6}$$

By computing the first screw product at point O_i and the second screw product at T, we obtain

$$\dot{s} = (H(O_i), u) \bullet (T_{i0}(O_i), \omega_{i0}) - (H(T), u) \bullet (T_{T0}(T), \omega_{T0}). \tag{7.1.7}$$

To find $(\partial s / \partial q)(q, t)$ and $(\partial s / \partial t)(q, t)$, it remains to write equation (7.1.7) in matrix form. Let us express all the coordinates of the vectors in the basis of F_0. Thus, (7.1.7) becomes

$$\dot{s} = (u^T \quad H^T(O_i))\begin{pmatrix} V_i \\ \omega_i \end{pmatrix} - (u^T \quad H^T(T))\begin{pmatrix} V_T \\ \omega_T \end{pmatrix}\bigg|_{F_0} \tag{7.1.8}$$

where we know that

$$H(P) = H(Q) + As(u)QP|_{F_0}. \tag{7.1.9}$$

Since O_i belongs to link i of the robot, we have (cf. Chapter 2)

$$\begin{pmatrix} V_i \\ \omega_i \end{pmatrix} = J_i(q)\dot{q}|_{F_0} \tag{7.1.10}$$

where $J_i(q)$ is the basic Jacobian matrix for link i in the basis of F_0. On the other hand, we have

$$\dot{s}(q, t) = \frac{\partial s}{\partial q}(q, t)\dot{q} + \frac{\partial s}{\partial t}(q, t). \tag{7.1.11}$$

After using (7.1.10) in (7.1.8), the comparison of (7.1.8) with (7.1.11) finally leads to

$$\frac{\partial s}{\partial q}(q, t) = (u^T \quad H^T(O_i))(q, t)J_i(q)|_{F_0} \tag{7.1.12}$$

$$\frac{\partial s}{\partial t}(q, t) = -(u^T \quad H^T(T))(q, t)\begin{pmatrix} V_T(t) \\ \omega_T(t) \end{pmatrix}\bigg|_{F_0}. \tag{7.1.13}$$

When the task function $e(q, t)$ depends on the signal $s(q, t)$, relation (7.1.12) can in turn be used to derive the task Jacobian $\partial e/\partial q$, while relation (7.1.13) can be used to derive $\partial e/\partial t$. We may already make two observations.

1. The matrix $J_i(q)$ involved in the derivation of $(\partial s/\partial q)(q,t)$ is generally well-known, as long as the kinematics of the robot are known.

2. The velocity of the object, $\begin{pmatrix} V_T(t) \\ \omega_T(t) \end{pmatrix}$, which may be zero in many cases, is often unknown in other applications, such as target tracking. Recall however that the control analysis of Chapter 6 takes this fact into account through the *model* J_t of $\partial e/\partial t$ utilized in the control. Without loss of generality, we may thus assume from now on and for the sake of simplicity that this velocity is zero.

7.1.2 The interaction screw

7.1.2.1 Definition

The meaning of the screw H requires careful interpretation. As seen previously, this screw depends not only on the characteristics of both the sensor and the object, but also on the relative position of the sensor with respect to the target. While being independent of the robot itself, it is an important characteristic of the *interaction*, between the sensor and the robot's environment. For this reason we propose the following.

Definition D7.1 *Given a sensor* (S), *located on a fixed link* L_i *of the robot, which provides a one-dimensional signal s, and an object* (T) *motionless in* F_0, *the* **interaction screw** *associated with* (S) *and* (T) *is the screw* $H = (H(.), u)$ *defined by the equation*

$$\dot{s} = H \bullet T_{S0} \qquad (7.1.14)$$

where T_{S0} is the velocity screw of the frame $F_i(O_i, R_i)$ linked to L_i with respect to F_0.

The concept of interaction screw is fundamental for modelling systems utilizing exteroceptive sensors because it contains most of the information required to design and analyse sensor-based control schemes. Moreover, it is general since it applies to any kind of sensor satisfying the assumption H1.

In some cases, when *models* of the environment and the sensor are available and simple, it is possible to know H fairly precisely. However, in the most frequently encountered situation, the characteristics of the sensor and/or the object are only approximately known. This fact will now be illustrated by considering the typical case of local sensors of the *optical* type, such as those described in Appendix A7.1.

7.1 MODELLING OF THE INTERACTIONS

7.1.2.2 Explicit computation of the interaction screw: two examples in proximity sensing

Example 7.1: Case of a pin-point target. Let us consider the case of an optical sensor and a pin-point target and compute the associated screw H. The situation is that of Fig. 7.2, where the sensor's signal is assumed to depend on both the distance δ and the angle α. Denoting $\delta = \|ST\|$, we have

$$s = f(\alpha, \delta). \tag{7.1.15}$$

With the change of variable:

$$\beta = \cos \alpha \tag{7.1.16}$$

we also have

$$\dot{s} = \frac{\partial f}{\partial \delta} \dot{\delta} + \frac{\partial f}{\partial \beta} \dot{\beta} \tag{7.1.17}$$

with

$$\dot{\delta} = -\left\langle \frac{ST}{\|ST\|}, V_{S/F_T} \right\rangle = -\langle n, V_{S/F_T} \rangle. \tag{7.1.18}$$

$\dot{\beta}$ can be computed by differentiating

$$\beta = \cos \alpha = \langle x_s, n \rangle \tag{7.1.19}$$

which yields

$$\dot{\beta} = \langle \dot{x}_s, n \rangle + \langle x_s, \dot{n} \rangle \tag{7.1.20}$$

with

$$\langle \dot{x}_s, n \rangle = \langle (\omega_{F_S/F_T} \times x_s), n \rangle = \langle \omega_{F_S/F_T}, (x_s \times n) \rangle \tag{7.1.21}$$

$$\dot{n} = \frac{d}{dt} \frac{ST}{\delta} = -\frac{1}{\delta} \dot{\delta} n - \frac{1}{\delta} V_{S/F_T} = -\frac{1}{\delta} (\dot{\delta} n + V_{S/F_T}). \tag{7.1.22}$$

Use of (7.1.18) in (7.1.22) leads to

$$\langle x_s, \dot{n} \rangle = -\frac{1}{\delta} (-\langle n, V_{S/F_T} \rangle \langle x_s, n \rangle + \langle x_s, V_{S/F_T} \rangle) \tag{7.1.23}$$

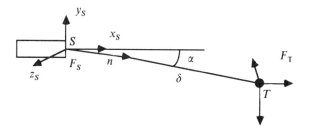

Fig. 7.2 Optical reflectance sensor and pinpoint target

Thus:

$$\dot{\beta} = \left(\langle (x_s \times n), \omega_{F_s/F_T} \rangle + \frac{1}{\delta} \langle x_s, n \rangle \langle n, V_{S/F_T} \rangle - \frac{1}{\delta} \langle x_s, V_{S/F_T} \rangle \right). \quad (7.1.24)$$

From (7.1.17), (7.1.18) and (7.1.24) we finally obtain

$$\dot{s} = \left\langle \left(\left(-\frac{\partial f}{\partial \delta} + \frac{\beta}{\delta} \frac{\partial f}{\partial \beta} \right) n - \frac{1}{\delta} \frac{\partial f}{\partial \beta} x_s \right), V_{S/F_T} \right\rangle + \left\langle \frac{\partial f}{\partial \beta} (x_s \times n), \omega_{F_s/F_T} \right\rangle$$
$$(7.1.25)$$

which shows that the elements of the interaction screw are in this case

$$u = \left(-\frac{\partial f}{\partial \delta} + \frac{\beta}{\delta} \frac{\partial f}{\partial \beta} \right) n - \frac{1}{\delta} \frac{\partial f}{\partial \beta} x_s \quad (7.1.26)$$

$$H(S) = \frac{\partial f}{\partial \beta} (x_s \times n). \quad (7.1.27)$$

In the above equations, the field of the interaction screw is expressed at the point S, linked to the *sensor*. Knowing that $H(T) = H(S) - \delta n \times u$, it is possible to compute the value of H at the point T. Using (7.1.26) and (7.1.27) we find easily:

$$H(T) = 0. \quad (7.1.28)$$

Therefore H is a slider through T.

Remark. An example of a function f is derived in Appendix A7.1 when the target is diffusely reflecting. Its general form is

$$f(\alpha, \delta) = \frac{h(\alpha)}{\delta^4} = \frac{v\beta^\mu}{\delta^4} \quad (7.1.29)$$

with

$$\mu = 2p + n_1 + n_2. \quad (7.1.30)$$

For this particular model of the sensor, we have

$$\frac{\partial f}{\partial \delta} = -\frac{4h(\alpha)}{\delta^5} \quad (7.1.31)$$

$$\frac{\partial f}{\partial \beta} = \frac{\mu v \beta^{\mu-1}}{\delta^4} \quad (7.1.32)$$

which, according to (7.1.26) and (7.1.27), leads to

$$u = \frac{h(\alpha)}{\delta^5} \left((4 + \mu)n - \frac{\mu}{\cos \alpha} x_s \right) \quad (7.1.33)$$

$$H(S) = \frac{\mu h(\alpha)}{\delta^4 \cos \alpha} (x_s \times n). \quad (7.1.34)$$

7.1 MODELLING OF THE INTERACTIONS

Notice that, if the target is on the axis of the sensor, we then have $n = x_s$, and (7.1.26), (7.1.27) become

$$u = \left(-\frac{\partial f}{\partial \delta} + \frac{\beta - 1}{\delta}\frac{\partial f}{\partial \beta}\right) n \qquad (7.1.35)$$

$$H(P) = 0 \qquad (7.1.36)$$

for any P belonging to the axis of the sensor. When f does not depend on the angle α, (7.1.26) and (7.1.27) give in this case

$$u = -\frac{\partial f}{\partial \delta} n \qquad (7.1.37)$$

$$H(P) = 0 \qquad (7.1.38)$$

for any P belonging to the axis of the sensor.

This last situation occurs for example when, instead of providing a signal proportional to the amount of the received flux, the sensor computes the time of flight (or measures the phase shift, in the case of optical sensors) from S to T.

Example 7.2: Case of a thin-field range finder. Let us now consider the situation represented in Fig. 7.3, where (S) is assumed to be narrow-field and n_S and n_T are unitary vectors, n_T being the normal to the surface of the target at point T. We assume that the sensor signal is a function of the distance δ between S and T:

$$s = f(\delta) \qquad (7.1.39)$$

where

$$ST = \delta n_S. \qquad (7.1.40)$$

From (7.1.39):

$$\dot{s} = \frac{\partial f}{\partial \delta} \dot{\delta} \qquad (7.1.41)$$

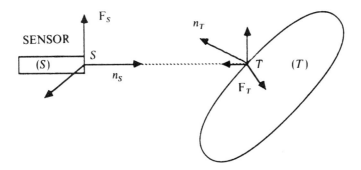

Fig. 7.3 A narrow-field sensor

and from (7.1.40):

$$\dot{ST} = \dot{\delta} n_S + \delta \dot{n}_S \qquad (7.1.42)$$

with

$$\dot{n}_S = \omega_{F_S/F_T} \times n_S \qquad (7.1.43)$$

Therefore

$$\dot{ST} = \dot{\delta} n_S + \delta(\omega_{F_S/F_T} \times n_S). \qquad (7.1.44)$$

On the other hand:

$$\dot{ST} = V_{T/F_T} - V_{S/F_T}. \qquad (7.1.45)$$

As T slides on the surface of the target, we have

$$\langle n_T, V_{T/F_T} \rangle = 0 \qquad (7.1.46)$$

Equations (7.1.45) and (7.1.46) yield

$$\langle n_T, V_{S/F_T} \rangle + \langle n_T, \dot{ST} \rangle = 0 \qquad (7.1.47)$$

and, from (7.1.44) and (7.1.47):

$$\dot{\delta} = -\frac{1}{\langle n_T, n_S \rangle} (\langle n_T, V_{S/F_T} \rangle + \delta \langle (n_S \times n_T), \omega_{F_S/F_T} \rangle). \qquad (7.1.48)$$

Finally, by using (7.1.48) in (7.1.41) we obtain the following interaction screw H associated with (S) and (T):

$$\left.\begin{array}{l} u = -\dfrac{\partial f}{\partial \delta} \dfrac{n_T}{\langle n_T, n_S \rangle} \\[1em] H(S) = -\dfrac{\partial f}{\partial \delta} \dfrac{\delta}{\langle n_T, n_S \rangle} (n_T \times n_S) \end{array}\right\} \qquad (7.1.49)$$

Again, using the property $H(T) = H(S) + TS \times u$, we find that $H(T) = 0$ and therefore that the interaction screw is a slider through T. Moreover, when the surface of the object is orthogonal to the sensor axis, we have $H(P) = 0$ for any P on the sensor axis.

Lessons from the examples. The above two examples point out some important aspects concerning sensory interaction screws. In particular, if we consider equations (7.1.26)–(7.1.27) or (7.1.49), the origin of possible uncertainties is clearly seen.

1. *Uncertainties due to the sensor itself.* As explained in Appendix A7.1, sensor models, like (7.1.29)–(7.1.30) are only approximations of reality. Furthermore, some characteristics of these models may be difficult to determine in practice. For example, the actual values of $\partial f/\partial \delta$ or $\partial f/\partial \beta$ are usually not known precisely. Then the scalar parameters which multiply n, x_s, and $x_s \times n$ in (7.1.26), (7.1.27) and n_T, $n_T \times n_S$ in (7.1.49) are also unknown.

7.1 MODELLING OF THE INTERACTIONS

2. *Uncertainties due to the environment.* In the case of Example 7.1, where the object is simply a pin-point target, the direction n in which the object is seen from the sensor is not directly provided by the sensor. Therefore, if the sensor field is wide, not only the *norms* but also the *directions* of the vectors u and $H(S)$ are not known precisely. In Example 7.2, the target is characterized by the vector n_T normal to the surface. If this vector is unknown, then the norms and directions of u and $H(S)$ are unknown again.

However, despite these uncertainties, the situation is not hopeless because control schemes can be made robust, at least to some extent, to modelling errors. The elements, in the interaction screws, which are really important for the control of the robot, will be emphasized later. Furthermore, the above examples have revealed the important fact that the interaction screw H often reduces to a slider. We will take advantage of this fact in a later section, where simplified models of the interaction screws will be proposed. Before that, we shall proceed with the analysis of the interaction screws by establishing a strong analogy with the classical kinematics of contacts between solids.

7.1.3 The analogy with the kinematics of contacts

In the remainder of this section, we shall consider only the case of non-contact sensors (for example proximity sensors or range finders). Force sensors will be studied in section 7.6.

In Definition D7.1 we defined the interaction screw associated with a sensor and an environment by writing the differential of the sensor signal s as a screw product. Let us now assume that the value of this signal is equal to a desired value, s_d, corresponding to a certain desired position $\bar{r}_d$. There may exist displacements of the sensing system with respect to the target which do not change this value. For example, in Fig. 7.4, any sliding motion of the plane target on itself leaves s unchanged.

Given the interaction screw H, the velocity screws T^* that leave s invariant are solutions of:

$$\dot{s} = H \bullet T^* = 0. \tag{7.1.50}$$

Since $\dot{s}$ is the screw product of H and T, (7.1.50) implies that T^* is a screw *reciprocal* to H in the six-dimensional vector space of screws.

Let us now consider a more general case where p sensors (S_j) belong to the *same* link L_i, with H_j denoting the related interaction screws and $\bar{s}$ the vector regrouping all sensors' outputs ($\bar{s} = (s_1, \ldots, s_p)^T$). The motions around an initial position $\bar{r}$ which leave $\bar{s}$ unchanged are characterized by the reciprocal subspace $\{T^*\}$ (cf. Definition D1.8) of the subspace spanned by the set of screws $\{H_1, \ldots, H_j, \ldots, H_p\}$.

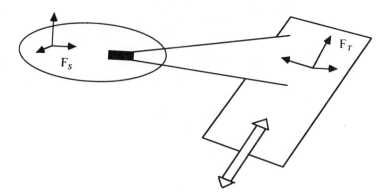

Fig. 7.4 Invariant motion

Now, as seen in Chapter 1, the formalism of screws can also be used to describe contacts between rigid bodies. This suggests the possibility of setting a parallel between true contacts and sensor-based tasks. The underlying idea is the following: imposing the condition $\dot{s}_j = 0$ on every sensor (S_j), $j = 1, \ldots, p$, is equivalent to introducing constraints in the configuration space SE_3 in order to achieve a *virtual contact* between the objects and the link carrying the sensors.

To proceed with this idea, we may state the following.

Definition D7.2 *A set of p compatible constraints:* $\bar{s}(\bar{r}) - \bar{s}_d = 0$ *determines a* **virtual linkage** *between the body L_i and the object (T). At a position $\bar{r}$ where the constraints are satisfied, the dimension of the reciprocal subspace in se_3 of the subspace* **S** *spanned by the set* $\{H_1, \ldots, H_j, \ldots, H_p\}$ *is called the* **class** *of the virtual linkage at $\bar{r}$.*

In this definition, the 'class' of the virtual linkage is the number of degrees of freedom of L_i, corresponding to the motions of L_i, which leave the outputs of all sensors unchanged. The velocities which are compatible with the virtual linkage are the elements of the screw subspace $\{T^*\}$ reciprocal to **S** (cf. Chapter 1, section 1.3.3.2). Note also that the concept of virtual linkage is introduced independently of the nature of the sensors. It thus applies to contact sensors as well. In this case, the virtual contact defined by the set $\{H_j\}$ may in fact be associated with a real contact, if some extra precautions are taken (see section 7.6).

Finally, due to the invariance property of the screw product, the reciprocal subspace $\{T^*\}$ is fully defined from a geometric point of view by the *null*

7.2 SIMPLIFIED MODELS OF INTERACTIONS

space of the ($p \times 6$) matrix operator:

$$L^T(P) = \begin{pmatrix} u_1^T & H_1^T(P) \\ \vdots & \vdots \\ u_p^T & H_p^T(P) \end{pmatrix} \qquad (7.1.51)$$

evaluated at some point P. Recall that all the vectors u_j and $H_j(p)$ must be expressed in the same basis. The matrix L, the transpose of which represents $\partial s/\partial \bar{r}$, will be named the *interaction matrix*. It will be seen later that L is a key element in the design and analysis of sensor-based task functions. According to Definition D7.2, the class of the virtual linkage is given by

$$N = 6 - \text{rank}(L) \qquad (7.1.52)$$

where $\text{rank}(L) \leq \min(6, p)$.

In this framework, each H_j can be interpreted as a force screw resulting from a (virtual) contact between the link L_i and the robot's environment, while $\dot{s}_j$ is the associated virtual power. By analogy with the case of real contacts between bodies, the two reciprocal screw spaces $\{T^*\}$ and $\mathbf{S}$ defined by (7.1.50) have the meaning of a *twist* space and a *wrench* space respectively.

7.2 SIMPLIFIED MODELS OF INTERACTIONS: A TOOL FOR THE USER

7.2.1 Introduction

In practice, the difficulty of designing sensor-based control schemes arises mainly from two problems.

1. *The uncertainties in $\{H_j\}$, which are responsible for uncertainties in the task Jacobian matrix.* The simple examples 7.1 and 7.2 have already illustrated the practical difficulty of knowing the interaction screws. More generally, the problems which appear in trying to form $\{H_j\}$ come from two sources.

    (a) *The sensors themselves.* We often have a fairly *qualitative* knowledge of their characteristics. For instance, in Example 7.1, this minimal knowledge can be summarized as follows: due to the shape of the field of an optical sensor, the unitary vector n belongs to a cone with known axis x_s and angle depending on the practical field width; and $f(\alpha, \delta)$ is a positive function decreasing with δ, and bell-shaped in α around $\alpha = 0$. This qualitative knowledge is usually complemented in practice by a calibration stage performed on dedicated targets.

(b) *The environment*. Even if an accurate model of sensory outputs with regard to known objects is available, the determination of the interaction model at each instant in a real environment remains difficult because of uncertainties concerning this environment.

When many sensors are used, it may also be difficult to determine the actual dimension of the interaction screw space. Furthermore, this dimension, and the structure itself of the virtual linkage, may change with the time, for example when the characteristics of the environment change. Finally, since the true $\{H_j\}$ is usually not known exactly, there is little chance of determining the task Jacobian matrix precisely. The first obvious consequence is that exact feedback linearization is usually not possible when using exteroceptive sensors.

2. *The diversity of possible sensors and tasks*. Given an application which needs the use of local sensors, the control designer has to take several decisions: which sensors to use; where they should be located; and how to elaborate an adequate task function.

The answers to these questions must integrate two main criteria: adequacy with regard to the application requirements and control robustness.

The previous comments stress the necessity of proposing simple methodological tools to the practitioner. Since technological aspects are of prime importance, it is certainly useful to classify and analyse the existing sensors. This is of course beyond the scope of this book, and we shall only give some references in the bibliographic note at the end of this chapter. A second aspect is to try to derive simple tools which facilitate the design of sensor-based tasks: location of sensors, choice of the task function, and models of the interaction screws. A natural method consists of proposing a set of basic 'canonical' structures corresponding to the most usual tasks, which may be combined and adapted to each specific application. The methodology proposed in the following takes advantage of the concept of virtual linkage presented above and may be considered as an extension of the classification of simple linkages between bodies based on the so-called *elementary contacts* (point on plane, line on plane, etc).

The basic idea is *to exploit simple but relevant* **models** *of interaction screws*. 'Simple' implies that the dependency on specific sensory features has to be minimal. A consequence of the aimed simplicity in these basic structures is to ease the analysis of the model of the related virtual linkage. As shown later, this in turn facilitates the constitution of a task vector, and helps to exhibit particular linear combinations of sensors, the regulation of which can be achieved in a robust manner.

7.2.2 A basic model of interaction

In the following development the carets over the variables are to remind us that the model of interaction that is derived does not necessarily represent the reality exactly. In Example 7.2, we have examined the generic case of a thin-field sensor, the signal of which depends on the distance between the sensor and the target in the direction of the axis of the sensor. Starting from this model, we will borrow the following two features.

1. The interaction screw, $\hat{H}$, is a slider through the point P which is at the intersection of the surface of the target and the axis of the sensor. Note that the existence of a point through which the interaction screw is a slider is a property which is not specific of the single case of Example 7.2; let us also recall (cf. Chapter 1) that six sliders in different points may span the whole of a screw space. This choice is thus not too restrictive.

2. The vector, $\hat{u}$, of this slider is normal to the surface of the target at the point P.

Since the corresponding interaction screw subspace $\hat{S}$ does not depend on the norm of $\hat{u}$, we may as well assume, for the sake of simplicity, that $\hat{u}$ is unitary. Finally, the 'direction' of $\hat{u}$ is determined from the assumption that the output of the sensor is a monotonically decreasing function of the distance. The third feature of our basic model is therefore:

3. $\hat{u} = -n_T$, where n_T is the unitary vector normal to the surface of the target at P.

This basic model of interaction is represented in Fig. 7.5. The corresponding interaction screw $\hat{H}$ is:

$$\left.\begin{array}{l}\hat{u} = -n_T \\ \hat{H}(P) = 0\end{array}\right\}. \tag{7.2.1}$$

In order to calculate an interaction matrix $\hat{L}$ associated with this screw, we must choose a point Q where the screw is evaluated and a frame F in the basis of which its elements are expressed.

Thus let Q denote a point linked to the body on which the sensor is fixed and F denote a frame linked to this body. The interaction matrix associated

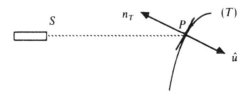

Fig. 7.5 A basic model of interaction

with the interaction screw $\hat{H}$ is then

$$\hat{L}(Q) = \begin{pmatrix} \hat{L}_1 \\ \hat{L}_2(Q) \end{pmatrix}\bigg|_F \qquad (7.2.2)$$

with

$$\hat{L}_1 = \hat{u}|_F \qquad (7.2.3)$$

$$\hat{L}_2(Q) = \Delta\hat{u}|_F \qquad (7.2.4)$$

$$\Delta = AS([PQ])|_F \qquad (7.2.5)$$

We will see in the following, and also through the examples of section 7.5, that this model is not so restrictive as it may seem at first glance.

7.2.3 Standard sensor-based primitives

7.2.3.1 Introduction

Having determined a basic model of interaction for a single sensor, we may now combine several of these elementary models so as to form more elaborate structures, named sensor-based primitives, where several sensors are involved and different (simple) models of targets are considered. The design of each primitive is guided by the objective of simply achieving a certain virtual linkage characterized by a given interaction screw subspace $\hat{S}$. In this way, it is possible to constitute a 'directory' of standard primitives representative of commonly encountered sensor-based tasks. The way of using this catalogue in practice will thus normally consist of selecting a primitive which seems to fit the application in relation to the virtual linkage that the user wishes to achieve. The utilization of primitives will be further considered later.

With $N = 6 - m$, where m is the dimension of $\hat{S}$ (equal to the rank of the interaction matrix $\hat{L}$), being the class of the related virtual linkage, we will say that the primitive is of class CN. If $p = m$, where p is the number of involved sensors, we will say that the primitive is *canonical*. If $p > m$, we will say that the primitive is *redundant* because the same virtual contact could be obtained with fewer sensors. Obviously, since $m \leq 6$, the primitive is necessarily redundant as soon as more than six sensors are used. The reasons for using redundant primitives will be explained later.

The primitives listed below are ordered by decreasing number of the class N. For every primitive we indicate:

(1) the usual name of the related contact;

(2) the corresponding interaction matrix $\hat{L} = \begin{pmatrix} \hat{u}_1 & \cdots & \hat{u}_p \\ \Delta_1 \hat{u}_1 & \cdots & \Delta_p \hat{u}_p \end{pmatrix}$, with $\Delta_j = AS([P_j O])$, evaluated at the origin O of some frame rigidly linked to the sensing system and expressed in the basis of this frame;

7.2 SIMPLIFIED MODELS OF INTERACTIONS

(3) a $(6 \times m)$ full rank matrix $\hat{W}^T$, the columns of which form an orthonormal basis (in $\mathbf{R}^6$) of the interaction screw space $\hat{\mathbf{S}}$.

By inspection of this matrix $\hat{W}^T$, it is easy to determine the motions which are 'constrained' by the virtual linkage and those which remain 'free'. The role of this matrix in the practical determination of sensor-based task functions will be explained in the following sections.

7.2.3.2 Some usual canonical primitives

Primitive P1 (Fig. 7.6): the basic model, for a particular position of the sensor.
  Class $C5$; 1 sensor
  (Point/plane contact without friction.)
  $P_1 = 0$

$$\hat{L}_1 = \begin{pmatrix} 0 \\ 0 \\ 1 \end{pmatrix}; \quad \hat{L}_2 = \begin{pmatrix} 0 \\ 0 \\ 0 \end{pmatrix}$$

$\hat{W} = (w_1^T) = (0\ 0\ 1\ 0\ 0\ 0)$.
All rotational motions around the axis Ox, Oy, and Oz are free. All translational motions are free except along the axis Oz.

Primitive P2.1 (Fig. 7.7)
  Class $C4$; 2 sensors
  (Line/plane contact without friction.)

$$P_1 = \begin{pmatrix} 0 \\ 1 \\ 0 \end{pmatrix}; \quad P_2 = \begin{pmatrix} 0 \\ -1 \\ 0 \end{pmatrix}$$

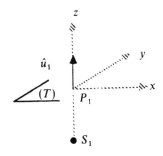

Fig. 7.6 Point/plane contact without friction

282   7 SENSOR-BASED TASKS

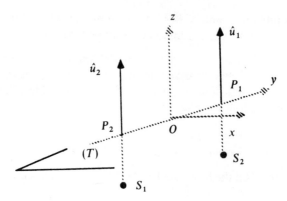

Fig. 7.7 Line/plane contact without friction

$$\hat{L}_1 = \begin{pmatrix} 0 & 0 \\ 0 & 0 \\ 1 & 1 \end{pmatrix}; \quad \hat{L}_2 = \begin{pmatrix} -1 & 1 \\ 0 & 0 \\ 0 & 0 \end{pmatrix}$$

$$\hat{W} = \begin{pmatrix} w_1^T \\ w_2^T \end{pmatrix} = \begin{pmatrix} 0 & 0 & 1 & 0 & 0 & 0 \\ 0 & 0 & 0 & 1 & 0 & 0 \end{pmatrix}.$$

Translational motions along the axis Ox and Oy and rotational motions around the axis Oy and Oz are free.

Primitive P2.2 (Fig. 7.8)
  Class $C4$; 2 sensors
  (Line/sphere contact without friction.)

$$S_1 = \begin{pmatrix} 0 \\ -a \\ \dfrac{\sqrt{2}}{2} \end{pmatrix}; \quad S_2 = \begin{pmatrix} 0 \\ -a \\ -\dfrac{\sqrt{2}}{2} \end{pmatrix}; \quad a > 1$$

$$P_1 = \begin{pmatrix} 0 \\ -\dfrac{\sqrt{2}}{2} \\ \dfrac{\sqrt{2}}{2} \end{pmatrix}; \quad P_2 = \begin{pmatrix} 0 \\ -\dfrac{\sqrt{2}}{2} \\ -\dfrac{\sqrt{2}}{2} \end{pmatrix}$$

7.2 SIMPLIFIED MODELS OF INTERACTIONS

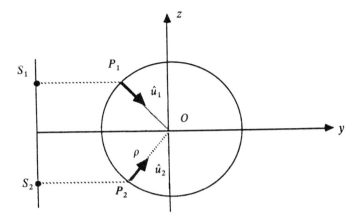

Fig. 7.8 Line/sphere contact without friction

$$\hat{L}_1 = \begin{pmatrix} 0 & 0 \\ \frac{\sqrt{2}}{2} & \frac{\sqrt{2}}{2} \\ -\frac{\sqrt{2}}{2} & \frac{\sqrt{2}}{2} \end{pmatrix}; \quad \hat{L}_2 = \begin{pmatrix} 0 & 0 \\ 0 & 0 \\ 0 & 0 \end{pmatrix}$$

$$\hat{W} = \begin{pmatrix} w_1^T \\ w_2^T \end{pmatrix} = \begin{pmatrix} 0 & 1 & 0 & 0 & 0 & 0 \\ 0 & 0 & 1 & 0 & 0 & 0 \end{pmatrix}$$

Translational motion along the axis Ox and all rotational motions are free.

Primitive P3 (Fig. 7.9)
  Class $C3$; 3 sensors
  (Plane/plane contact without friction.)

$$P_1 = \begin{pmatrix} 1 \\ 0 \\ 0 \end{pmatrix}; \quad P_2 = \begin{pmatrix} -\frac{1}{2} \\ \frac{\sqrt{3}}{2} \\ 0 \end{pmatrix}; \quad P_3 = \begin{pmatrix} -\frac{1}{2} \\ -\frac{\sqrt{3}}{2} \\ 0 \end{pmatrix}$$

$$\hat{L}_1 = \begin{pmatrix} 0 & 0 & 0 \\ 0 & 0 & 0 \\ 1 & 1 & 1 \end{pmatrix}; \quad \hat{L}_2 = \begin{pmatrix} 0 & \frac{\sqrt{3}}{2} & -\frac{\sqrt{3}}{2} \\ -1 & \frac{1}{2} & \frac{1}{2} \\ 0 & 0 & 0 \end{pmatrix}$$

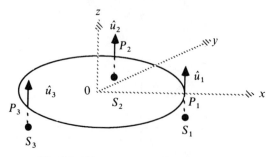

Fig. 7.9 Plane/plane contact without friction

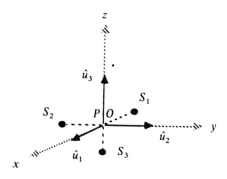

Fig. 7.10 Point/plane contact with friction

$$\hat{W} = \begin{pmatrix} w_1^T \\ w_2^T \\ w_3^T \end{pmatrix} = \begin{pmatrix} 0 & 0 & 1 & 0 & 0 & 0 \\ 0 & 0 & 0 & 1 & 0 & 0 \\ 0 & 0 & 0 & 0 & 1 & 0 \end{pmatrix}.$$

Translational motions along the axis Ox and Oy, and rotational motions around the axis Oz are free.

Primitive P4 (Fig. 7.10)
  Class $C3$; 3 sensors
  (Point/plane contact with friction.)
  $P_j = O; \quad j = 1, 2, 3$

$$\hat{L}_1 = \begin{pmatrix} 1 & 0 & 0 \\ 0 & 1 & 0 \\ 0 & 0 & 1 \end{pmatrix}; \quad \hat{L}_2 = \begin{pmatrix} 0 & 0 & 0 \\ 0 & 0 & 0 \\ 0 & 0 & 0 \end{pmatrix}$$

$$\hat{W} = \begin{pmatrix} w_1^T \\ w_2^T \\ w_3^T \end{pmatrix} = \begin{pmatrix} 1 & 0 & 0 & 0 & 0 & 0 \\ 0 & 1 & 0 & 0 & 0 & 0 \\ 0 & 0 & 1 & 0 & 0 & 0 \end{pmatrix}.$$

All rotational motions are free.

7.2 SIMPLIFIED MODELS OF INTERACTIONS

Primitive P5.1 (Fig. 7.11)
  Class $C2$; 4 sensors ('soft finger')

$$P_1 = -P_2 = \begin{pmatrix} 0 \\ -1 \\ 0 \end{pmatrix}; \quad P_3 = P_4 = 0$$

$$\hat{L}_1 = \begin{pmatrix} 1 & 1 & 0 & 0 \\ 0 & 0 & 1 & 0 \\ 0 & 0 & 0 & 1 \end{pmatrix}; \quad \hat{L}_2 = \begin{pmatrix} 0 & 0 & 0 & 0 \\ 0 & 0 & 0 & 0 \\ 1 & -1 & 0 & 0 \end{pmatrix}$$

$$\hat{W} = \begin{pmatrix} w_1^T \\ w_2^T \\ w_3^T \\ w_4^T \end{pmatrix} = \begin{pmatrix} 1 & 0 & 0 & 0 & 0 & 0 \\ 0 & 1 & 0 & 0 & 0 & 0 \\ 0 & 0 & 1 & 0 & 0 & 0 \\ 0 & 0 & 0 & 0 & 0 & 1 \end{pmatrix}.$$

Rotational motions around the axis Ox and Oy are free.

Primitive P5.2 (Fig. 7.12)
  Class $C2$; 4 sensors

$$S_1 = \begin{pmatrix} -a \\ -\frac{\sqrt{2}}{2} \\ 1 \end{pmatrix}; \quad S_2 = \begin{pmatrix} -a \\ -\frac{\sqrt{2}}{2} \\ -1 \end{pmatrix}; \quad S_3 = \begin{pmatrix} -a \\ -\frac{\sqrt{2}}{2} \\ 1 \end{pmatrix};$$

$$S_4 = \begin{pmatrix} -a \\ \frac{\sqrt{2}}{2} \\ -1 \end{pmatrix}$$

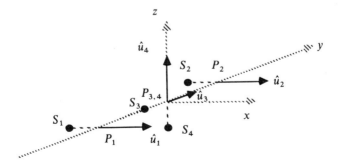

Fig. 7.11 'Soft finger'

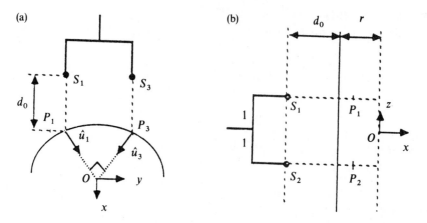

Fig. 7.12 Association of primitives for positioning with respect to a cylinder (a) plan view (b) side view

$$P_1 = \begin{pmatrix} -\frac{\sqrt{2}}{2} \\ -\frac{\sqrt{2}}{2} \\ 1 \end{pmatrix}; \quad P_2 = \begin{pmatrix} -\frac{\sqrt{2}}{2} \\ -\frac{\sqrt{2}}{2} \\ -1 \end{pmatrix}; \quad P_3 = \begin{pmatrix} -\frac{\sqrt{2}}{2} \\ \frac{\sqrt{2}}{2} \\ 1 \end{pmatrix};$$

$$P_4 = \begin{pmatrix} -\frac{\sqrt{2}}{2} \\ \frac{\sqrt{2}}{2} \\ -1 \end{pmatrix}$$

$a = d_0 + r$

$$\hat{L}_1 = \frac{\sqrt{2}}{2}\begin{pmatrix} 1 & 1 & 1 & 1 \\ 1 & 1 & -1 & -1 \\ 0 & 0 & 0 & 0 \end{pmatrix}; \quad \hat{L}_2 = \frac{\sqrt{2}}{2}\begin{pmatrix} -1 & 1 & 1 & -1 \\ 1 & -1 & 1 & -1 \\ 0 & 0 & 0 & 0 \end{pmatrix}$$

$$\hat{W} = \begin{pmatrix} w_1^T \\ w_2^T \\ w_3^T \\ w_4^T \end{pmatrix} = \begin{pmatrix} 1 & 0 & 0 & 0 & 0 & 0 \\ 0 & 1 & 0 & 0 & 0 & 0 \\ 0 & 0 & 0 & 1 & 0 & 0 \\ 0 & 0 & 0 & 0 & 1 & 0 \end{pmatrix}.$$

7.2 SIMPLIFIED MODELS OF INTERACTIONS

7.2.3.3 Redundant primitives

One of the main advantages in using redundant primitives is that, by increasing the number of sensors, it is possible to augment the size of the sensed local environment and introduce averaging (or filtering) effects by regrouping several sensors and adding their outputs.

Some examples of redundant primitives, derived from canonical primitives, are now given.

Primitive P6 (Fig. 7.13)
   Class $C5$; 2 sensors
   (Bilateral point contact without friction.)
   $P_1 = P_2 = O$

$$\hat{L}_1 = \begin{pmatrix} 1 & -1 \\ 0 & 0 \\ 0 & 0 \end{pmatrix}; \quad \hat{L}_2 = \begin{pmatrix} 0 & 0 \\ 0 & 0 \\ 0 & 0 \end{pmatrix}$$

$$\hat{W} = (w_1^T) = (0\ 0\ 1\ 0\ 0\ 0).$$

Only translational motions along the axis Ox are constrained.

Primitive P7 (Fig. 7.14)
   Class $C4$; p sensors, $p > 2$
   (Line/plane contact without friction.)

$$P_j = \begin{pmatrix} 0 \\ \alpha_j \\ 0 \end{pmatrix}; \quad j = 1, \ldots, p; \quad \alpha_j \neq \alpha_k \text{ for } j \neq k$$

$$\hat{L}_1 = \begin{pmatrix} 0 & \ldots & 0 \\ 0 & \ldots & 0 \\ 1 & \ldots & 1 \end{pmatrix}; \quad \hat{L}_2 = \begin{pmatrix} \alpha_1 & \ldots & \alpha_p \\ 0 & \ldots & 0 \\ 0 & \ldots & 0 \end{pmatrix}$$

$$\hat{W} = \begin{pmatrix} w_1^T \\ w_2^T \end{pmatrix} = \begin{pmatrix} 0 & 0 & 1 & 0 & 0 & 0 \\ 0 & 0 & 0 & 1 & 0 & 0 \end{pmatrix}.$$

Translational motions along the axis Ox and Oy and rotational motions around the axis Oy and Oz are free.

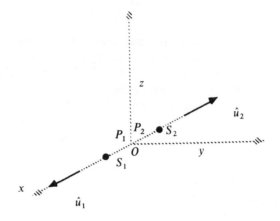

Fig. 7.13 Bilateral point contact without friction

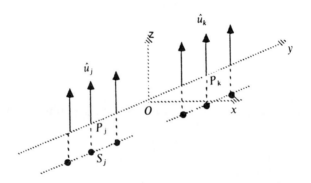

Fig. 7.14 Line/plane contact without friction

Primitive P8 (Fig. 7.15)
   Class $C4$; p sensors, $p > 2$
   (Ball joint on a slipper guide)

$$P_j = \begin{pmatrix} 0 \\ \rho\, c\theta_j \\ \rho\, s\theta_j \end{pmatrix}; \quad j = 1, \ldots, p; \quad \theta_j \neq \theta_k \text{ for } j \neq k$$

$$\hat{L}_1 = \begin{pmatrix} 0 & \ldots & 0 \\ c\theta_1 & \ldots & c\theta_p \\ s\theta_1 & \ldots & s\theta_p \end{pmatrix}; \quad \hat{L}_2 = \begin{pmatrix} 0 & \ldots & 0 \\ 0 & \ldots & 0 \\ 0 & \ldots & 0 \end{pmatrix}$$

$$\hat{W} = \begin{pmatrix} w_1^T \\ w_2^T \end{pmatrix} = \begin{pmatrix} 0 & 1 & 0 & 0 & 0 & 0 \\ 0 & 0 & 1 & 0 & 0 & 0 \end{pmatrix}.$$

Translational motions along Ox and all rotational motions are free.

7.2 SIMPLIFIED MODELS OF INTERACTIONS

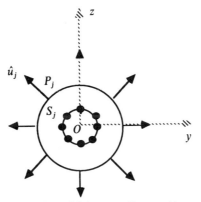

Fig. 7.15 Ball joint on a slipper guide

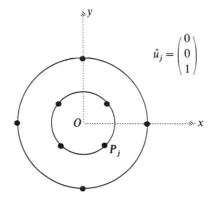

Fig. 7.16 Plane/plane contact without friction

Primitive P9 (Fig. 7.16)
  Class $C3$; p sensors, $p > 3$
  (Plane/plane contact without friction.)

$$P_j = \begin{pmatrix} -\beta_j \\ \alpha_j \\ 0 \end{pmatrix}; \quad j = 1, \ldots, p$$

$$\hat{L}_1 = \begin{pmatrix} 0 & \cdots & 0 \\ 0 & \cdots & 0 \\ 1 & \cdots & 1 \end{pmatrix}; \quad \hat{L}_2 = \begin{pmatrix} \alpha_1 & \cdots & \alpha_p \\ \beta_1 & \cdots & \beta_p \\ 0 & \cdots & 0 \end{pmatrix}$$

$$\hat{W} = \begin{pmatrix} w_1^T \\ w_2^T \\ w_3^T \end{pmatrix} = \begin{pmatrix} 0 & 0 & 1 & 0 & 0 & 0 \\ 0 & 0 & 0 & 1 & 0 & 0 \\ 0 & 0 & 0 & 0 & 1 & 0 \end{pmatrix}.$$

Translational motions along the axis Ox and Oy, and rotational motions around the axis Oz are free.

7.2.4 Constitution of a sensor-based task vector

7.2.4.1 Determination of the sensing system

Prior to the synthesis of a sensor-based task vector, it is necessary to determine a sensing system which is adapted to the application that the user has in mind. This stage involves the choice of the sensors, their number, their location (not necessarily on the robot) and their direction of sensing. Although crucial, this stage hardly lends itself to a deep and self-contained analysis. The reason is that the solutions, when they exist, are never unique, are limited to the available or technically feasible sensors, and critically depend on experimental conditions which may vary greatly, even within the same class of applications (for example, the shape, surface characteristics and motion of the sensed object). As a consequence, the practical training and experience of the designer is particularly important at this level.

However, in many cases, simple and idealized models of sensors/environment interactions, such as the standard primitives which were previously introduced, can effectively be used to determine the number of required sensors and their relative positioning on a robot link. Indeed, the user often has a prior idea of the directions along which he would like to control the motion of the (let us say) robot's end-effector from sensor-based information. In other words, he has an idea of the kind of virtual contact that he would like to achieve during the application. A possibility then consists in finding a corresponding primitive which can be associated with a virtual contact of the same type. When this is done, the number of sensors, their relative positions on the end-effector and their directions of sensing are automatically given by the geometry of the primitive. It should however be emphasized that the use of a primitive is theoretically justified when the application requires the invariance of the interaction screw subspace with respect to the sensing system.

Once the basic features of a sensing system have been determined from the selection of a primitive (or a set of primitives), the designer has the possibility of utilizing the knowledge of the geometry of the target and the characteristics of the sensors in order to derive a more refined model of the physical interactions between the sensing system and the target. This more accurate (and usually more complex) model will also consist of a set of interaction screws gathered into an interaction matrix $L(P)$. A first application of this model will be to verify that it is compatible with the structure of the primitive which has been used. This notion of compatibility between the physical

7.2 SIMPLIFIED MODELS OF INTERACTIONS

system and a primitive will be made explicit a little later. It will also be possible to use this model in the subsequent stages of control design and analysis, as explained later in this chapter.

From the user's point of view, primitives are therefore a methodological tool which allows him to quickly select a sensory structure corresponding to his application, provided that his objective is expressed in terms of a virtual linkage to be achieved. It is however useful to be aware of two important aspects concerning the use of primitives. A first point is that some primitives are not self-contained in the sense that the satisfaction of the underlying constraints do not ensure the invariance of the interaction screw subspace with respect to the sensing system. For example, the linkages associated with primitives P1 or P7 do not bind the axis of the sensor to stay orthogonal to the surface of the target. In this case additional constraints must be specified in order to enforce the invariance property of the interaction screw subspace. A second aspect is that it is not possible to achieve any virtual linkage through the use of exteroceptive sensors. The reason is that, for a given target, a given type of sensor and some positions of the sensing system with respect to the target, there may exist motions which do not change the outputs of the sensors. In other words, there may exist a lower bound to the class of the virtual linkages that can be achieved. For example, with sensor models of the form (7.1.29) or (7.1.39), the minimal class is $C1$ when the target is a dihedron. It is $C0$ for a trihedron, $C2$ for the outer or the inner of an infinite cylinder, etc.

The two aspects that we have just described show that the choice of a primitive must be preceded by a careful analysis of the application. Once the linkage is defined, the task of implementing it falls on the control. But, prior to the control design, a relevant task function has to be determined. Primitives may also be used at this stage by providing us with an intuitive mechanical approach to the problem of sensor-based task design.

7.2.4.2 Design of the sensor-based task: a simplified mechanical point of view

Let us conceptually assume that the selected primitive and its corresponding model of interaction represent the physical reality exactly (i.e. $\hat{\mathbf{S}} = \mathbf{S}$ and $\partial \bar{s}/\partial \bar{r} = \hat{H}$). By analogy with physical contacts, and anticipating section 7.6, we may imagine that the virtual linkage to be achieved is elastic in the sense that the resultant F of the virtual forces of contact applied to the sensing system derives from a potential function $V(\bar{r})$. We may choose for example

$$V(\bar{r}) = \tfrac{1}{2} \|\bar{s}(\bar{r})\|^2 \qquad (7.2.6)$$

where $\bar{r}$ represents as usually the relative position of the sensing system with respect to the target. Then, the force (wrench) F is given by

$$F = -dV = -\sum_{j=1}^{p} F_j \tag{7.2.7}$$

with

$$F_j = \frac{\partial s_j}{\partial \bar{r}} s_j = \hat{H}_j s_j \tag{7.2.8}$$

where $\hat{H}_j$ is the interaction screw associated with the sensor (S_j).

Since $\hat{H}_j$ is, by assumption, a unitary slider through the point P_j at the intersection of the axis of the sensor and the surface of the target, the relation (7.2.8) shows that the output s_j may, in this case, be interpreted as the measure of the intensity of the elementary force of contact F_j at the point P_j. Furthermore, the vector $\hat{u}_j$ of the screw $\hat{H}_j$ being orthogonal to the surface of the target, we see that the force of contact F_j is also orthogonal to the surface of the target at the point P_j. In order to complete the analogy with true forces of contact, we may imagine, for example, that this force F_j results from the contact between a spring and the target, this spring being on the axis of the sensor, with one extremity tied to the sensor (S_j) and the other touching the target at the point P_j (see Fig. 7.17).

Following this interpretation, the wrench F associated with the forces $\{F_j\}, j = 1, \ldots, p$ represents the *global* force (wrench) resulting from the virtual contact between the sensing system and the target. Therefore, this screw, expressed in the frame F with origin O is:

$$F = -\sum_{j=1}^{p} \begin{pmatrix} s_j \hat{u}_j \\ s_j \Delta_j \hat{u}_j \end{pmatrix} = -\hat{L}\bar{s}\bigg|_F \tag{7.2.9}$$

with $\bar{s}^T = (s_1, \ldots, s_p)$ and $\Delta_j = AS(P_j O)$.

With this interpretation, the constraint $\{\bar{s} - \bar{s}_d = 0\}$ is equivalent to the mechanical constraint of having the force F equal to the desired force $F_d = -\hat{L}\bar{s}_d$. In order to characterize this constraint with a minimal set of independent equations we may express both forces F and F_d in a basis of the screw subspace $\hat{S}$ to which they belong. Thus, let $\hat{W}^T$ denote a $(6 \times m)$ matrix, the columns of which form an orthonormal basis of $\hat{S}$, i.e. a matrix such that $R(\hat{W}^T) = R(\hat{L})$ and $\hat{W}\hat{W}^T = I_m$. Let f and f_d denote the coordinates of F and F_d in this basis:

$$F = \hat{W}^T f|_F \tag{7.2.10}$$

$$F_d = \hat{W}^T f_d|_F. \tag{7.2.11}$$

The constraint

$$F - F_d = 0|_F \tag{7.2.12}$$

is equivalent to the m equations

$$f - f_d = 0|_F \tag{7.2.13}$$

7.3 IMMERSION OF SENSOR-BASED TASKS

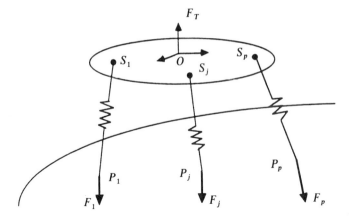

Fig. 7.17 Mechanical interpretation of virtual contact forces

which, according to (7.2.10) (7.2.11), can also be written

$$\hat{W}(F - F_d) = 0|_F \qquad (7.2.14)$$

A sensor-based task vector which characterizes the task to be achieved is therefore

$$e_1 = \hat{W}(F - F_d) = \hat{W}\hat{L}(\bar{s} - \bar{s}_d)|_F \qquad (7.2.15)$$

Remarks. This mechanical interpretation based on the analogy with the kinematics of contacts shows how a set of exteroceptive sensors may conceptually be considered as a (virtual) force contact sensor, and how the related sensor-based task may be expressed as a force control problem.

By differentiating (7.2.9), and still assuming that the model of interaction $\hat{L}$ represents the reality exactly, we obtain

$$\dot{F} = \hat{L}\hat{L}^T \begin{pmatrix} V \\ \omega \end{pmatrix} \bigg|_F. \qquad (7.2.16)$$

By analogy with the case of elastic linkages (which will be studied in section 7.6), the relation (7.2.16) shows that the (6 × 6) matrix $\hat{L}\hat{L}^T$ may also be interpreted as a *stiffness matrix*. This is the stiffness matrix of the 'virtual generalized spring' composed of the sensing system and the target.

7.3 IMMERSION OF SENSOR-BASED TASKS IN THE TASK REDUNDANCY FRAMEWORK

7.3.1 Introduction

As already evoked, in many applications the regulation of sensor outputs has to be combined with other objectives such as, for example, trajectory tracking. Such tasks are often referred to as *hybrid tasks*. Usually, a partial specification of the global task leads to the derivation of a sensor-based task $e_1(q,t)$ made of $m \leq n$ independent components, the regulation of which summarizes the part of the task in which exteroceptive sensors are involved. The methodology based on the use of sensor-based primitives, which was presented above, may be used to define both the sensing system and a vector e_1. This vector may also be obtained by other means, when, for example, no primitive can be associated with the problem. The specification of a second objective (for example the displacement of the end-effector along a given direction) would in the same way lead to a second task vector $e_2(q,t)$. The main concern at this point is to make sure that the two objectives are compatible so that the two partial task vectors e_1 and e_2 can be regrouped into a single global admissible task-function $e(q,t)$. In terms of virtual linkages, this means for example that the second objective should be achievable only by taking advantage of the free motions of the virtual linkage associated with the sensor-based task. More generally, it appears that, whenever the regulation of sensor signals constitutes the main task which is to be complemented by a secondary objective, we are in fact dealing with a problem of redundancy which can be treated within the general approach developed in Chapter 4. Before going on with this treatment, we recall and summarize in the next section some results of Chapter 4 which will be applied later to the specific case of sensor-based tasks.

7.3.2 Recall of some results about redundant tasks

Let

(1) e_1 denote the primary task vector, with dim $(e_1) = m(\leq n)$; for the sake of simplicity, we will assume that the task $\{e_1(q,t); q_0\}$ is a canonical redundant task in the sense of Definition D4.6 of Chapter 4;

(2) h_s be the secondary cost function to be minimized (or maximized) under the constraint $e_1 = 0$;

(3) x be a regular system of robot coordinates; usually, $x = q$, or, by a breach of notation, $x = \bar{r}$;

(4) $J_1^x = \partial e_1 / \partial x$ denote the $(m \times n)$ full rank Jacobian matrix associated with e_1;

(5) $g_s^x = \partial h_s / \partial x$ denote the n-dimensional gradient vector associated with the secondary cost;

(6) α be a strictly positive (or negative when h_s is to be maximized) scalar function given by the user; usually, α is chosen constant.

It is assumed that the values taken by e_1 and g_s^x along any trajectory of the robot are either directly measured (through the use of adequate sensors) or

7.3 IMMERSION OF SENSOR-BASED TASKS

computed from the measurement of the joint coordinates $q(t)$. In order to be able to minimize h_s under the constraint $e_1 = 0$ one must be able to determine the subspace of motions which are left free by this constraint. This means that one has to know the null space, $N(J_1^x)$, of J_1^x, (or equivalently the range space, $R(J_1^{xT})$, of (J_1^{xT})), along the ideal trajectory of the robot. In practice, this knowledge is equivalent to the knowledge of a matrix function W with the property

P_1:
$$R(W^T) = R(J_1^{xT}) \quad (\text{or } N(W) = N(J_1^x))$$

along the ideal trajectory of the robot.

Notice that, J_1^x being by assumption of a full rank matrix, the property P_1 is also equivalent to the following matrix characterizations:

$P_1^{(1)}$: there exists a regular matrix C such that $W = C J_1^x$;

$P_1^{(2)}$:
$$\left.\begin{array}{l} \text{rank}(W) = m \\ (I - W^+ W) J_1^{xT} = 0 \end{array}\right\};$$

$P_1^{(3)}$:
$$\left.\begin{array}{l} \text{rank}(W) = m \\ (I - J_1^{x+} J_1^x) W^T = 0 \end{array}\right\}.$$

Once such a matrix is known, a possible global task function that can be associated with the problem is

$$e = W^+ e_1 + \alpha(I - W^+ W) g_s^x. \tag{7.3.1}$$

As explained in Chapter 4, other task functions could also be considered. For the sake of simplicity, and because there is no clear advantage of using other task functions in practice, we will restrict our analysis to the class (7.3.1).

If, in addition to the property P_1, the chosen matrix W also satisfies the property

P_2: $J_1^x W^T > 0$ *along the ideal trajectory of the robot,*

then, under 'normal circumstances' (see Lemma 4.4 of Chapter 4 for more precision), the global task Jacobian matrix $J_T^x = \partial e/\partial x$ is such that, along the ideal trajectory $q_r(t)$

$$J_T^x(I_n + \gamma(I_n - W^+ W)) > 0 \tag{7.3.2}$$

for any positive scalar function γ equal to, or larger than some positive function γ_m which depends on α. If α is small enough, then $\gamma_m = 0$ and J_T^x is positive.

As explained in Chapter 6, the reason for trying to satisfy the positivity property (7.3.2) is that, in that case, it is no longer necessary to know the Jacobian matrix J_T^x. Indeed, the important control stability condition $J_T J_q^{-1} > 0$ (with $J_T = J_T^x(\partial x/\partial q)$) is then satisfied by simply choosing

$$J_q^{-1} = \left(\frac{\partial x}{\partial q}\right)^{-1}(I_n + \gamma(I_n - W^+ W)). \qquad (7.3.3)$$

For small values of α, one can also take $J_q = \partial x/\partial q$. Therefore property P_2 is particularly important when the exact computation of J_T is not possible (the case of most sensor-based tasks) or when it is desired to simplify the control expression without creating instability.

A particular case, evoked in Chapter 4, is when the subspace $N(J_1^x(q,t))$ is invariant, at any time t, on the set $\{q^{**}\}_{|t}$ of solutions of $e_1(q,t) = 0$. In the case of sensor-based tasks, this situation occurs when the subspace S spanned by the interaction screws (this subspace being evaluated in a frame linked to the sensing system) is invariant on the set of positions of the sensing system satisfying the constraint of virtual linkage $e_1 = 0$. The importance of this case results from the fact that the subspace $N(J_1^x)$, along $q_r(t)$, then does not depend on the choice of the secondary objective. As a consequence, the knowledge of the subspace $S^*(t)$ spanned by the interaction screws when the constraint of virtual linkage is satisfied is, in this case, sufficient to determine a matrix $W(t)$ satisfying the property P_1. Furthermore, if this subspace is also time-invariant, then it is possible to choose W as a *constant* matrix. Some examples illustrating this case will be given later on.

Let us comment further on the practical issues related to the satisfaction of the properties P_1 and P_2. Although the property P_1 does not imply the exact knowledge of J_1^x, it is rarely possible to *fully* comply with it in practice. This simply results from the fact that W is usually obtained from a model of the system that we wish to control and that such a model, however good it is, never represents the physical system exactly. One can easily imagine that the case of sensor-based tasks is particularly sensitive to this kind of problem because of the necessity of modelling, besides the robot's kinematics, the responses of the sensors and the characteristics of the environment in order to obtain a reasonable model of their interaction. This problem has already been considered. It is therefore important to understand, at least qualitatively, the consequences of transgressing the property P_1 when choosing the matrix W and using the task function (7.3.1).

To do so, it is useful to mathematically characterize the 'amount' of mismatch between the two subspaces $R(J_1^{x^T})$ and $R(W^T)$ so as to be able to distinguish the case when the property P_1 is 'slightly' transgressed from the cases when it is 'much' transgressed. A possible characterization is the notion of the angle $\Theta(\mathbf{A}, \mathbf{B})$ between two m-dimensional vector subspaces $\mathbf{A}$ and $\mathbf{B}$,

7.3 IMMERSION OF SENSOR-BASED TASKS

which we define as follows:

$$\Theta(\mathbf{A},\mathbf{B}) = \arccos\left(\inf_{\substack{x \in \mathbf{A} \\ \{\|x\|=1\}}} \|P_{\mathbf{B}}x\|\right); \quad \Theta(\mathbf{A},\mathbf{B}) \in \left[0, \frac{\pi}{2}\right] \quad (7.3.4)$$

where $P_{\mathbf{B}}$ denotes the operator of orthogonal projection on $\mathbf{B}$.

From this definition, one can readily deduce the following propositions:

$$\Theta(\mathbf{A},\mathbf{B}) = \Theta(\mathbf{B},\mathbf{A}) \qquad (7.3.5)$$

$$\Theta(\mathbf{A},\mathbf{B}) = 0 \Leftrightarrow \mathbf{A} = \mathbf{B} \qquad (7.3.6)$$

$$\Theta(\mathbf{A},\mathbf{B}) = \frac{\pi}{2} \Leftrightarrow \dim(P_{\mathbf{B}}\mathbf{A}) < m \qquad (7.3.7)$$

and the property P_1 can also be written

$P_1^{(4)}$:
$$\left.\begin{array}{r}\text{rank}(W) = m \\ \Theta(R(J_1^{x^T}), R(W^T)) = 0\end{array}\right\}.$$

We will say that this property is 'slightly' transgressed when the angle $\Theta(R(J_1^{x^T}), R(W^T))$ remains 'small' along $q_r(t)$. It is now possible to distinguish and roughly examine three cases, according to the size of this angle.

Case 1: $\Theta(R(J_1^{x^T}), R(W^T)) = 0$ along $q_r(t)$. This is the ideal case. Then, assuming that the secondary cost function h_s belongs to the set of functions for which the condition 4 of Lemma 4.3 is fulfilled, the task associated with the task function (7.3.1) is admissible and the regulation of e to zero ensures the constrained minimization of h_s.

The choice of W can be further restricted to the set on which the property P_2 is satisfied, so as to confer positivity properties to J_T^x.

Case 2: $0 < \Theta(R(J_1^{x^T}), R(W^T)) < \pi/2$ along $q_r(t)$. This is the usual case encountered in practice. By arguments of continuity, one may still expect the task to be admissible when $\Theta(\mathbf{A},\mathbf{B})$ remains small. The matrix W being of rank m, the regulation of the global task vector (7.3.1) to zero still implies that the primary task vector e_1 is regulated to zero. However, since the property P_1 is no longer satisfied, the constraint $e = 0$ no longer yields the minimization of h_s systematically. In general, one can only expect h_s to be *almost* minimized when the angle Θ remains small. An exception to this rule is when the minimization of h_s is not affected by the constraint $e_1 = 0$, which corresponds to the case where the equations $e_1 = 0$ and $g_s^x = 0$ are compatible. Then the regulation of e to zero may still yield the unconstrained minimization of h_s, even if the angle Θ is not small. However, the admissibility of the task will be lost if, at some point along $q_r(t)$, the Hessian $\partial^2 h_s/\partial x^2$ is no longer positive on $N(W)$.

Notice that the property P_2 may still be satisfied while P_1 is not, and that the satisfaction of P_2 will usually continue to confer positivity properties to the task Jacobian J_T^x when Θ remains small.

Case 3: $\Theta(R(J_1^{x^T}), R(W^T)) = \pi/2$ at some point on $q_r(t)$. In this case, $J_1^x W^T$ is no longer a full rank matrix and J_T^x is singular. The task is not admissible and, clearly, this case must be avoided by any means.

To summarize this paragraph, some results of Chapter 4 about task redundancy led us to propose a set of task-functions $e(q, t)$ in the form (7.3.1) which can be utilized whenever the regulation of sensor outputs constitutes the main objective. Ideally, the matrix W which appears in the expression of e should be chosen so as to satisfy the property P_1. Then, the regulation of e to zero normally yields the (constrained) minimization of the secondary cost function h_s. By further restricting the choice of W so as to satisfy the property P_2, it is possible to provide the global task Jacobian matrix J_T^x with positivity properties which can in turn be exploited at the control level. Obviously, when the primary task Jacobian matrix is known, both properties P_1 and P_2 are automatically satisfied by choosing $W = J_1^x$. However, the satisfaction of P_1 only requires the knowledge of $R(J_1^{x^T})$ along $q_r(t)$. This fact, complemented with other considerations given below, provides a theoretical justification to the study and utilization of simple models of this subspace, such as the sensor-based primitives which were previously introduced. Invariance of this subspace on the manifold defined by the primary constraint $e_1 = 0$ indicates that W can be calculated independently of the secondary objective. Furthermore, even when P_1 is slightly transgressed in practice, the regulation of the task function (7.3.1) may still constitute a well-conditioned control problem and generate an acceptable behaviour of the robot, although usually the secondary cost function is no longer minimized.

The application of these general results to the specific case of sensor-based tasks is discussed in the next section.

7.3.3 Application to sensor-based tasks

It will first be assumed that the choice '$x = \bar{r}$' can be made. This corresponds, for example, to the practical situation of a six-joint robot manipulator equipped with a sensing system rigidly linked to its last body. The case $n \neq 6$ will be discussed briefly at the end of this section. In order to simplify the notation, it will also be assumed that all screws and related matrices are evaluated at a point Q_i and expressed in a well chosen frame F_i, with Q_i and F_i rigidly linked to the sensing system carried on the robot.

Recall that $\bar{s}(\bar{r}, t)$ denotes the p-dimensional vector regrouping the sensor outputs that we wish to regulate, and that $L(\bar{r}, t)(-(\partial \bar{s}/\partial \bar{r})^T)$ is the $6 \times p$

7.3 IMMERSION OF SENSOR-BASED TASKS

interaction matrix associated with the sensing system and the observed environment. The rank m ($m \leq \min(p, 6)$) of this matrix along the ideal trajectory of the robot, $q_r(t)$, is assumed to be constant and known. Recall that $N = 6 - m$ was previously defined as the class of the 'virtual linkage' resulting from the interaction between the sensing system and the target.

We will restrict the discussion to the following set of sensor-based functions:

$$e_1(q, t) = D(t)\bar{s}(\bar{r}, t) - \sigma(t) \tag{7.3.8}$$

where $D(t)$ is a ($m \times p$) full rank combination matrix, defined by the user and such that the ($m \times 6$) matrix DL^T is also of full rank m along $q_r(t)$. This rank condition is needed for the components of e_1 to be compatible and independent. $\sigma(t)$ is an m-dimensional reference vector chosen by the user which determines the ideal trajectory for the vector $D\bar{s}$.

It is possible to convey more generality to the form (7.3.8) by allowing $\bar{s}$ to represent a vector of regular functions of the sensor outputs. Notice that (7.3.8) can also be written

$$e_1(q, t) = D(t)\tilde{s}(\bar{r}, t) \tag{7.3.9}$$

with

$$\tilde{s}(\bar{r}, t) = \bar{s}(\bar{r}, t) - s_r(t) \tag{7.3.10}$$

and

$$s_r(t) = D^+(t)\sigma(t). \tag{7.3.11}$$

Therefore, when D is a square matrix (i.e. when $p = m$), the task associated with e_1 is equivalent to having each sensor output s_j regulated around the ideal value $s_{r,j}$.

The Jacobian matrix associated with the sensor-based task is:

$$J_1^r = \frac{\partial e_1}{\partial \bar{r}} = D\frac{\partial \tilde{s}}{\partial \bar{r}} = DL^T. \tag{7.3.12}$$

From (7.3.12), we readily see that $R((J_1^r)^T) \subseteq R(L)$. Since DL^T is of rank m, we also have $\dim\{R((J_1^r)^T)\} = m = \dim\{R(L)\}$. Therefore

$$R((J_1^r)^T) = R(L). \tag{7.3.13}$$

This last relation reminds us that the knowledge of the vector subspace $R((J_1^r)^T)$ is equivalent to the knowledge of the screw subspace **S** spanned by the interaction screws.

The sensor-based task characterized by the vector e_1 being redundant (when $m < n$), it must be complemented by a secondary objective, the minimization of a cost function h_s for example. According to (7.3.1) and (7.3.9), one may then consider the following global task function

$$e = W^+ D\tilde{s} + \alpha(I_6 - W^+ W)g_s^r \tag{7.3.14}$$

where

$$g_s^r = \frac{\partial h_s}{\partial \bar{r}}. \tag{7.3.15}$$

As discussed before, the matrix function W should ideally be chosen so as to satisfy property P_1 along $q_r(t)$. According to (7.3.13), this is equivalent to having

$$R(W^T) = R(L). \tag{7.3.16}$$

Because of the dimensions of W, this property is also equivalent to

$$W^T = (w_1, \ldots, w_m) \tag{7.3.17}$$

where the set $\{w_j, j = 1, \ldots, m\}$ is any basis of the interaction screw subspace S. Besides P_1, the satisfaction of the property P_2, which in this case becomes (because of (7.3.12))

$$DL^T W^T > 0 \tag{7.3.18}$$

constitutes an additional constraint which prevents the matrices D and W from being chosen independently.

For example, if the choice of W is the following:

$$W = DL^T \tag{7.3.19}$$

or if D is determined from the prior knowledge of a basis of S according to

$$D = WLR \tag{7.3.20}$$

where R is any positive $(p \times p)$ matrix, then both properties P_1 and P_2 are satisfied.

In practice, the 'true' interaction matrix L usually has to be replaced by an estimate $\hat{L}(t)$ evaluated along the trajectory actually followed by the robot. In some cases this estimate is determined off-line, or provided by a model of a primitive which seems adequate. In other cases, this off-line estimate may be complemented or replaced by an on-line estimation scheme.

Then, by extension of (7.3.19), a possible choice for W is:

$$W = D\hat{L}^T. \tag{7.3.21}$$

Notice that in this case the property P_1 is satisfied if

$$R(\hat{L}) = R(L) \tag{7.3.22}$$

while the property P_2 is satisfied if

$$DL^T \hat{L} D > 0. \tag{7.3.23}$$

A second possibility consists of determining, prior to the choice of D, a basis $\{\hat{w}_j, j = 1, \ldots, m\}$ of the estimated interaction screw subspace $\hat{S}$ asso-

7.3 IMMERSION OF SENSOR-BASED TASKS

ciated with $\hat{L}$. Then a possible choice for W is

$$W^T = (\hat{w}_1, \ldots, \hat{w}_m), \tag{7.3.24}$$

and, by extension of (7.3.20), a possible choice of D is

$$D = W\hat{L}. \tag{7.3.25}$$

The choices (7.3.24) and (7.3.25) call for a few remarks.

1. An intuitive interpretation of these choices, by analogy with the kinematics of contacts, has previously been given when discussing the utilization of primitives. In this case the estimate $\hat{L}$ is a constant matrix directly provided by the model of the chosen primitive.

2. The properties P_1 and P_2 are both satisfied if and only if the matrices L and $\hat{L}$ can be decomposed as follows

$$L = W^T Q \tag{7.3.26}$$

$$\hat{L} = W^T \hat{Q} \tag{7.3.27}$$

with

$$\hat{Q}Q^T > 0. \tag{7.3.28}$$

The normalization of the models of interaction screws (unitary sliders) associated with the primitives can then be justified in relation to the freedom of choice of $\hat{Q}$ allowed by (7.3.28), as illustrated by several examples in section 7.5. Notice that a sufficient condition for the property P_2 alone to be satisfied is the positivity of the matrix $\hat{L}L^T$ on the subspace of constrained motions.

3. By replacing D in (7.3.14) by the expression (7.3.25), the global task function can be written:

$$e = S\tilde{F} + (I - S)\alpha g_s^r \tag{7.3.29}$$

where $S = W^+ W$ is the orthogonal projection operator on the subspace of constrained motions, and where

$$\tilde{F} = \hat{L}\tilde{s} \tag{7.3.30}$$

The writing of the task function in this way shows how the formulation of hybrid tasks involving force control can formally be extended to the case of tasks based on the use of non-contact sensors.

It should be also emphasized that the matrix S defined above is a generalization of the 'selection matrix' encountered in the common so-called 'hybrid control'. Clearly, the case when this matrix is diagonal with entries equal to 1 or 0 is not the most general case, even though it often occurs in practice.

To conclude this section, let us briefly comment upon the case $n \neq 6$. The previous analysis may still be applicable to the case $n < 6$ by working in a reduced configuration space. This possibility is illustrated in section 7.3.5, where the robot moves in a plane.

On the other hand, when $n > 6$, the choice $x = \bar{r}$ is no longer possible and one is bound to work in the space of joint coordinates. However, it is sometimes possible to split the problem into two stages. Indeed, when the sensor-based vector e_1 and the secondary cost function h_s depend only on $\bar{r}$ and the time t, one may still work, at this level, in the configuration space, and consider a task function of the form:

$$e = W^+ e_1 + \alpha(I_6 - W^+ W)\frac{\partial h_s}{\partial \bar{r}} \qquad (7.3.31)$$

with $\dim(e) = 6$.

This function characterizes a task which is still redundant, with a redundancy rate equal to $n - 6$. It may therefore be associated with the problem of minimizing a *second* cost function $h'_s(q, t)$ in order to define a global task which, finally, is admissible. By applying the general approach of redundancy once again, we then obtain a global n-dimensional task function of the form (cf. Chapter 4):

$$e' = B^T e + \alpha'(I_n - W'^T(AW'^T)^{-1} A)\frac{\partial h'_s}{\partial q}. \qquad (7.3.32)$$

7.3.4 About the secondary objective in hybrid tasks

In many applications, the secondary objective associated with a sensor-based task can be expressed in the form of a set of (holonomous) constraints:

$$f_i(\bar{r}, t) = 0; \quad i = 1, \ldots, l \qquad (7.3.33)$$

that the user wishes to satisfy as closely as possible. Then, a possible secondary cost function to be minimized is

$$h_s = \frac{1}{2}\sum_{k=1}^{l} f_k^2 \qquad (7.3.34)$$

the 'gradient' of which is the screw

$$g_s^r(Q) = \left(\frac{\partial f_1}{\partial \bar{r}}(Q), \ldots, \frac{\partial f_l}{\partial \bar{r}}(Q)\right)\begin{pmatrix} f_1 \\ \vdots \\ f_1 \end{pmatrix}. \qquad (7.3.35)$$

From the knowledge of g_s^r, a task function associated with the hybrid task can be determined along the lines of section 7.3.3. Clearly, when $n = 6$, the number l of constraints should at least be equal to the redundancy rate $6 - m$ of the sensor-based task for the hybrid task to have a chance of

7.3 IMMERSION OF SENSOR-BASED TASKS

being admissible. Notice also that, when $l > 6$, all constraints cannot be independent.

Some examples of such constraints, which can be found in applications, like surface inspection, tracking of the seam in arc welding, contour deburring, or other force/motion control problems, will now be discussed on the basis of the generic situation represented in Fig. 7.18.

In this figure, the frame F_6, with origin O_6, is associated with the end-effector. F_0 is a fixed reference frame. The intersection of the sensed target with a given plane (P) is the curve (C). The plane (P) is completely defined by the normal unitary vector n and by the distance d to the origin O_0 of the frame F_0. Some possible constraints are now listed and discussed, independently of the sensor-based task.

Constraint 1: The point O_6 belongs to the plane (P).

This constraint can be expressed in the form

$$f_1 = \langle n, OO_6 \rangle - d = 0. \qquad (7.3.36)$$

Then

$$\frac{\partial f_1}{\partial \bar{r}}(O_6) = \begin{pmatrix} n \\ 0 \end{pmatrix}. \qquad (7.3.37)$$

Constraint 2: the plane $(O_6; i_6, j_6)$ remains parallel to (P).

This constraint is equivalent to having the two following constraints satisfied simultaneously:

$$\left. \begin{array}{l} f_2 = \langle i_6, n \rangle = 0 \\ f_3 = \langle j_6, n \rangle = 0 \end{array} \right\} \qquad (7.3.38)$$

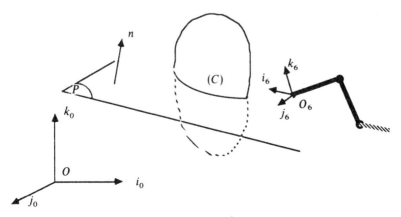

Fig. 7.18 A generic constrained situation

Then
$$\frac{\partial f_1}{\partial \bar{r}}(O_6) = \begin{pmatrix} 0 \\ i_6 \times n \end{pmatrix}$$
$$\frac{\partial f_2}{\partial \bar{r}}(O_6) = \begin{pmatrix} 0 \\ j_6 \times n \end{pmatrix}. \tag{7.3.39}$$

Constraint 3: the point O_6 tracks a point D, the 3 coordinates of which in F_0 are regrouped in the vector $x_d(t)$.

This constraint is equivalent to
$$\begin{pmatrix} f_4 \\ f_5 \\ f_6 \end{pmatrix} = x - x_d(t) = 0 \tag{7.3.40}$$

where x denotes the coordinates of O_6 in F_0. Then
$$\left(\frac{\partial f_4}{\partial \bar{r}}(O_6) \frac{\partial f_5}{\partial \bar{r}}(O_6) \frac{\partial f_6}{\partial \bar{r}}(O_6) \right) \bigg|_{F_0} = \begin{pmatrix} I_3 \\ 0 \end{pmatrix} \tag{7.3.41}$$

and therefore
$$\left(\frac{\partial f_4}{\partial \bar{r}}(O_6) \frac{\partial f_5}{\partial \bar{r}}(O_6) \frac{\partial f_6}{\partial \bar{r}}(O_6) \right) \bigg|_{F_6} = \begin{pmatrix} R_{06}^T \\ 0 \end{pmatrix} \tag{7.3.42}$$

where R_{06} is the rotation matrix which characterizes the orientation of F_6 with respect to F_0.

Notice that the desired trajectory $x_d(t)$ may be obtained in various ways. For example, it may be a trajectory in the plane (P), determined in advance from the knowledge of an explicit model of the curve (C). This possibility will be illustrated by the example of a hybrid task discussed in the next section. It may also be obtained by integration of a desired velocity $V_d(t)$. Then

$$x_d(t) = x_d(0) + \int_0^t V_d(s) ds \tag{7.3.43}$$

with the initial condition chosen for example as:
$$x_d(0) = x(0). \tag{7.3.44}$$

However, if $V_d(t)$ is determined independently of the actual position $\hat{x}(t)$ of the end-effector and if the constraint of perfect tracking is not compatible with the fulfilment of other constraints (the primary sensor-based task, for example), there is a risk of having the distance $\|x - x_d\|$ between O_6 and D grow without bound. In order to avoid this problem, a possibility, among others, consists of correcting the initial desired velocity $V_{d_1}(t)$ as follows

$$V_d(t) = \alpha_1(x(t) - x_d(t)) + V_{d_1}(t); \quad \alpha_1 > 0. \quad (7.3.45)$$

This is in fact the equation of a first-order linear filter with gain α_1. The correction $\alpha_1(x(t) - x_d(t))$ can also be interpreted as a velocity vector pointing in the direction of O_6. Obviously, the better the tracking, the smaller the correction is.

In practice, the choice of $V_{d_1}(t)$ may be related to the desire of moving along the surface of the target at a constant speed. This leads for example to

$$V_{d_1} = v\hat{l}(t) \quad (7.3.44)$$

where $\hat{l}(t)$ is a unitary vector orthogonal to a current estimate $\hat{n}(t)$ of the normal to the surface. In some cases, this estimate may be obtained by processing the information issued from the sensor's outputs. Precautions concerning the utilization of this estimate in the construction of $V_{d_1}(t)$ have to be taken in order to ensure the differentiability of $V_d(t)$ as required by the admissibility.

Remark. A numerically stable implementation of (7.3.43), (7.3.44) is:

$$x_d((k+1)\Delta) = (1-\beta)x_d(k\Delta) + \beta x(k\Delta) + V_d(k\Delta)\Delta; \quad 0 < \beta < 1; \quad k \in N \quad (7.3.47)$$

where Δ is the control sampling interval.

Constraint 4: the orientation r_{06} of the end-effector tracks a desired orientation $r_d(t)$.

In order to express this objective with a minimal set of constraints, a parametrization of the orientations is needed. One may for example choose the parametrization PL defined in section 1.2.5.4 of Chapter 1. The objective is then equivalent to the satisfaction of the following three constraints:

$$\begin{pmatrix} f_7 \\ f_8 \\ f_9 \end{pmatrix} = PL(r_d^{-1} r_{06})|_{F_0} = 0 \quad (7.3.48)$$

with

$$\left(\frac{\partial f_7}{\partial \bar{r}} \frac{\partial f_8}{\partial \bar{r}} \frac{\partial f_9}{\partial \bar{r}}\right)\bigg|_{F_0} = \begin{pmatrix} 0 \\ \dfrac{\partial PL}{\partial r}(r_d^{-1} r_{06})|_{F_0} \end{pmatrix} \quad (7.3.49)$$

where $\partial PL/\partial r$ is given in Proposition P1.9 of Chapter 1 (note that $\left(\frac{\partial f_7}{\partial \bar{r}} \frac{\partial f_8}{\partial \bar{r}} \frac{\partial f_9}{\partial \bar{r}}\right)\bigg|_{F_0} \simeq \begin{pmatrix} 0 \\ I_3 \end{pmatrix}$ and $\left(\frac{\partial f_7}{\partial \bar{r}} \frac{\partial f_8}{\partial \bar{r}} \frac{\partial f_9}{\partial \bar{r}}\right)\bigg|_{F_6} \simeq \begin{pmatrix} 0 \\ R_{06}^T \end{pmatrix}$ when $r_{06} \approx r_d$).

As for $x_d(t)$, the desired trajectory $r_d(t)$ may be determined in various ways, either in advance or on-line. The rotation matrix R_d associated with r_d may for example be obtained by integration of a desired angular velocity $\omega_d(t)$:

$$R_d(t) = R_d(0) + \int_0^t As(\omega_d(s))R_d(s)ds \qquad (7.3.50)$$

with the initial condition chosen, as for example:

$$R_d(0) = R_{06}(0). \qquad (7.3.51)$$

Notice that, due to the specific structure of rotation matrices, some precautions have to be taken when implementing (7.3.50) numerically. A parametrization of orientations may also be used at this level. As in the case of translations, in order to prevent $\|PL(r_d^{-1}r_{06})\|$ from growing large when the initial desired angular velocity $\omega_{d_1}(t)$ is determined independently of the actual orientation of the end-effector, it is possible to correct this velocity according to

$$\omega_d(t) = \omega_{d_1}(t) + \omega_{d_2}(t) \qquad (7.3.52)$$

with

$$\omega_{d_2}(t) = \alpha_2 PL(r_d^{-1}(t)r_{06}(t)); \quad \alpha_2 > 0 \qquad (7.3.53)$$

or

$$\omega_{d_2}(t) = \alpha_3 \phi(t); \quad \alpha_3 > 0 \qquad (7.3.54)$$

where $\phi(t)$ is the rotation vector of the rotation $r_d^{-1}(t)r_{06}(t)$.

7.3.5 An example of hybrid task

Let us consider the situation of Fig. 7.19 where a three-joint planar robot has to follow the surface of a part, the intersection, (C), of which with the robot's plane is convex and roughly circular. The two sensors are used to keep the gripper at a given distance from the object and maintain the axis $O_3 x_3$ normal to (C). The related configuration space may be parametrized by $\bar{r}_3 = (x, y, \theta)$ where $\begin{pmatrix} x \\ y \end{pmatrix}$ stands for $\begin{pmatrix} x_{O_3} \\ y_{O_3} \end{pmatrix}\bigg|_{F_0}$.

Since the surface of the part observed by the sensors can locally be approximated by its tangent plane, we model the interaction between the sensors and the part by a primitive P2.1 as represented in the figure. The corresponding (3×2) interaction matrix model evaluated at O_3 and expressed in the frame F_3 is

$$\hat{L} = \begin{pmatrix} -1 & -1 \\ 0 & 0 \\ -1 & 1 \end{pmatrix}. \qquad (7.3.55)$$

7.3 IMMERSION OF SENSOR-BASED TASKS

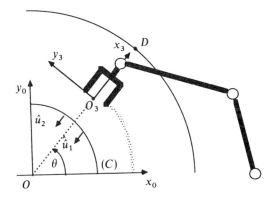

Fig. 7.19 Sensor-based following of a quasi-circular part in a plane

By choosing $W^T = \hat{L}$, we obtain

$$W^+ W = \begin{pmatrix} 1 & 0 & 0 \\ 0 & 0 & 0 \\ 0 & 0 & 1 \end{pmatrix} \quad (7.3.56)$$

$$\hat{L}\tilde{s} = \begin{pmatrix} -(\tilde{s}_1 + \tilde{s}_2) \\ 0 \\ \tilde{s}_1 - \tilde{s}_2 \end{pmatrix} \quad (7.3.57)$$

and the task function is, using (7.3.14) and (7.3.25)

$$e = \begin{pmatrix} 1 & 0 & 0 \\ 0 & 0 & 0 \\ 0 & 0 & 1 \end{pmatrix} \begin{pmatrix} -(\tilde{s}_1 + \tilde{s}_2) \\ 0 \\ \tilde{s}_1 - \tilde{s}_2 \end{pmatrix} + \alpha \begin{pmatrix} 0 & 0 & 0 \\ 0 & 1 & 0 \\ 0 & 0 & 0 \end{pmatrix} g_s^{\bar{r}_3}|_{F_3}. \quad (7.3.58)$$

In order to specify the hybrid task and the corresponding task function completely, there remains the determination of the secondary objective.

Since the intersection (C) of the sensed part with the robot's plane is known to be circular as a first approximation, a way of monitoring the motion of the end-effector along (C) consists of minimizing the distance between O_3 and a point D moving, at a chosen speed, on a circle concentric to (C). Due to the convexity of (C), the optimal position of O_3 is unique if the radius ρ of the circle is large enough (so that (C) lies entirely inside the circle). Furthermore, in the ideal case where (C) is also a circle centered on O, the velocity of this ideal point is proportional to that of D.

Let us therefore consider the following cost function

$$h_s = \tfrac{1}{2}((x - x_d(t))^2 + (y - y_d(t))^2 \quad (7.3.59)$$

where

$$\left.\begin{array}{l}x_d(t) = \rho \cos \phi_d(t) \\ y_d(t) = \rho \sin \phi_d(t)\end{array}\right\} \quad (7.3.60)$$

are the coordinates of the point D in the plane $(x_0 O y_0)$.

The gradient of h_s expressed in F_0 is then

$$g_s^{r_3} = \left.\frac{\partial h_s}{\partial \bar{r}_3}\right|_{F_0} = \begin{pmatrix} x - \rho \cos \phi_d(t) \\ y - \rho \sin \phi_d(t) \\ 0 \end{pmatrix} \quad (7.3.61)$$

and, in F_3

$$g_s^{r_3} = \left.\frac{\partial h_s}{\partial \bar{r}_3}\right|_{F_3} = \begin{pmatrix} \cos\theta(x - \rho\cos\phi_d(t)) + \sin\theta(y - \rho\sin\phi_d(t)) \\ -\sin\theta(x - \rho\cos\phi_d(t)) + \cos\theta(y - \rho\sin\phi_d(t)) \\ 0 \end{pmatrix}. \quad (7.3.62)$$

Using (7.3.62) in (7.3.58) finally yields (with $\alpha = 1$):

$$e = \begin{pmatrix} \tilde{s}_1 + \tilde{s}_2 \\ -\sin\theta(x - \rho\cos\phi_d(t)) + \cos\theta(y - \rho\sin\phi_d(t)) \\ \tilde{s}_1 - \tilde{s}_2 \end{pmatrix}. \quad (7.3.63)$$

7.4 SUMMARY

A methodology for the design and analysis of sensor-based tasks has been proposed. At the modelling level, it was first established that the differential, with respect to a relative configuration space, of a sensor signal was a particular screw, called the interaction screw. This screw is an important characteristic of the interaction between a sensor and its environment. When several sensors are used, the corresponding interaction screws span a subspace which determines the motions which can be controlled with the sensors. At this stage, the notion of virtual linkage associated with a set of sensors was defined by analogy with the kinematics of contacts. Then, in the same way as the concept of ideal mechanical linkage is used to solve complex mechanical problems, the notion of a sensor-based primitive, that reduces the set of interaction screws to unitary sliders, was introduced in order to simplify the modelling and the utilization of the interaction screws. These sliders, expressed at the same point and in the same frame rigidly linked to the sensing system, were regrouped in the so-called interaction matrix, $\hat{L}$. A vector, representative of the sensor-based task and the dimension of which depends on the *class* of the virtual linkage (the complement to six of the dimension of the subspace spanned by the interaction screws) was then proposed in the

form:
$$e_1 = W\hat{L}\tilde{s}, \qquad (7.4.1)$$

where the dimension of $\hat{L}$ is $(6 \times p)$, $\tilde{s}$ is the difference between the vector of the p sensor outputs and a reference vector, and W^T is a $(6 \times m)$ full rank matrix whose columns form a basis of the subspace spanned by the interaction screws. When the sensor-based task is redundant (i.e. when $m < n$), it has to be complemented by a secondary objective. It was shown how the approach of redundancy, developed in Chapter 4, can be applied to derive a global task function $e(q, t)$, the regulation of which constitutes the final control objective.

This suggests the following methodology, that the user may adapt to the particularities of the application.

1. Starting from a more or less formalized specification of the sensor-based task to be performed, and an approximated model of the environment, choose adequate primitives and mount them on the relevant link. The non-exhaustive Table 7.1 may provide a few guidelines for some basic applications. The characteristics and positioning of the sensors should of course be consistent with the chosen primitives and the underlying model of interaction. In some cases, a refined model will be needed.

2. Choose a frame, rigidly linked to the sensing system, and compute the interaction matrix $\hat{L}$ associated with the selected primitives. Determine a basis of the subspace spanned by the interaction screws to form a matrix W.

3. When the sensor-based task is redundant (in the sense of Chapter 4), choose a secondary cost function to be minimized (or maximized).

4. Determine the task function associated with the global task. For example, when $n = 6$, take

$$e = (W^+ W)\hat{L}\tilde{s} + \alpha(I_6 - W^+ W)g_s^r. \qquad (7.4.2)$$

Then, if the task is admissible, and if the properties P_1 and P_2 are satisfied (at least in the first approximation), a possible matrix J_q to be used in the control expression is:

$$J_q = (I_6 + \gamma(I_6 - W^+ W))\frac{\partial \bar{r}}{\partial q}. \qquad (7.4.3)$$

7.5 EXAMPLES

We will end this part of the chapter devoted to the use of non-contact sensors by discussing a few examples of sensor-based tasks. In each case we will focus our attention on the sensor-based part of the task (secondary objectives will

Table 7.1 *Typical Classes of Application of Standard Primitives*

Basic tasks	Elementary obstacle avoidance	Line/plane positioning	Centering towards a cylindrical target	Centering with full orientation control towards a cylindrical target	Plane surface following	Guidance inside a cylinder	Frame calibration	Complete obstacle avoidance
Some possible primitives	P1	P2 P7	P6	P9	P3 P9	P8	P4	P7 P8

7.5 EXAMPLES

not be discussed). In the first three examples, the utilization of primitives will be compared to the use of a refined model of the interaction screw subspace. The specific case of sensory obstacle avoidance will be considered last.

7.5.1 Surface following: the case of a plane

Let F_S be a frame linked to the end-effector of a robot, with origin S and basis vectors $\{i_S, j_S, k_S\}$. Let us consider the problem of keeping, through the use of sensors, the plane $\{S; i_S, j_S\}$ parallel to the planar surface, (P), of some object, while maintaining a certain distance δ_0 between the two planes (see Fig. 7.20). Since translations along the plane (P) as well as rotations around the axis normal to the surface are not constrained, the class of the virtual linkage to be achieved is $C3$. The minimal number of required sensors is thus three.

Let us therefore consider the case of three sensors, each of the type studied in Example 7.2 equally distributed and positioned on the circle (C) as shown in the figure. Along the ideal trajectory of the end-effector, the unitary vector n_s, which points in the direction of observation of the sensors, is equal to $-n_T$, where n_T denotes the unitary vector normal to (P). By application of the relation (7.1.49), we find that, along the ideal trajectory, the interaction screws are

$$\left. \begin{array}{l} u_j = \dfrac{\partial f}{\partial \delta}(\delta_0) n_T \\ \\ H_j(S_j) = 0 \end{array} \right\} \quad (7.5.1)$$

for $j = 1, 2, 3$.

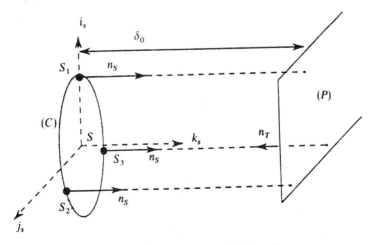

Fig. 7.20 A plane surface following task

The corresponding interaction matrix, L, evaluated at S and expressed in F_S, can then be computed and we obtain

$$L = \frac{\partial f}{\partial \delta}(\delta_0) \begin{pmatrix} 0 & 0 & 0 \\ 0 & 0 & 0 \\ 1 & 1 & 1 \\ 0 & \frac{\sqrt{3}}{2}\rho & -\frac{\sqrt{3}}{2}\rho \\ -\rho & \frac{\rho}{2} & \frac{\rho}{2} \\ 0 & 0 & 0 \end{pmatrix} \qquad (7.5.2)$$

where ρ is the radius of the circle (C). From (7.5.2), it is easy to verify that the columns of the following matrix

$$W^T = \begin{pmatrix} 0 & 0 & 0 \\ 0 & 0 & 0 \\ 1 & 0 & 0 \\ 0 & 1 & 0 \\ 0 & 0 & 1 \\ 0 & 0 & 0 \end{pmatrix} \qquad (7.5.3)$$

form an orthonormal basis of the interaction screw space S. Then, according to (7.3.9) and (7.3.20), a possible sensor based task vector is

$$e_1 = WL\tilde{s}. \qquad (7.5.4)$$

We could, in fact, make the economy of computing L explicitly by considering the primitive P3 of section 7.2.3.2 instead. The interaction matrix associated with it is

$$\hat{L} = \begin{pmatrix} 0 & 0 & 0 \\ 0 & 0 & 0 \\ 1 & 1 & 1 \\ 0 & \frac{\sqrt{3}}{2} & -\frac{\sqrt{3}}{2} \\ -1 & \frac{1}{2} & \frac{1}{2} \\ 0 & 0 & 0 \end{pmatrix} \qquad (7.5.5)$$

7.5 EXAMPLES

and a possible sensor-based task vector is, according to (7.3.25)

$$e_1 = W\hat{L}\tilde{s} = \begin{bmatrix} \tilde{s}_1 + \tilde{s}_2 + \tilde{s}_3 \\ \dfrac{\sqrt{3}}{2}(\tilde{s}_2 - \tilde{s}_3) \\ \tfrac{1}{2}(\tilde{s}_2 + \tilde{s}_3) - s_1 \end{bmatrix}. \qquad (7.5.6)$$

Let us quickly verify that properties P_1 and P_2 are satisfied when using this primitive. From (7.5.2) and (7.5.3), we see that L can be decomposed as follows:

$$L = W^T Q \qquad (7.5.7)$$

with

$$Q = -\dfrac{\partial f}{\partial \delta}(\delta_0) \begin{bmatrix} 1 & 1 & 1 \\ 0 & \dfrac{\sqrt{3}}{2}\rho & -\dfrac{\sqrt{3}}{2}\rho \\ -\rho & \dfrac{\rho}{2} & \dfrac{\rho}{2} \end{bmatrix} \qquad (7.5.8)$$

while, from (7.5.5) and (7.5.3), $\hat{L}$ can be decomposed as follows

$$\hat{L} = W^T \hat{Q} \qquad (7.5.9)$$

with

$$\hat{Q} = \begin{bmatrix} 1 & 1 & 1 \\ 0 & \dfrac{\sqrt{3}}{2} & -\dfrac{\sqrt{3}}{2} \\ -1 & \dfrac{1}{2} & \dfrac{1}{2} \end{bmatrix}. \qquad (7.5.10)$$

Since $\hat{Q}$ is a full rank matrix, we already see that $R(\hat{L}) = R(W^T)$. Therefore, the property P_1 is satisfied when using $\hat{L}$ instead of the 'true' interaction matrix L. Furthermore, from (7.5.8) and (7.5.10):

$$\hat{Q} Q^T = -3 \dfrac{\partial f}{\partial \delta}(\delta_0) \begin{bmatrix} 1 & 0 & 0 \\ 0 & \rho & 0 \\ 0 & 0 & \dfrac{\rho}{2} \end{bmatrix}. \qquad (7.5.11)$$

If $(\partial f/\partial \delta)(\delta_0)$ is negative, the product matrix $\hat{Q} Q^T$ is positive and P_2 is also satisfied. This condition on the sign of $(\partial f/\partial \delta)(\delta_0)$ is not surprising because it was implicitly assumed, when choosing the directions of the unitary sliders composing the primitive, that the signals of the sensors are monotonically decreasing functions of the distance to the target. In the case where $(\partial f/\partial \delta)(\delta_0)$ is positive, the opposite of $\hat{L}$ must be used.

For the same task, a redundant primitive of type P9 (Fig. 7.21), involving for example eight sensors instead of three, could also be used. The interested reader will verify, by computing the related matrix $\hat{L}$ and using (7.3.25) again, that this possibility would have led to a sensor-based vector of the form

$$e_1 = \begin{bmatrix} \sum_{j=1}^{8} \tilde{s}_j \\ \tilde{s}_2 - \tilde{s}_4 + \tfrac{1}{2}((\tilde{s}_5 + \tilde{s}_6) - (\tilde{s}_7 + \tilde{s}_8)) \\ \tilde{s}_1 - \tilde{s}_3 + \tfrac{1}{2}((\tilde{s}_5 + \tilde{s}_8) - (\tilde{s}_6 + \tilde{s}_7)) \end{bmatrix}. \quad (7.5.12)$$

7.5.2 Positioning with respect to a cylinder

Let us consider the problem of positioning an end-effector with respect to a cylindrical part, depicted in Fig. 7.22. Sensors of the same type as in the previous example are used to keep the gripper at a given distance, d_0, from the cylinder and to ensure the centring of the gripper with respect to the cylinder. Because of the particular geometry of the target, it already appears that motions along and around the axis z_T cannot be constrained by using sensors. The virtual linkage to be achieved is thus of class $C2$, and at least four sensors are required. Let us therefore assume that four sensors are utilized. Their locations on the end-effector are the points S_j and their

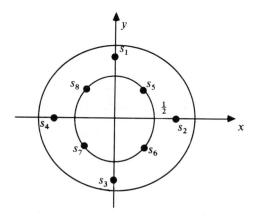

Fig. 7.21 Plane surface following with eight sensors

7.5 EXAMPLES

common direction of observation is given by the unitary vector n_s. When the gripper is correctly positioned (along the ideal trajectory of the robot), the corresponding interaction screws evaluated at the points P_j are, according to equation (7.1.49):

$$\left. \begin{array}{l} u_j = -\dfrac{\partial f}{\partial \delta}(\delta_0) \dfrac{n_j}{\langle n_j, n_s \rangle} \\ H_j(P_j) = 0 \end{array} \right\}. \tag{7.5.13}$$

These screws, evaluated, at the same point O_T, on the axis of the cylinder, are

$$\left. \begin{array}{l} u_j = -\dfrac{\partial f}{\partial \delta}(\delta_0) \dfrac{n_j}{\langle n_j, n_s \rangle} \\ H_j(O_T) = -\dfrac{\partial f}{\partial \delta}(\delta_0) P_j O_T \times \dfrac{n_j}{\langle n_j, n_s \rangle} \end{array} \right\}. \tag{7.5.14}$$

Expressing every screw in the frame F_T and regrouping them in the interaction matrix $L(O_T)$, we get

$$L(O_T) = -\dfrac{\partial f}{\partial \delta}(\delta_0) \begin{bmatrix} 1 & 1 & 1 & 1 \\ \tan\alpha & \tan\alpha & -\tan\alpha & -\tan\alpha \\ 0 & 0 & 0 & 0 \\ -l_1 \tan\alpha & l_1 \tan\alpha & l_1 \tan\alpha & -l_1 \tan\alpha \\ l_1 & -l_1 & l_1 & -l_1 \\ 0 & 0 & 0 & 0 \end{bmatrix}. \tag{7.5.15}$$

From (7.5.15), it is easy to verify that the columns of the matrix:

$$W^T = \begin{bmatrix} 1 & 0 & 0 & 0 \\ 0 & 1 & 0 & 0 \\ 0 & 0 & 0 & 0 \\ 0 & 0 & 1 & 0 \\ 0 & 0 & 0 & 1 \\ 0 & 0 & 0 & 0 \end{bmatrix} \tag{7.5.16}$$

form a basis of the interaction screw space **S**.

A frame F_S, linked to the sensor system, i.e. to the end-effector, and with respect to which all screws must be expressed, has to be chosen. The above development suggests choosing the frame which coincides with the frame F_T

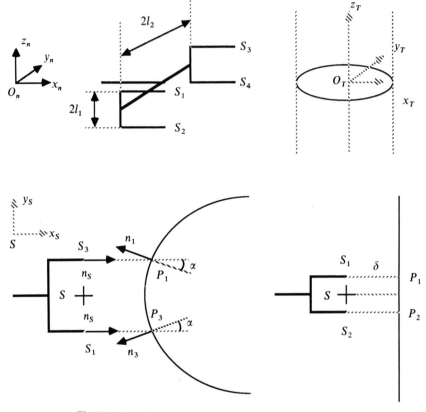

Fig. 7.22 Positioning of a gripper with respect to a cylinder

when the end-effector is ideally positioned with respect to the cylinder. However, the determination of the origin of the frame requires the knowledge of the radius of the cylinder, r, since the distance $(d_0 + r)$ between the point O_T and the gripper has to be calculated. This fact is not anodyne because it means that it is not possible in reality to determine **S** precisely, unless r is known exactly. Some consequences, related to the satisfaction of the decoupling property P_1 have been discussed in section 7.3.2. Let us however assume here, for the sake of simplicity, that this particular frame F_S can be determined and is the one chosen. Then, a possible sensor-based task vector is

$$e_1 = WL\tilde{s} \qquad (7.5.17)$$

with W and L given by (7.5.16) and (7.5.15) respectively.

7.5 EXAMPLES

As in the previous example, we could have avoided the computation of L by considering predetermined primitives instead. For example, the primitive P5.2 would have led to the following interaction matrix, expressed in F_S.

$$\hat{L} = \frac{\sqrt{2}}{2} \begin{bmatrix} 1 & 1 & 1 & 1 \\ 1 & 1 & -1 & -1 \\ 0 & 0 & 0 & 0 \\ -1 & 1 & 1 & -1 \\ 1 & -1 & 1 & -1 \\ 0 & 0 & 0 & 0 \end{bmatrix}. \quad (7.5.18)$$

A basis of the corresponding interaction screw space, $\hat{S}$, is the set of columns of the matrix W^T given by (7.5.16), and a possible sensor-based task vector is

$$e_1 = W\hat{L}\tilde{s} = \frac{\sqrt{2}}{2} \begin{bmatrix} \tilde{s}_1 + \tilde{s}_2 + \tilde{s}_3 + \tilde{s}_4 \\ (\tilde{s}_1 + \tilde{s}_2) - (\tilde{s}_3 + \tilde{s}_4) \\ (\tilde{s}_2 + \tilde{s}_3) - (\tilde{s}_1 + \tilde{s}_4) \\ (\tilde{s}_1 + \tilde{s}_3) - (\tilde{s}_2 + \tilde{s}_4) \end{bmatrix}. \quad (7.5.19)$$

As in the previous example, by inspection of the matrices L and $\hat{L}$ given by (7.5.15) and (7.5.18) respectively, it is easy to verify that the properties P_1 and P_2 are satisfied when $(\partial f/\partial \delta)(\delta_0)$ is negative.

Remark. When reflectance sensors like the ones presented in Appendix A7.1 are used, the output signal may often be considered as linear in reflectivity. In some cases, this last parameter may play the same role as a physical distance if needed. For example, the above setting of sensors would still give rise to a virtual linkage of class $C2$ when positioned in front of a target like the one represented in Fig. 7.23. In this case, the missing third dimension is 'simulated' by the drawing on the target.

Fig. 7.23 Drawing on a plane target

7.5.3 Tracking of a pointwise target

We consider a two-jointed robot and a pin-point target moving in the same plane (see Fig. 7.24). We will take advantage of the simplicity of this example to have a closer look at the conditions of admissibility of the related task. The task consists of moving the gripper so as to maintain the target T motionless with respect to the gripper. More precisely, the objective is to have the point D, the position of which in the frame F_2 is characterized by the predetermined distance l and angle α (see Fig. 7.25), always coincide with the target T. The configuration space of the gripper may be parametrized by $\bar{r}_3 = (x, y, \theta)$ where (x, y) are the coordinates of O_2 in F_0, and $\theta = q_1 + q_2$ is the angle $(O_0 x_0, O_2 x_2)$.

Since the orientation of the gripper is the only variable which is not constrained by the task, the virtual linkage to be achieved is of class $C1$, and thus at least two sensors are needed. They are located at S_1 and S_2. The output of each sensor is assumed to be a function of the distance δ_j between S_j

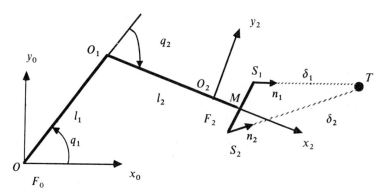

Fig. 7.24 Tracking a pointwise target with a two degrees-of-freedom plane robot

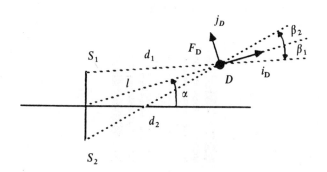

Fig. 7.25 Target tracking

and the target:
$$s_j = f(\delta_j); \quad j = 1, 2. \tag{7.5.20}$$

Using (7.1.37) and (7.1.38), we find that the corresponding interaction screws are

$$\left. \begin{array}{l} u_j = -\dfrac{\partial f}{\partial \delta}(\delta_j) n_j \\ H_j(T) = 0 \end{array} \right\} \tag{7.5.21}$$

where $n_j, j = 1, 2$ are the unitary vectors defined in Fig. 7.24.

The relation (7.5.21) tells us that the dimension of S is 2, unless n_1 and n_2 are collinear. Obviously, this last situation occurs only when T is on the axis $S_1 S_2$. This is to be avoided since the motion of the gripper is then constrained in only one direction. Therefore the point D should not be chosen on $S_1 S_2$. This first restriction upon the choice of D will appear again when discussing the admissibility of the task later. When the gripper is correctly positioned (i.e. when T and D coincide), the interaction matrix $L(D)$, deduced from (7.5.21), and expressed in the frame F_D is (in the reduced configuration space)

$$L(D) = \begin{bmatrix} -\dfrac{\partial f}{\partial \delta}(d_1) \cos \beta_1 & -\dfrac{\partial f}{\partial \delta}(d_2) \cos \beta_2 \\ -\dfrac{\partial f}{\partial \delta}(d_1) \sin \beta_1 & -\dfrac{\partial f}{\partial \delta}(d_2) \sin \beta_2 \\ 0 & 0 \end{bmatrix} \tag{7.5.22}$$

where the frame F_D, the distance d_1, d_2, and the angles β_1, β_2 are defined in Fig. 7.25.

From (7.5.22), we easily verify that a basis of S along the ideal trajectory is composed of the two column vectors of

$$W^T = \begin{bmatrix} 1 & 0 \\ 0 & 1 \\ 0 & 0 \end{bmatrix}. \tag{7.5.23}$$

A possible choice for the sensor-based task vector thus is

$$e_1 = WL\tilde{s} = \begin{bmatrix} -\dfrac{\partial f}{\partial \delta}(d_1) \cos \beta_1 & -\dfrac{\partial f}{\partial \delta}(d_2) \cos \beta_2 \\ -\dfrac{\partial f}{\partial \delta}(d_1) \sin \beta_1 & -\dfrac{\partial f}{\partial \delta}(d_2) \sin \beta_2 \end{bmatrix} \begin{pmatrix} \tilde{s}_1 \\ \tilde{s}_2 \end{pmatrix} \tag{7.5.24}$$

where
$$\tilde{s}_j = s_j - s_j^*; \quad j = 1, 2 \tag{7.5.25}$$

and the s_j^*'s are premeasured values of the sensor signals when the gripper is correctly positioned with respect to the target. Notice that, in this example, the sensor-based task is not redundant, since the dimension of e_1 is equal to the number of degrees of freedom of the robot ($n = 2$).

In order to analyse the conditions of admissibility of the task, let us derive the task Jacobian matrix along the ideal trajectory of the robot (i.e. when $D = T$). By differentiating (7.5.24) with respect to q, we obtain

$$J_T = \frac{\partial e_1}{\partial q} = WL \frac{\partial s}{\partial \bar{r}_3} \frac{\partial \bar{r}_3}{\partial q} = WLL^T \bar{R}_{OD}^T J_D \qquad (7.5.26)$$

with $\bar{R}_{OD} = \begin{pmatrix} R_{OD} & 0 \\ 0 & 1 \end{pmatrix}$, R_{OD} being the (2×2) rotation matrix between the frames F_0 and F_D; and J_D the (3×2) basic Jacobian matrix of the robot evaluated at the point D.

From (7.5.26), we see that the admissibility condition $\det(J_T) \neq 0$ implies:

(1) rank $(L) = 2$;

(2) rank $(J_D) = 2$.

As discussed earlier, the first condition is related to the dimension of S, and means that the point D should not lie on the axis $S_1 S_2$. The second condition means that the tracking of the target should not bind the robot to cross geometrical singularities. In the case represented by Fig. 7.24, we already know that these singularities correspond to configurations where the points O_0, O_1 and D are aligned. The satisfaction of this condition thus depends on the trajectory of the target. For example, when $\alpha = 0$, the target must stay within the ring delimited by the circles centered in O and with radius $l_1 - (l_2 + l)$ and $l_1 + (l_2 + l)$.

Let us now comment further on the particular case $\alpha = 0$. Then, $d_1 = d_2(=d)$ and $\beta_1 = -\beta_2(=-\beta)$, and the sensor-base task vector (7.5.24) becomes

$$e_1 = -\frac{\partial f}{\partial \delta}(d) \begin{pmatrix} \cos \beta(\tilde{s}_1 + \tilde{s}_2) \\ \sin \beta(\tilde{s}_2 - \tilde{s}_1) \end{pmatrix}. \qquad (7.5.27)$$

If, instead of the 'true' model (7.5.22) of L, we had used the simpler model $\hat{L}$ provided by the combination of two primitives P1 as shown in Fig. 7.26, we would have obtained

$$\left. \begin{array}{r} \hat{u}_j = n_j \\ \hat{H}_j(D) = 0 \end{array} \right\} \qquad (7.5.28)$$

$$\hat{L}(D)|_{F_D} = \frac{\sqrt{2}}{2} \begin{bmatrix} 1 & 1 \\ -1 & 1 \\ 0 & 0 \end{bmatrix} \qquad (7.5.29)$$

7.5 EXAMPLES

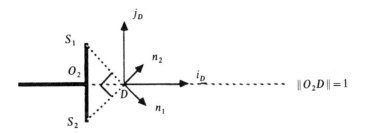

Fig. 7.26 Primitives for the tracking of a pointwise target

and the obtained sensor-based task vector would have been

$$e_1 = W\hat{L}\tilde{s} = \frac{\sqrt{2}}{2}\begin{pmatrix} \tilde{s}_1 + \tilde{s}_2 \\ \tilde{s}_2 - \tilde{s}_1 \end{pmatrix}. \tag{7.5.30}$$

Here, the coefficient $\sqrt{2}/2$ is of no importance and can be dropped. As in the previous examples, the interested reader may verify that this model is compatible with the satisfaction of the properties P_1 and P_2.

7.5.4 Obstacle avoidance: some principles

The ability of avoiding obstacles is primordial for manipulators working in complex environments. When the world is not known, is poorly known or may change in an unpredictable way, sensors are usually required. However, the avoidance of obstacles is not a well-defined objective in the sense that it cannot be formulated as a regulation problem: the underlying constraint is an inequality (the distance to the obstacles should be positive or greater than a given value) while an equality would be necessary to define a reference trajectory. For this reason, it cannot be treated, within the approach of redundancy, as a primary objective. On the other hand, it can theoretically be chosen as a secondary objective, since a way of avoiding obstacles obviously consists of maximizing the distance to the obstacles. However, the avoidance of obstacles cannot be granted independently of the primary objective in this case. A safer solution may therefore consist of considering a potential-like approach, the principles of which will now be briefly presented.

Let us assume that every sensory signal s_j is equal to zero when no obstacle is close, and that their output monotonically tends to infinity when their distance to the observed obstacle tends to zero. The proximity of an obstacle is detected when one of the signals, let us say s_j, passes a given threshold τ_j. In order to prevent a collision, one may try to minimize a potential function of the form

$$h = h_1 + \delta_j(s_j - \tau_j)^2 \tag{7.5.31}$$

with

$$\left.\begin{array}{l}\delta_j = 0 \quad \text{when} \quad s_j \leq \tau_j \\ \delta_j = 1 \quad \text{when} \quad s_j > \tau_j\end{array}\right\} \quad (7.5.32)$$

and where $h_1(q, t)$ denotes the cost function to be minimized in order to achieve the user's other objectives.

In order to smooth the transition between the two cases $\delta_j = 0$ and $\delta_j = 1$, it is possible to replace δ_j by a differentiable function and consider, instead of (7.5.31):

$$h = h_1 + \beta_j(s_j - \tau_j)^2 \quad (7.5.33)$$

with, for example

$$\beta_j(s_j) = \begin{cases} 0, & \text{when } s_j \leq \tau_j \\ \frac{1}{2}\left(\sin\left(\frac{\pi}{\varepsilon}(s_j - \tau_j) - \frac{\pi}{2}\right) + 1\right), & \text{when } \tau_j \leq s_j \leq \tau_j + \varepsilon \quad (7.3.34) \\ 1 & \text{when } s_j \geq \tau_j + \varepsilon. \end{cases}$$

The generalization of (7.5.33) to the other sensors is

$$h = h_1 + \sum_{j=1}^{p} \beta_j(s_j - \tau_j)^2. \quad (7.5.35)$$

A corresponding task function is then obtained by considering the gradient of h, which gives

$$e = \frac{\partial h_1}{\partial q} + \sum_{j=1}^{p} (s_j - \tau_j)\gamma_j \frac{\partial s_j}{\partial q} \quad (7.5.36)$$

with

$$\gamma_j = \beta_j + (s_j - \tau_j)\frac{d\beta_j}{ds_j} \quad (7.5.37)$$

$$\frac{d\beta_j}{ds_j} = \begin{cases} 0, & \text{for } s_j \leq \tau_j \text{ and } s_j \geq \tau_j + \varepsilon \\ \frac{\pi}{2\varepsilon}\cos\left(\frac{\pi}{\varepsilon}(s_j - \tau_j) - \frac{\pi}{2}\right), & \text{for } \tau_j \leq s_j \leq \tau_j + \varepsilon \end{cases} \quad (7.5.38)$$

$$\frac{\partial s_j}{\partial q} = J_j^T(O_j)\begin{pmatrix} R_{oj} & 0 \\ 0 & R_{oj} \end{pmatrix}\begin{pmatrix} u_j \\ H_j(O_j) \end{pmatrix}\bigg|_{F_j} \quad (7.5.39)$$

where $H_j = (H(O_j), u_j)$ is the interaction screw associated with the jth sensor, F_j is a frame rigidly linked to the body on which the jth sensor is fixed, and J_j is the $(6 \times n)$ basic Jacobian matrix associated with this body. Notice that $\gamma_j = \beta_j$ when s_j does not belong to $]\tau_j, \tau_j + \varepsilon[$.

When all sensors are fixed on the same body, and all screws are evaluated at the same point S and expressed in the basis of a frame F_S linked to this body, the relation (7.5.36) can also be written

$$e = \frac{\partial h_1}{\partial q} + J^{\mathrm{T}}(S)\begin{pmatrix} R_{F_S} & 0 \\ 0 & R_{F_S} \end{pmatrix} L(S)|_{F_S} \Gamma \tilde{s} \qquad (7.5.40)$$

where J is the Jacobian matrix associated with the frame F_S, L the $(6 \times p)$ interaction matrix associated with the sensors and the environment, and

$$\Gamma = \mathrm{diag}(\gamma_1, \ldots, \gamma_p)$$
$$\tilde{s}^{\mathrm{T}} = (s_1 - \tau_1, \ldots, s_p - \tau_p).$$

However, the task function (7.5.40) has only a theoretical value because the interaction screws, which also depend on the obstacles' characteristics, are not known precisely. In practice, L will thus be replaced by some estimate $\hat{L}$. For example, each interaction screw H_j will be replaced by a slider $\hat{H}_j$ pointing in the average direction of observation of the corresponding sensor (see Fig. 7.27). This approximation is usually best justified when the direction of observation of the sensor is normal to the surface of the obstacle.

The resulting task function is then

$$e = \frac{\partial h_1}{\partial q} + J^{\mathrm{T}}(S)\begin{pmatrix} R_{F_S} & 0 \\ 0 & R_{F_S} \end{pmatrix} \hat{L}(S)|_{F_S} \Gamma \tilde{s}. \qquad (7.5.41)$$

Notice that it can also be written

$$e = \frac{\partial h_1}{\partial q} + J^{\mathrm{T}}(S) F(S)|_{F_0} \qquad (7.5.42)$$

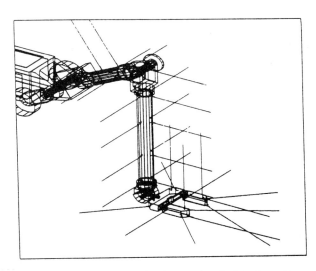

Fig. 7.27 Sensor axis in an example of sensor-based obstacle avoidance task

where

$$F(S) = \hat{L}(S)\Gamma\tilde{s} \qquad (7.5.43)$$

can be interpreted as a virtual wrench resulting from a virtual contact with obstacles.

Remark. It should be emphasized that an obstacle avoidance task is *a priori* less constrained that the realization of a virtual linkage: in fact, with the sensor model chosen in this section, any motion allowing the sensor outputs to decrease (for example a motion opposite to the direction of observation of a sensor) can be considered as correct. On the other hand, as in all potential-like methods, a problem which may occur during the task achievement is to have the robot trapped in a local minimum where the primary user's objective cannot be satisfied.

7.6 TASKS BASED ON THE USE OF FORCE SENSORS

7.6.1 Introduction

The control of the forces exerted between a robot and its environment is one of the most important challenges in production automation. The two main classes of manufacturing application of force control are as follows.

1. *Assembly* tasks (Fig. 7.28). The aim is to monitor the relative positioning of two or more workpieces with generally high accuracy constraints, while controlling the exerted forces in order to prevent the parts from being damaged and avoid failure of the assembly task (because of wedging or jamming for example). Usually the motions are rather slow, with low amplitude, and dynamic effects are not very important.

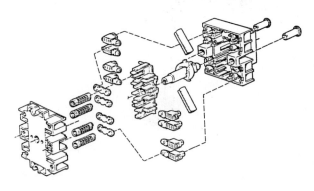

Fig. 7.28 An example of an assembly task

7.6 TASKS BASED ON THE USE OF FORCE SENSORS

2. *Contouring* tasks (deburring, milling, etc.). The goal here is to follow a contour, the geometry of which is not always known precisely, while exerting given forces on the environment.

Let us briefly review a few technical issues. *Exertion* of forces with a robot can be performed in various ways: the first idea is to use the actuators of the robot, and determine the joint torques needed to exert given contact forces on the environment. When fine control is required, it is often more efficient to use dedicated actuators mounted on the end-effector (tool or gripper). Especially for assembly tasks, another approach consists of using a special external device ('left hand'), in order to improve the accuracy and the efficiency of force monitoring (see example in Fig. 7.29).

Also it must be pointed out that, although the method is beyond the scope of the present analysis, passive compliant devices are frequently used in industrial assembly tasks. Of course in practice all these methods may be

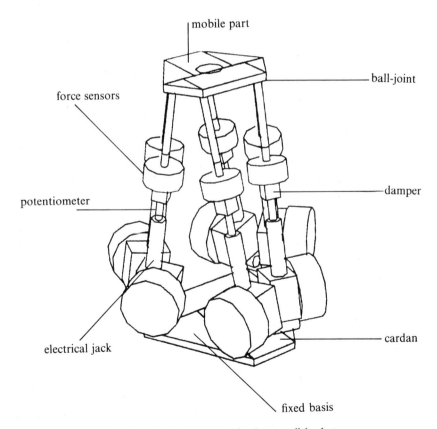

Fig. 7.29 A six degrees-of-freedom parallel robot

mixed and, for example, semi-active compliant wrists constitute a good alternative solution.

Measurement of forces is another problem. We may distinguish two kinds of measurement methods.

1. *Through actuators* (measurement of current, pressure, etc.). In many cases, inaccuracies and noise in the measurements may degrade the efficiency of the approach significantly.

2. *Through the elastic deformation of a material*. To avoid amplitudes of deformation becoming too large and the use of cumbersome devices, measurements are generally made by strain gauges. Existing devices are mainly wrist sensors and finger sensors. Other sensing technologies are based on the use of calibrated springs or inductive sensors. The important point to be emphasized is that forces are never measured *directly*, but only through their *effects*. In all cases the measurement of some motion (often small) is involved.

Although the analysis performed in the previous sections does apply to the problem of force control, some aspects specific to this class of sensor-based tasks motivate a complementary study. For example, the fact that force exertion as well as force measurement are generally performed through a *displacement*, is related to the existence of an 'elastic' medium which can be characterized by its stiffness or its compliance. This means also that, from a control point of view, it is necessary to clarify the differences between position control and force control. Some other questions naturally arise in this framework. What is a force really? What is elasticity and where should it be located? What are the consequences of a physical contact and of compliance on the overall system stability? How are the concepts of virtual contacts and real linkages connected? All these questions are addressed in the following. We do not however intend to treat these issues in an exhaustive manner. We plan, rather, on the one hand to give a few guidelines concerning the modelling aspects specifically linked to the existence of elasticities, and on the other hand to give some insight into how these elasticities affect the structure and the performances of the control schemes.

7.6.2 Generalized springs and force sensors

7.6.2.1 Background

We shall implicitly refer, in this section, to some basic modelling concepts given in Chapter 1. In the following the term 'force' will stand for both 'linear' force and torque, i.e. the resultant and the value of the momentum field of the associated screw evaluated in some point.

7.6 TASKS BASED ON THE USE OF FORCE SENSORS

Let us briefly recall that we are concerned with the special Euclidean group of rigid body displacements, SE_3. When a reference working frame (origin and basis) is defined, an element $\bar{r}$ of SE_3 may be represented by a homogeneous 4×4 matrix. Since SE_3 is a Lie group, it is possible to consider small displacements at the identity only. The space of vectors tangent to SE_3 at the identity is se_3 and a force F is an element of its dual se_3^*. It is also a screw in the usual sense. It will be denoted F, with, when considered as a vector of $\mathbf{R}^6$:

$$F = \begin{pmatrix} f \\ m(.) \end{pmatrix}$$

where f is the resultant and $m(.)$ the momentum field, or torque, in some point.

Now, when a contact occurs between two rigid bodies, the motions become constrained. As seen in Chapter 1, the subspace of static contact forces is locally the reciprocal of the subspace of allowed velocities. The problem with this theoretical model is that it does not describe the physical nature of the contact. In fact, it predicts that away of the contact the motions are free and that, at the contact, reaction forces appear, so as to forbid *any* motions in a certain subspace. Fortunately, the reality is different, and there is always some 'elasticity' in the contact which allows motions in the constrained directions and connects these motions to the reaction forces in a certain way. Among possible models of this phenomenon, one of the most useful is the concept of *generalized spring*, that we now recall.

Consider a rigid body immersed in a medium which exerts constraints on the body depending on its position. The set {body, medium} constitutes a generalized spring. By definition, an elastic medium is characterized by a potential function, $V(\bar{r})$, from SE_3 to $\mathbf{R}$, such that the resulting force F exerted by the medium on the body derives from this potential. We thus have, at $\bar{r} \in SE_3$:

$$F_{\bar{r}} = -\mathrm{d}V_{\bar{r}}. \tag{7.6.1}$$

Let $id \in SE_3$ denote the initial configuration of the body in the medium, and $\tau = \begin{pmatrix} \omega \\ V(.) \end{pmatrix}$ a small displacement ($\tau \in se_3$). Let $\bar{r}$ denote the new configuration of the body after this displacement took place. A first-order expansion of $\mathrm{d}V$ gives us

$$\mathrm{d}V_{\bar{r}} = \mathrm{d}V_{id} + (\mathrm{d}^2 V_{id})(\tau) + \|\tau\|\varepsilon_{id}(\tau) \tag{7.6.2}$$

with: $\varepsilon_{id}(\tau) \to 0$ when $\|\tau\| \to 0$. $\mathrm{d}^2 V_{id}$ is a linear mapping from se_3 to se_3^*. Therefore it is a symmetric bilinear form (cf. Chapter 1, section 1.3.2.6).

From (7.6.1) and (7.6.2), the force $F_{\bar{r}}$ applied to the body in the position $\bar{r}$ is related to the force F_{id} applied to the body in its initial position, according to

$$F_{\bar{r}} \approx F_{id} - (\mathrm{d}^2 V_{id})(\tau). \tag{7.6.3}$$

Expressing all screws in a frame F with origin O, (7.6.3) can also be written

$$\begin{pmatrix} f \\ m(O) \end{pmatrix}_{\bar{r}} \approx \begin{pmatrix} f \\ m(O) \end{pmatrix}_{id} - K_{id}(O) \begin{pmatrix} \omega \\ V(O) \end{pmatrix} \bigg|_F \qquad (7.6.4)$$

In this relation, K_{id} is the 6×6 symmetric matrix called the *stiffness matrix at id*, associated with the bilinear form $d^2 V_{id}$. Notice that, expressed in another frame, this matrix would be different. Relations of passage from one frame to another are given in Appendix A7.2. The stiffness matrix is also defined for a given configuration of the body in the medium. Another configuration of the body is usually characterized by a different stiffness matrix.

If the preconstraint F_{id} is equal to zero, then the system is in natural equilibrium at *id*. If K_{id} is a (semi)positive matrix, the medium is said to be *elastic at id*. The medium is *purely elastic at id* if K_{id} is positive definite. When the medium is elastic without being purely elastic, there are motions which do not produce any force. These 'free' motions are given by the eigenvectors of K_{id} associated with zero eigenvalues.

When the matrix K_{id} is regular, its inverse K_{id}^{-1} is called the *compliance matrix*. For both stiffness and compliance, there usually exists a set of frames in which decoupling (between constrained and free motions) is maximal in some sense. The origins of these frames are called centres of stiffness and centres of compliance respectively. These two sets of points do not usually coincide.

7.6.2.2 Linkages and generalized springs

We may distinguish two classes of linkages.

1. *Non-dissipative linkages.* In this case, the energy of the system remains constant when no external force is applied. This was the case of the virtual linkages associated with non-contact sensors where the virtual contacts are elastic and the forces applied to the sensors derive from a potential of the form $\frac{1}{2} \| \bar{s}(\bar{r}) \|^2$.
2. *Dissipative linkages.* In this case, energy is dissipated. Usually dissipation of energy occurs when the system is in motion and is due to friction phenomena. Classically, two main classes of friction can be distinguished: *Coulomb* (dry) friction and *viscous* friction (damping). Viscous friction acts as a linear damper and has a stabilizing effect. Without much loss of generality we shall not consider them at this level. Note also that accurate models of Coulomb friction are difficult to obtain in general.

Let us consider the case of a robot which is in contact with its environment through a force sensor. The force sensor may be modelled as an elastic generalized spring with a symmetric and (at least) semi-positive stiffness

7.6 TASKS BASED ON THE USE OF FORCE SENSORS

matrix. Now, the elasticity of the force sensor may not (or should not) be the main source of elasticity of the contact. Other possible sources are the robot itself (structural flexibility, elastic transmissions, etc.) and the environment. Consider therefore the situation depicted in Fig. 7.30 where each element of elasticity is modelled by a generalized spring. It is possible to show (see Appendix A7.2) that the serially linked generalized springs are equivalent statically to a single generalized spring. However, unless the masses of the interconnected intermediary bodies can be neglected (as it may be the case for the constitutive elements of the force sensor), or some springs are much stiffer than the others (in which case they may be considered as rigid), this synthetic representation cannot be used for the analysis of the dynamics of the system because of the inertial forces appearing when the bodies are in motion.

7.6.2.3 Modelling of a force sensor

Most force sensors encountered in robotics are mechanical devices equipped with a set of elementary sensors (strain gauges, for example) and designed so that they can be modelled as elastic linear generalized springs (see Fig. 7.31). They are made of two rigid parts connected by one or more bodies, some of which are flexible.

Two important characteristics of a force sensor are the *stiffness matrix* and the *calibration matrix* which are associated with it around its natural equilibrium configuration. Let F denote the internal force (wrench) applied to one of the rigid parts and associated with a small deformation of the flexible bodies. As seen before, the stiffness matrix expressed in some frame F is defined as

$$K = -\frac{\partial F}{\partial \bar{r}}\bigg|_F . \qquad (7.6.5)$$

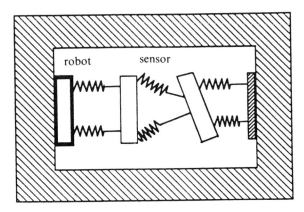

Fig. 7.30 Robot, sensor, and environment as generalized springs

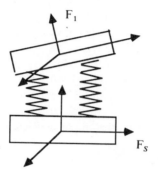

Fig. 7.31 Simplified representation of a force sensor

This matrix, which is symmetric and (semi-)positive, relates the force to the deformation of the sensor. Usually a force sensor is very stiff (the positive eigenvalues of K are large) so that small deformations can produce large forces. This is, in particular, the case of devices based on the use of piezo-electrical sensors.

As to the calibration matrix C, it relates the force to the p sensor signals regrouped in the vector $\bar{s}$ and can be defined as

$$C = -\left.\frac{\partial F}{\partial \bar{s}}\right|_F. \qquad (7.6.6)$$

In the range where the sensor is linear, we have

$$F(\bar{r}) = -C\bar{s}(\bar{r})|_F. \qquad (7.6.7)$$

As in the case of the stiffness matrix, the calibration matrix depends on the frame in which the force screw is expressed. More precisely, we have

$$C_{|F_i} = \Theta_{ik} C_{|F_k} \qquad (7.6.8)$$

where the transformation matrix Θ_{ik} is given by (A7.2.6).

To illustrate how the stiffness and calibration matrices may relate to each other, let us consider the particular case, represented in Fig. 7.32, where each elementary sensor measures the local force component f_j along a direction u_j resulting from the deformation of a one-dimensional linear spring with stiffness k_j:

$$\frac{\partial s_j}{\partial q_j} = -k_j \qquad (7.6.9)$$

$$f_j = s_j u_j. \qquad (7.6.10)$$

7.6 TASKS BASED ON THE USE OF FORCE SENSORS

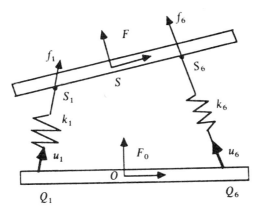

Fig. 7.32 A type of force sensor

The resultant force, expressed in the frame F with origin S is

$$F = \begin{pmatrix} f \\ m \end{pmatrix} = \sum_{i=1}^{6} \begin{pmatrix} s_j u_j \\ s_j \Delta_j u_j \end{pmatrix} \bigg|_F \qquad (7.6.11)$$

where $\Delta_j = As(S_j S)|_F$ as usual. From (7.6.7) and (7.6.11), we see that the calibration matrix of the sensor is

$$C = -J^{-T}|_F \qquad (7.6.12)$$

with

$$J^{-T} = \begin{pmatrix} u_1 & \ldots & u_6 \\ \Delta_1 u_1 & \ldots & \Delta_6 u_6 \end{pmatrix}. \qquad (7.6.13)$$

We leave to the interested reader, as an exercise, the task of verifying that J is nothing other than the matrix associated with the Jacobian operator $\partial \bar{r}/\partial \bar{q}$, i.e. the matrix such that

$$T = J\dot{\bar{q}} \qquad (7.6.14)$$

where T is the velocity screw of the frame F with respect to F_0, and $\bar{q}^T = (q_1, \ldots, q_6)$ the natural system of generalized coordinates associated with the system (in Fig. 7.32, we have $Q_j S_j = q_j u_j$). This may be shown, either by a direct calculation of J^{-1} or by applying the principle of virtual powers.

Let us now calculate the stiffness matrix of the sensor. We have

$$K = -\frac{\partial F}{\partial \bar{r}} = -\frac{\partial F}{\partial \bar{s}}\frac{\partial \bar{s}}{\partial \bar{q}}\frac{\partial \bar{q}}{\partial \bar{r}} = C\frac{\partial \bar{s}}{\partial \bar{q}} J^{-1}\bigg|_F \quad (7.6.15)$$

Taking (7.6.9) and (7.6.12) into account, we finally obtain

$$K = J^{-T}\mathrm{diag}(k_j)J^{-1}|_F \quad (7.6.16)$$

which is symmetric positive definite.

If the sensor is well designed, there may exist a frame in which the stiffness matrix is diagonal. Usually, this is obtained by exploiting structural symmetries. The normal use of a force sensor is to measure the external force F_{ext} applied to one of the rigid parts. When a static equilibrium is reached, we have

$$F_{ext} = -F \quad (7.6.17)$$

and therefore

$$F_{ext} = C\bar{s}|_F. \quad (7.6.18)$$

This last relation is also often utilized to identify the calibration matrix experimentally. Note that relations (7.6.17) and (7.6.18) are not valid during dynamic motions, unless the inertia effects within the force sensor can be neglected.

In practice, the calibration matrix is the most important characteristic of a force sensor. We shall see that, when the force sensor is very stiff, the stability of the control demands the existence of another source of elasticity (the robot itself, or the environment) so as to increase the overall compliance of the system. For this reason the stiffness matrix of a stiff sensor is usually not a characteristic of much relevance for the control problem. There are, however, exceptions when the force sensor is not very stiff and when it is the main source of elasticity. Then a good modelling of the sensor may be important.

7.6.3 Control aspects

It is known that among the many problems related to the active control of the forces exerted by a robot on its environment, the stability issue is of particular importance. Even without considering the problems linked to the numerical implementation of the control, the sources of poor working of the system, and the very nature of the phenomena involved, are extremely varied. In the first place, the non-linear phenomena occurring in the contact linkages, generally poorly modelled, give rise to various annoyances, unstabilities or limit cycles: Coulomb friction, stiction, and even discontinuities associated with impact forces. The stiffness of the environment is also an important factor. When the environment and the force sensor are altogether very stiff, the robot's behaviour is likely to be underdamped or even locally unstable. Other

7.6 TASKS BASED ON THE USE OF FORCE SENSORS

parameters, linked to the physical structure of the system, are also significant: non-collocation of the sensors and the actuators, limited bandwidth and power of the actuators, existence of drive-train backlash or unmodelled high frequency dynamics are potential factors of instability. Finally, within the control loops themselves, incautious use of low-band filters or integral correcting terms may degrade the performance. In this brief section, we shall not reconsider these aspects. This would necessitate long developments based on experimental results. On the other hand, and so as to maintain a certain coherence with the aspects developed in the previous chapters, we shall try to point out a few facts related to the numerical implementation of the control such as the effects of discretization. This aspect will be treated from the study of a very simple example.

7.6.3.1 A basic stability problem

Consider the simple system of Fig. 7.33, with an ideal spring of stiffness K. The robot is a cart with one degree of freedom, and γ is the control force applied to the cart. The force measured by the sensor is

$$s = Kq. \tag{7.6.19}$$

and the equation of motion of the system is

$$m\ddot{q} + Kq = \gamma \tag{7.6.20}$$

which may also be written

$$\ddot{s} + \omega^2 s = \omega^2 \gamma \tag{7.6.21}$$

where

$$f_0 = \frac{1}{2\pi}\omega = \frac{1}{2\pi}\sqrt{\frac{K}{m}} \tag{7.6.22}$$

is the *natural oscillation frequency* of the homogeneous (free) system (when $\gamma = 0$).

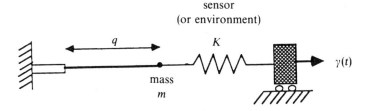

Fig. 7.33 A one-dimensional force control problem

7 SENSOR-BASED TASKS

Let s_d be the desired sensor value corresponding to the contact force that we would like to achieve. A continuous feedback control of the form

$$\gamma = -mg\left(\frac{\mu}{K}(s - s_d) + \dot{q}\right) \qquad (7.6.23)$$

would lead to a stable behaviour for any positive set $\{g, \mu\}$. Let us now apply a discretized control law, such as

$$\gamma(t) = -mg\left(\frac{\mu}{K}(s(t_n) - s_d) + \dot{q}(t_n)\right), \qquad \text{for } t_n < t \leq t_n + h \qquad (7.6.24)$$

where h is the sampling period of the control. Then, the equation of the closed-loop system, on the time interval $]t_n, t_n + h[$, is

$$\ddot{q} + \omega^2 q = -g(\mu(q(t_n) - q_d) + \dot{q}(t_n)) \qquad (7.6.25)$$

where $q_d = s_d/K$.

The related state equation is

$$\dot{X} = AX + \begin{pmatrix} 0 \\ -1 \end{pmatrix} g(\mu(q(t_n) - q_d) + \dot{q}(t_n)) \qquad (7.6.26)$$

where

$$A = \begin{pmatrix} 0 & 1 \\ -\omega^2 & 0 \end{pmatrix}; \quad X = \begin{pmatrix} q \\ \dot{q} \end{pmatrix}. \qquad (7.6.27)$$

Integrating along one period h gives the discrete-time equation of evolution

$$X_{t_n+h} = e^{Ah}X_{t_n} + \int_0^h e^{A\tau}\begin{pmatrix} 0 \\ -1 \end{pmatrix} g(\mu(q(t_n) - q_d) + \dot{q}(t_n))\,d\tau \qquad (7.6.28)$$

Let p denote the Laplace variable. We have

$$e^{At} = \mathbf{L}^{-1}\{(pI - A)^{-1}\} \qquad (7.6.29)$$

where $\mathbf{L}$ is the Laplace transform. Thus, with A given in (7.6.27)

$$(pI - A)^{-1} = \frac{1}{p^2 + \omega^2}\begin{pmatrix} p & 1 \\ -\omega^2 & p \end{pmatrix}. \qquad (7.6.30)$$

Knowing that $\mathbf{L}(\cos \omega t) = \dfrac{p}{p^2 + \omega^2}$ and $\mathbf{L}(\sin \omega t) = \dfrac{\omega}{p^2 + \omega^2}$, we get

$$e^{A\tau} = \begin{pmatrix} \cos \omega\tau & \dfrac{1}{\omega}\sin \omega\tau \\ -\omega \sin \omega\tau & \cos \omega\tau \end{pmatrix} \qquad (7.6.31)$$

7.6 TASKS BASED ON THE USE OF FORCE SENSORS

and finally the state equation of the controlled system is

$$X_{t_n+h} = \begin{pmatrix} \cos\omega h + \dfrac{g\mu}{\omega^2}(\cos\omega h - 1) & \dfrac{1}{\omega}\sin\omega h + \dfrac{g}{\omega^2}(\cos\omega h - 1) \\ -\omega\sin\omega h - \dfrac{g\mu}{\omega}\sin\omega h & \cos\omega h - \dfrac{g}{\mu}\sin\omega h \end{pmatrix} X_{t_n}$$

$$- \begin{pmatrix} g\dfrac{\mu}{\omega^2}(\cos\omega h - 1) \\ -g\dfrac{\mu}{\omega}\sin\omega h \end{pmatrix} q_d \qquad (7.6.32)$$

The corresponding characteristic polynomial is

$$P(\lambda) = \lambda^2 + a_1\lambda + a_2 \qquad (7.6.33)$$

with

$$a_1 = 2\cos\omega h + g\dfrac{\mu}{\omega^2}(\cos\omega h - 1) - \dfrac{g}{\omega}\sin\omega h \qquad (7.6.34)$$

$$a_2 = 1 + g\dfrac{\mu}{\omega^2}(1 - \cos\omega h) - \dfrac{g}{\omega}\sin\omega h. \qquad (7.6.35)$$

The application of Jury's criteria (cf. Chapter 0) provides us with three stability conditions:

(C1): $a_2 < 1$

(C2): $a_2 > a_1 - 1$

(C3): $a_2 > -a_1 - 1$.

Let us assume that g and μ are both chosen positive, and examine these three conditions.

Condition (C1) leads to:

$$\dfrac{\mu}{\omega}(1 - \cos\omega h) < \sin\omega h. \qquad (7.6.36)$$

Since the first term of the inequality is positive, a first requirement is the positivity of $\sin\omega h$. If we restrict the analysis to the first permitted frequency band (i.e. $\omega h \in]0, \pi[$), we thus find that the following relation must be satisfied:

$$\dfrac{\omega}{2\pi} < \dfrac{1}{2h}. \qquad (7.6.37)$$

This means that the natural oscillation frequency, f_0, *of the system has to be smaller than half the control sampling frequency* $f_s(=1/h)$. *Once this condition* is fulfilled, (7.6.36) gives also

$$\mu < \frac{\omega}{\sin \omega h}(1 + \cos \omega h) \qquad (7.6.38)$$

which defines the maximal possible value for μ, plotted in Fig. 7.34 as a function of ω.

Condition (C2) yields

$$g \frac{\mu}{\omega^2}(1 - \cos \omega h) > \cos \omega h - 1 \qquad (7.6.39)$$

which is always satisfied.

As for condition (C3), since $\sin \omega h$ is positive, it is easy to see that this condition is equivalent to

$$g < \frac{\omega}{\sin \omega h}(1 + \cos \omega h) \qquad (7.6.40)$$

which is similar to the condition (7.6.38) on μ, obtained from (C1).

To summarize the analysis of this example, we have found that the stability conditions are of two kinds.

1. *Gain conditions*, given by:

$$g \text{ and } \mu < \frac{\omega}{\sin \omega h}(1 + \cos \omega h). \qquad (7.6.41)$$

These conditions show that the maximum allowable control gains are $2f_s$ when the natural oscillation frequency tends to zero. They show also that the maximal size of the gains decreases when the natural frequency (i.e. the stiffness) increases.

2. *A frequency condition*:

$$f_0 < \tfrac{1}{2} f_s \qquad (7.6.42)$$

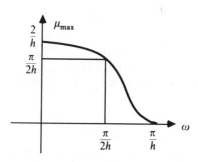

Fig. 7.34 Maximal size of the gains

7.6 TASKS BASED ON THE USE OF FORCE SENSORS

which may also be written

$$K < \frac{m}{4h^2}. \tag{7.6.43}$$

This constraint, which is analogous to Shannon's condition, is very important. It tells us that the control sampling frequency must be bounded, from below, by twice the natural oscillation frequency. It also tells us that this lower bound is proportional to the square root of the stiffness K. In other words, the stiffer the contact, the larger the control sampling frequency must be.

Let us now consider that K is in fact a scalar *equivalent* stiffness, given for example by

$$\frac{1}{K} = \frac{1}{K^1} + \frac{1}{K^2} + \frac{1}{K^3} \tag{7.6.44}$$

where K^1, K^2, and K^3 represent for example the stiffness coefficients of the robot's gripper, sensor and environment respectively. In the case where these coefficients are very different, we see that:

$$\frac{1}{K} \approx \frac{1}{K_{\min}}. \tag{7.6.45}$$

This expression, together with condition (7.6.43) gives an indication of how the smallest stiffness in the system is related to the control sampling frequency. Since sensors are generally very stiff, the overall stability of the system will often require the introduction of compliant elements in the system, either in the robot arm or in the environment itself.

Remark. These results may be extended to the linear multivariable case where (7.6.20) is replaced by $M\ddot{q} + Kq = U$ and q is a vector. It is easy to show that in the basis of the eigenvectors of $M^{-1}K$, and with a single control sampling rate, h, we are led back to a set of single-variable problems in which conditions (C1), (C2) and (C3) have to be satisfied for every eigenvalue. In particular, we would find that the control sampling frequency has to be greater than twice the highest natural eigenfrequency of the system.

7.6.3.2 Compensation of the steady-state contact force

1. *Case of a simple spring.* Let us return to equations (7.6.21) and (7.6.23). The equation of the controlled system, without precompensation of the steady-state contact force, is

$$\ddot{s} + \omega^2 s = -mg\omega^2 \left(\frac{\mu}{K}(s - s_d) + \dot{q} \right). \tag{7.6.46}$$

If the system is stable (see the analysis above), the steady-state value is obtained by setting all velocities and accelerations to zero. This gives

$$s_{eq} = \frac{mg\frac{\mu}{K} s_d}{1 + mg\frac{\mu}{K}}. \qquad (7.6.47)$$

Therefore, the absolute value of the steady-state error $e_{eq} = s_{eq} - s_d$ is

$$|e_{eq}| = \frac{|s_d|}{1 + mg\frac{\mu}{K}} = \frac{\omega^2}{\omega^2 + g\mu} |s_d|. \qquad (7.6.48)$$

This relation shows that it is possible to reduce this error by increasing the feedback gain $g\mu$. However, when the control is sampled, this gain is limited by the stability condition (7.6.41) which gives

$$(g\mu)_{max} = \frac{\omega^2}{\sin^2 \omega h} (1 + \cos \omega h)^2. \qquad (7.6.49)$$

Thus

$$|e_{eq}|_{min} = \frac{1}{1 + \left(\frac{1 + \cos \omega h}{\sin \omega h}\right)^2} |s_d| \qquad (7.6.50)$$

and the static error tends to zero only when ω (or K) tends to zero also. The maximal value of $|e_{eq}|_{min}$ is reached when $(1 + \cos \omega h) = 0$, i.e. when the control sampling frequency tends to the critical value $f_s = 2f_0$.

A complementary way (independent of stability issues) of reducing the static error consists of adding a precompensation of the steady-state contact force, s_d to the control expression. The control then becomes

$$\gamma(t) = -mg\left(\frac{\mu}{K}(s - s_d) + \dot{q}\right) + s_d \qquad (7.6.51)$$

and it is easy to verify that the static error is then eliminated (at least in the ideal case treated here). Another well-known possibility consists of complementing the feedback part of the control with an integral action. However this solution may also introduce additional problems (stability and others) and is not investigated here.

2. *Precompensation in the general case.* The dynamic model of a rigid robot derived in Chapter 2 and used in the control analysis is

$$\Gamma = M(q)\ddot{q} + N(q, \dot{q}, t). \qquad (7.6.52)$$

7.6 TASKS BASED ON THE USE OF FORCE SENSORS

The non-linear term N now has to include the effect of the contact force at the joint level. At the equilibrium, i.e. in the absence of motion, (7.6.52) reduces to

$$\Gamma_{eq} = G(q_{eq}) + \Gamma_c \tag{7.6.53}$$

where $G(q_{eq})$ is the gravity compensation term and Γ_c is the control torque associated with the steady-state contact force F_{eq}, sometimes called the 'interface torque'. At the equilibrium, the virtual power principle can be applied and used to establish the relationship between Γ_c and F_{eq}. This well-known relation is

$$\Gamma_c = J_n^T(q_{eq}) F_{eq|F_0} \tag{7.6.54}$$

where J_n is the Jacobian matrix associated with the gripper.

Let us recall that this expression is valid only *statically*. In practice, it can be used to derive a precompensation term which, when added to the control expression, reduces the static error $F_{eq} - F_d$ (where F_d is the desired force at the contact with the environment). For example, this term may be of the form

$$\Gamma_{pre} = J_n^T(q) F_{d|F_0} \tag{7.6.55}$$

or

$$\Gamma_{pre} = J_n^T(q_{eq}) F_{d|F_0} \tag{7.6.56}$$

if q_{eq} is known in advance.
Note that, in the previous one-dimensional example, we had simply $J_n = 1$ and $F_{d|F_0} = s_d$.

7.6.3.3 Force control

1. Introduction. The aim of a force control scheme is to produce an efficient control of the forces exerted by a robot on the environment. It is first important to realize that the transformation of driving torques from joint space to another space, through an expression like (7.6.54) obtained by application of the virtual power principle, is a purely mathematical operation, and that it does not necessarily relate to the forces actually exerted on the environment. Indeed, we have seen that this expression was valid only in the static case. This is an important limitation. Moreover, the force screw F in the generic equation $\Gamma = J_n^T F$ is generally different from the screw F_m measured by contact sensors. For example, when important friction phenomena occur at the contact, the difference between F and F_m may be large, even at low velocities. For these reasons, this equation should only be used at the control level to derive (as indicated before) a precompensation term for reducing static errors. It is not to be confused with the expression of the control itself.

A more reliable solution to the initial force control problem consists of exploiting the notion of generalized spring, which allows the mapping of forces into displacements through the notion of stiffness. *The original force control problem is in this way implicitly transformed into a position control problem.* This is why it is possible to claim that most of the force control schemes are in fact position control schemes. This does *not* mean however that the design of the control requires transforming force errors into position errors *explicitly*. Nevertheless, several elements have to be taken into account at the control design level.

1. In section 7.6.3.1 it was shown that for a single one-dimensional linear spring the control sampling frequency has to be greater than twice the natural frequency of the system. Generalization of this result to the multidimensional case shows that the equivalent stiffness of the system and the sampling frequency cannot be considered independently.

2. Force sensors may be so stiff that their natural frequency may not be compatible with usual control sampling frequencies. In this case, the robot's structure may be more 'compliant' than the sensor. The control problem then becomes all the more difficult that the compliance of the robot is distributed along the structure.

3. A solution would be to use sensors with low stiffness. However, in most applications, the displacements of a tool mounted on the end-effector have to be controlled in addition to the contact forces. Then, because of the compliance of the sensor (for example a wrist sensor), it is not possible to derive the position of the tool from the joint position accurately. Even if it was possible to obtain an estimation of this position using a model of the geometry of the sensor and its instantaneous output, the problem of *controlling* the extremity would remain difficult, as is the case of a *flexible* system.

From the above considerations, we may deduce that in order to stay within the rigid robot control framework, the compliance which is needed has to be transferred towards the environment. This remark justifies the idea of 'left hand', briefly evoked in the introduction to this section. In some cases, this may be an active robot which manipulates a tool. In other applications, this robot may handle the workpiece and be controlled so as to simulate the needed compliance, while the tool is mounted on another robot.

Active simulation of a compliance (or a complete *mechanical impedance*) can be obtained for example by implementing the computed torque method in Cartesian space (cf. Chapter 5) and tuning the control gains so as to create the desired stiffness (position gain) and damping (velocity gain). For this use, a parallel robot seems to be particularly well-suited.

2. Task Functions. The general framework is the same as in the case of non-contact sensors. The reader may therefore refer to section 7.3. for an

7.6 TASKS BASED ON THE USE OF FORCE SENSORS

exposition of the basic ideas concerning the derivation of 'hybrid task' functions.

Let F denote the contact force (wrench) to be controlled. We may assume that F is measured through the use of a force sensor characterized by its matrix of calibration $C_{|F}$ (usually, the frame F in which the contact force is expressed is linked to one of the rigid parts of the force sensors):

$$F = C\bar{s}|_F. \qquad (7.6.57)$$

Let us also assume that the contact can be modelled as an ideal elastic generalized spring characterized by the stiffness operator K:

$$\frac{\partial F}{\partial \bar{r}} = K|_F. \qquad (7.6.58)$$

As explained earlier, some compliance is needed at the contact, and the source of compliance may be either the force sensor, the robot, the environment or a combination of these three elements.

Notice that the assumption according to which the contact can be modelled as an ideal generalized spring does not necessarily reflect the physical reality accurately, especially when contact friction forces appear. In fact, this assumption may be weakened because only the controlled part of the contact force is, in the forthcoming development, required to be a function of the relative position $\bar{r}$ of the robot's end-effector with respect to the sensed body. In this case the stiffness matrix relates to this part of the contact force only. This point will be illustrated by an example given later.

We hope that the reader is now convinced that many 'equivalent' task functions characterizing the same task may be considered. The basic reason for this diversity is that operational constraints may be expressed in many different ways. However, some choices are better than others in terms of admissibility and also because they may lead to simpler control expressions. The guidelines given in section 7.3 suggest, among other possibilities, the following methodology for the derivation of an adequate task function e.

The first step consists of expressing the force control objective as a set of m independent constraints $\{e_1(\bar{r}, t) = 0\}$ to be achieved, with a vector e_1 of the form:

$$e_1(\bar{r}, t) = D(t)F(\bar{r}) - \sigma(t) = D(t)C\bar{s}(\bar{r}) - \sigma(t)|_F \qquad (7.6.59)$$

where the $(6 \times m)$ 'combination' matrix $D(t)$ and the m-dimensional reference vector $\sigma(t)$ have to be determined. Notice that (7.6.59) can also be written:

$$e_1(\bar{r}, t) = D(t)\tilde{F}(\bar{r}, t)|_F \qquad (7.6.60)$$

with

$$\tilde{F}(\bar{r}, t) = F(\bar{r}) - F_r(t)|_F \qquad (7.6.61)$$

$$F_r(t) = D^+(t)\sigma(t). \qquad (7.6.62)$$

If the stiffness matrix associated with the contact is non-singular (pure elasticity), then the choice of $D(t)$ is entirely free and, if the user wishes to do it, it is possible to control all the components of the force F by monitoring the position of the robot's end-effector. Choose for example $D = I_6$ in this case. If $n = 6$, a possible task function then is $e = e_1 = F(\bar{r}) - \sigma(t) |_F$ where $\sigma(t)$ now represents the desired contact force. Notice that in this case the task Jacobian matrix $(\partial e/\partial \bar{r})|_F$ in SE_3, coincides with the stiffness matrix and thus is positive. If $n > 6$, then the sensor-based task is redundant and a secondary objective has to be specified in order to derive an admissible global task function.

In practice, however, the case where the stiffness matrix K is only semi-positive (rank$(K) = l < 6$) is much more common. In this case, it is not possible to control all the components of F independently of each other since $F|_F$ is bound to belong to the vector subspace $R(K|_F)$. This means that at most l components of $F|_F$ can be controlled independently. The selection of these components is done through the choice of the matrix D, which, in this case, is no longer entirely free. More precisely, D has to be an $(m \times 6)$ full rank matrix with $m \leq l$ and such that $\partial e_1/\partial \bar{r} = DK|_F$ is also of maximum rank m. To proceed with the derivation of a task function, we may at this state distinguish two cases.

Case 1: $n = 6$. In this case it is possible to keep working in SE_3. From section 7.3.2, we know that a possible task function is given by

$$e = W^+ e_1 + \alpha(I - W^+ W)g_s^r|_F \qquad (7.6.63)$$

where $g_s^r = \partial h_s/\partial \bar{r}$, $h_s(\bar{r}, t)$ being a cost function to be minimized under the constraint $e_1 = 0$. Examples of such functions were given in section 7.3.4. W is a full rank $(m \times 6)$ matrix function which has to be chosen so as to satisfy the decoupling property P_1, as well as possible along the robot's trajectory. It is also desirable to choose W so as to satisfy the positivity property P_2 along the robot's trajectory, in order to provide the task Jacobian matrix $(\partial e/\partial \bar{r})|_F$ with a positivity property that can be exploited at the control level (in relation to the calculation of the term J_q).

From (7.6.58) and (7.6.59) we have

$$J_1^r = \frac{\partial e_1}{\partial \bar{r}} = DK|_F \qquad (7.6.64)$$

and therefore the properties P_1 and P_2 become in this case

$$P_1: \quad (I - W^+ W)KD^T = 0|_F \qquad (7.6.65)$$

$$P_2: \quad DKW^T > 0|_F. \qquad (7.6.66)$$

7.6 TASKS BASED ON THE USE OF FORCE SENSORS

Obviously, a first possibility for the choice of W is

$$W = DK|_F. \tag{7.6.67}$$

However, this solution is of limited value in practice because it requires knowledge of the stiffness matrix associated with the contact. A more implementable solution is obtained by restricting the choice of D to the set of $m \times 6$ full rank matrices such that

$$D^T = K_1|_F E \tag{7.6.68}$$

where $K_1|_F$ is a $(6 \times m)$ matrix, the columns of which are a selection of eigenvectors of the stiffness matrix $K|_F$ associated with non-zero eigenvalues, and E is any nonsingular $(m \times m)$ matrix. The eigenvectors of $K|_F$ involved in the calculation of D determines the directions which are *force-controlled*.

Then the properties P_1 and P_2 are satisfied by choosing

$$W = D. \tag{7.6.69}$$

so that $R(W^T) = R(K_1|_F)$.

The advantage of this second solution is that it only requires knowledge of the range space of the stiffness operator (i.e. knowledge of the directions constrained by the contact) instead of the stiffness matrix itself. By using (7.6.60) and (7.6.69) in (7.6.63), we see that the resulting task function can be written:

$$e = S\tilde{F} + \alpha(I_6 - S)g_s^r|_F \tag{7.6.70}$$

where the (6×6) matrix

$$S = W^+ W \tag{7.6.71}$$

is obtained from the choice of a $(m \times 6)$ full rank matrix W such that $R(W^T)$ coincides with the *force-controlled* subspace $R(K_1|_F)$ spanned by m eigenvectors of $K|_F$ associated with non-zero eigenvalues. Thus

$$R(W^T) \subseteq R(K|_F) \tag{7.6.72}$$

and S has the classical meaning of a *selection* matrix.

Remarks.

1. S is in fact the matrix associated with an operator of orthogonal projection. The projection is on the subspace $R(W^T)$, itself contained in the subspace $R(K|_F)$ of the velocities constrained by the contact.

2. In practice, the subspace $R(K_1|_F)$ may not be known exactly even if the contact is truly elastic and the source of elasticity is well identified. For example, the geometry of the body in contact with the robot may be unknown. In this case, an on-line estimation of $R(K_1|_F)$ may have to be implemented in order to determine a matrix W such that the 'angle' (see

section 7.3.2) between $R(W^T)$ and $R(K_1|_F)$ remains small. This type of difficulty, which is generic as soon as exteroceptive sensors are used in the control loop, will hopefully motivate in the near future an important research effort towards the development of dedicated estimation algorithms.

Let us illustrate the above development with a simple example which may be adapted to various contouring and deburring applications. Let us consider the situation depicted in Fig. 7.35. In this example, n is the unitary vector normal to the surface of the object at the contact point, C, and F is a frame, with origin C, rigidly linked to the tool. Let us assume that the contact is elastic in the direction n. Then, the stiffness matrix, expressed in F, is:

$$K = k \begin{pmatrix} nn^T & 0 \\ 0 & 0 \end{pmatrix} \bigg|_F \quad (7.6.73)$$

where k is a coefficient of stiffness and n is the (3×1) matrix of coordinates of the unitary normal vector, here expressed in the basis of F.

A possible sensor-based task then consists of controlling the intensity of the normal force f_n so as to always have it equal to a given value σ. This suggests choosing

$$e_1 = \langle n, f \rangle - \sigma. \quad (7.6.74)$$

Since f is the vector of the contact screw F, (7.6.74) may also be written

$$e_1 = WF - \sigma |_F \quad (7.6.75)$$

with

$$W^T = \begin{pmatrix} n \\ 0 \end{pmatrix} \bigg|_F . \quad (7.6.76)$$

From (7.6.73) and (7.6.75), it is easy to verify that

$$R(W^T) = R(K|_F). \quad (7.6.77)$$

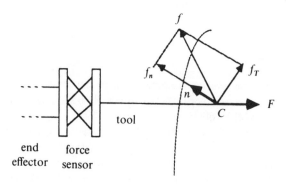

Fig. 7.35 A simple case of force control

7.6 TASKS BASED ON THE USE OF FORCE SENSORS

This choice of W is thus compatible with the aforementioned methodology. According to (7.6.70) and (7.6.71), a possible global task function then is

$$e = SF - \sigma \binom{n}{0} + \alpha(I_6 - S)g_s^r|_F \qquad (7.6.78)$$

with, in this case

$$S = \begin{pmatrix} nn^T & 0 \\ 0 & 0 \end{pmatrix}\bigg|_F. \qquad (7.6.79)$$

In practice, the normal vector $n(t)$ along the ideal trajectory of the robot has to be estimated. If the surface of the object is planar and its motion known with respect to the fixed frame F_0, then this vector may be directly deduced from the initial knowledge of $n(0)$, at time $t = 0$. Otherwise, this vector may be determined from the measurement of the wrench $F|_F$. Indeed, if the contact is frictionless, it is known that the vector f (equal to f_N in this case) of this wrench is collinear to n. A more difficult case is when a tangential friction force appears at the contact. Then

$$f = f_N + f_T. \qquad (7.6.80)$$

Notice that, if this friction force depends on the velocity of the end-effector with respect to the object, then our initial assumption according to which the sensor signals only depend on the relative position of the sensing system with respect to the target is not satisfied. In this case, the control of the component f_T is not directly covered by the approach. However, the control of the normal component f_N is still possible because this component remains a function of the position of the sensing system. Of course, the estimation of n is not as straightforward as in the frictionless case since f is no longer collinear to n. A possibility then consists of using a model of the friction force f_T in (7.6.80) so as to obtain an estimate of f_N from which an estimate of n can be deduced.

Case 2: $n > 6$. In this case, we have to work in the joint space directly. As before, a possible sensor-based task vector is

$$e_1 = W\tilde{F}|_F \qquad (7.6.81)$$

where W is chosen such that

$$R(W^T) = R(K_1|_F) \qquad (7.6.82)$$

$$\subseteq R(K|_F).$$

Then

$$\frac{\partial e_1}{\partial q} = WK|_F \begin{pmatrix} R^T & 0 \\ 0 & R^T \end{pmatrix} J \qquad (7.6.83)$$

where R is the rotation matrix associated with the passage from the fixed frame F_0 to the frame F linked to the sensing system, and J is the basic Jacobian matrix associated with the body to which the frame F is linked.

If the second objective consists of minimizing the cost function $h_s(q, t)$ under the constraint $e_1 = 0$, then a possible global task function is (cf. Chapter 4)

$$e = B^T e_1 + \alpha(I_n - B^+ B)\frac{\partial h_s}{\partial q} \quad (7.6.84)$$

where the matrix function B has to be chosen such that $R(B^T) = R((\partial e_1/\partial q)^T)$ in order to have the decoupling property P_1 satisfied. Recall that this property is also equivalent to the existence of a regular matrix C such that $\partial e_1/\partial q = CB$. Moreover, if C is positive then the property P_2 is satisfied and the global task Jacobian $J_T = \partial e/\partial q$ is associated with a positivity property that can be exploited at the control level (cf. Chapter 4).

Recalling that we have

$$(I - W^+ W)K|_F W^T = 0 \quad (7.6.85)$$

it is easy to verify that a possible choice for B is

$$B = (W^+)^T \begin{pmatrix} R^T & 0 \\ 0 & R^T \end{pmatrix} J. \quad (7.6.86)$$

Indeed, from (7.6.83), (7.6.85) and (7.6.86), we have

$$\frac{\partial e_1}{\partial q} = CB \quad (7.6.87)$$

with

$$C = WK|_F W^T. \quad (7.6.88)$$

If the contact is elastic, this is a positive matrix.

7.7 CONCLUDING REMARKS

In the light of the previous study, let us quickly summarize the similarities and differences between non-contact sensor-based tasks and force-based tasks. The first point is that the task-function approach applies to both problems. At a conceptual level, it has been shown how non-contact sensor-based tasks may be interpreted in the force framework through the notion of 'virtual linkage'. Namely, a sensor signal is the intensity of a local 'virtual force' of contact, the direction of which depends on the model of interaction between the sensor and the environment. For example, when the sensor signal is a function of the distance sensor/object evaluated along the direction of observation of the sensor, the direction of the virtual force is normal to the surface of the object at the 'virtual point of contact' located at the intersection of the direction of observation of the sensor with the surface of the object. This corresponds to the case of a physical contact without friction. The interaction matrix L associated with the sensing system has the significance of a calibration matrix, while the symmetric positive matrix LL^T may be interpreted as a virtual stiffness matrix characterizing the stiffness of the virtual contact.

The differences between the two classes of tasks come, in part, from the physical characteristics of *real* contacts. For example, 'parasitic' friction forces may appear in a real contact. These forces then have to be modelled so as to eliminate their contribution to the measured contact force. In this respect, we may say that virtual contacts are easier to control than real contacts. Also, the 'equivalent' stiffness associated with a real contact, which must not be confused with the stiffness of the force sensor itself, may be very high if no particular attention is paid to this issue. As explained before, this may yield control stability problems. This type of problem may even be amplified by the existence of distributed structural flexibilities. In the case of non-contact sensors, since the stiffness matrix LL^T depends on the characteristics of the sensor signals, it is usually not too difficult to provide the system with enough compliance. A counterpart of this facility is that some knowledge of the interaction matrix is needed in the case of non-contact sensors, especially with regard to the fulfilment of a useful positivity property.

In both cases, fine control of sensor-based tasks may require the use of dedicated estimation algorithms (estimation of the normal of the sensed object, for example) or filtering techniques, the design of which will motivate future studies.

7.8 BIBLIOGRAPHIC NOTE

The analysis of force control techniques has been done by several authors since about 1975. Basic tools issued from the configuration space have been studied for example by Mason (1981, 1982), and Raibert and Horn (1978); impedance control by Hogan (1981, 1984, 1987); and hybrid control schemes by Anderson and Spong (1987), Craig and Raibert (1981), Guinot et al. (1985), Merlet (1986, 1987), Paul (1987), Salisbury (1985), Reboulet and Robert (1985), Zhang and Paul (1985), Bidaud and Guinot (1987), among others. Studies of instabilities in force control may be found for example in Townsend and Salisbury (1987), An and Hollerbach (1987b), and Eppinger and Seering (1987). Proximity-based control problems have been analysed initially from the point of view of telemanipulation by Andre (1983), Andre and Boulic (1985a), Andre and Fournier (1985b), Bejczy (1980), Hirzinger (1985), Wampler (1984). The application to robotics through the approach presented in this chapter is new for it allows connection of several kinds of sensor-based tasks, and also because it lies in a coherent control framework. However, first steps in this direction may be found in the works of the authors [see Espiau (1984, 1985a), Leborgne and Espiau (1984) and Samson (1987a, b) for example]. An overview of optical sensing techniques is given in Espiau (1986b). A study of normal forms for stiffness and compliance matrices is presented in Loncaric (1987). Force control analysis and passive compliance ('RCC') have been studied in detail by Whitney (1982, 1985). Among examples of force sensors, the description of a representative device may be found in Gaillet and Reboulet (1983).

Figure 7.27 is taken from Boulic (1986). Figure 7.28 is from Merlet (1986) and Figure 7.29 was provided by J. P. Merlet (private communication). Another example of a parallel manipulator is given in Inoue (1985). Some basic mechanics of stiffness and linkages may be found, for example, in Bamberger (1981).

APPENDIX A7.1 A MODEL OF OPTICAL PROXIMITY SENSORS

Consider the simple sensory system shown in Fig. A7.1. The emitter is a light-emitting diode (LED) or a laser diode and the receiver is a photodiode or phototransistor. Whatever the driving mode of the emitter (continuous, pulsed, modulated), we may derive a simplified model of the received flux, on which the output signal s depends. With the classical underlying photometric assumptions, its general form is

$$s = \int_\Sigma c \frac{\cos^{n_1}\alpha_1 \cos^p\beta_1}{r_1^2} \frac{\cos^{n_2}\alpha_2 \cos^p\beta_2}{r_2^2} d\Sigma \qquad (A7.1.1)$$

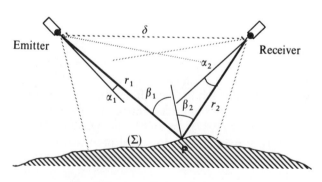

Fig. A7.1 An optical proximity sensor

APPENDIX A7.1 A MODEL OF OPTICAL PROXIMITY SENSORS

where the positive parameters n_1, n_2, p and c depend mainly on the photometric properties of emitter, receiver, and object. Σ depends on the shapes of the sensitive areas of the optical components, on the geometry of the object itself and on its position relative to the sensor. Obviously, in the general case, these parameters are not well known. Although it is sometimes possible to measure some of these related to the sensor, the characteristics of the object are often unknown in practice. This shows that the output, s, includes the contribution of poorly known phenomena. In practice simplified versions of the model (A7.1.1) may be considered instead.

Example A7.1. Figure A7.2 shows a model in which emitter and receiver are merged, while facing a plane target orthogonal to their common axis. The object is assumed to be perfectly diffusely reflecting, with coefficient λ. In this case (A7.1.1) reduces to

$$s = c'\lambda \int_{\Sigma} \frac{\cos^h \alpha}{d^4} \, d\Sigma \qquad (A7.1.2)$$

with

$$h = 2p + n_1 + n_2 + 4. \qquad (A7.1.3)$$

Σ is simply a disc determined by the maximum sensor aperture Φ and the distance d; thus

$$d\Sigma = 2\pi r \, dr = 2\pi d^2 \frac{\tan \alpha}{\cos^2 \alpha} \, d\alpha \qquad (A7.1.4)$$

and

$$s = \frac{2\pi c'\lambda}{d^2} \int_0^{\Phi} \sin \alpha \cos^{h-5} \alpha \, d\alpha \qquad (A7.1.5)$$

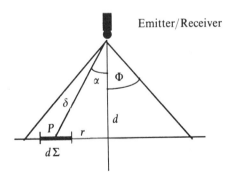

Fig. A7.2 Simplified model of optical reflectance sensors

which is of the form

$$s = \frac{\mu}{d^2}. \tag{A7.1.6}$$

Note that even in this simple case the sensory output still involves an unknown parameter, namely the diffusivity, λ, of the object.

Example A7.2 When the object reduces to a pin-point diffusely reflecting target located in P, the output signal, from equation (A7.1.2), simply takes the form

$$s = v \frac{\cos^{h-4}\alpha}{\delta^4} = \frac{h(\alpha)}{\delta^4}. \tag{A7.1.7}$$

APPENDIX A7.2 COMBINATION OF GENERALIZED SPRINGS

When p generalized springs are serially linked (cf. Fig. 7.30), it is possible to associate with this system a single equivalent generalized spring. The stiffness matrix associated with each generalized spring is denoted as K^i, and is here assumed to be non-singular. It is generally naturally expressed in frame F_i, thus written K_i^i). Let us also denote by $\tau_{i,j}$ the screw of small displacements (i.e. homogeneous to velocities as an element of se_3) of frame F_i with respect to frame F_j. We thus may write

$$F = -K_i^i \tau_{i,i-1}|_{F_i}, \quad i = 1, \ldots, p. \tag{A7.2.1}$$

The equivalent generalized spring between F_p and F_0 with matrix K_k expressed in a frame F_k is given by

$$F = -K_k \tau_{p,0}|_{F_p}. \tag{A7.2.2}$$

We have also

$$\tau_{p,0} = \sum_{i=1}^{p} \tau_{i,i-1} \tag{A7.2.3}$$

with the screw change of frames formulae

$$F|_{F_i} = \Theta_{ik} F|_{F_k} \tag{A7.2.4}$$

$$\tau_{i,i-1}|_{F_k} = \Theta_{ik}^* \tau_{i,i-1}|_{F_i} \tag{A7.2.5}$$

$$\Theta_{ik} = \begin{pmatrix} R_{ik} & 0 \\ As(O_k O_i)|_{F_i} R_{ik} & R_{ik} \end{pmatrix} \tag{A7.2.6}$$

where Θ^* is the adjoint of Θ in the sense of the bilinear form given in Chapter 1, i.e.

$$\Theta_{ik}^* = \begin{pmatrix} R_{ik}^T & R_{ik}^T As(O_i O_k)|_{F_i} \\ 0 & R_{ik}^T \end{pmatrix} \tag{A7.2.7}$$

APPENDIX A7.2 COMBINATION OF GENERALIZED SPRINGS

R_{ik} being the usual rotation matrix from F_i to F_k. Equation (A7.2.1) yields

$$\tau_{i,i-1} = -(K_i^i)^{-1} F |_{F_i}. \qquad (A7.2.8)$$

From equations (A7.2.4) to (A7.2.8)

$$\tau_{i,i-1} |_{F_k} = -\Theta_{ik}^* (K_i^i)^{-1} \Theta_{ik} F |_{F_k}. \qquad (A7.2.9)$$

From equations (A7.2.2), (A7.2.3) and (A7.2.9)

$$-(K_k)^{-1} F |_{F_k} = \sum_{i=1}^{p} \Theta_{ik}^* (K_i^i)^{-1} \Theta_{ik} F |_{F_k} \qquad (A7.2.10)$$

which shows that the equivalent compliance is given by

$$(K_k)^{-1} = \sum_{i=1}^{p} \Theta_{ik}^* (K_i^i)^{-1} \Theta_{ik}. \qquad (A7.2.11)$$

Moreover, it is easy to check that in a change of frame F_i to F_k, a stiffness matrix is transformed through

$$K_k = \Theta_{ik} K_i \Theta_{ik}^*. \qquad (A7.2.12)$$

Using (A7.2.12) in (A7.2.11), we obtain finally the expression of the equivalent stiffness matrix

$$(K_k)^{-1} = \sum_{i=1}^{p} (K_k^i)^{-1}. \qquad (7.2.13)$$

REFERENCES

Aldon, M. J. (ed.) (1986). *Outils mathématiques pour la modélisation et la commande des robots*. CNRS, Paris, France.

An, C. H., Atkeson, C. G., Griffiths, J. D., and Hollerbach, J. M. (1987a). Experimental evaluation of feedforward and computer torque control. *Proceedings of the IEEE conference on robotics and automation, Raleigh*, pp. 165–8.

An, C. H. and Hollerbach, J. M. (1987b). Dynamic stability issues in force control of manipulators. *Proceedings of the IEEE conference on robotics and automation, Raleigh*, pp. 890–6.

Anderson, R. J. and Spong, M. W. (1987). Hybrid impedance control of robotic manipulators. *IEEE conference on robotics and automation, Raleigh*, pp. 1073–80.

André, G. (1983). Conception et modélisation de systèmes de perception proximétrique. Application à la commande en téléopération. Thesis. University of Rennes.

André, G. and Boulic, R. (1985a). Système graphique et capteurs proximétriques pour la programmation de robots. *Proceedings of 4th European conference on CAD/CAM and computer graphics*, pp. 185–98. Hermès, Paris.

André, G. and Fournier, R. (1985b). Generalized end effector control in a computer aided teleoperation system with application to motion coordination of a manipulator arm on an oscillating carrier. *International Conference on Advanced Robotics, ICAR 85*, pp. 337–44. Tokyo.

André, P., Kauffmann, J. M., Lhote, F., and Taillard, J. P. (1983). *Les robots (Tome 4) constituants technologiques*. Hermès, Paris.

Asada, H. and Kanade, T. (1983a). Design of direct-drive mechanical arms. *ASME Journal of Vibrations, Acoustics, Stress and Reliability in Design*, 105, 312–6.

Asada, H., Kanade, T., and Takeyama, I. (1983b). Control of a direct-drive arm. *Journal of Dynamic Systems, Measurement, and Control*, 105, 136–42.

Asada, H. and Slotine, J. J. E. (1986). *Robot analysis and control*. Wiley, New York.

Astrom, K. J. and Wittenmark, B. (1984). *Computer controlled systems*. Prentice Hall, Englewood Cliffs.

Baillieul, J., Hollerbach, J., and Brockett, R. (1984). Programming and control of kinematically redundant manipulators. *Proceedings of the 23rd conference on decision and control, Las Vegas*, pp. 768–74.

Baillieul, J. (1986). Avoiding obstacles and resolving kinematic redundancy. *Proceedings of the IEEE conference on robotics and automation, San Francisco*, pp. 722–8.

Baillieul, J. (1987). A constrained approach to inverse problems for kinematically redundant manipulators. *Proceedings of the IEEE conference on robotics and automation, Raleigh*, pp. 1827–33.

Bamberger, F. (1981). *Mécanique de l'ingénieur: systèmes de corps rigides*. Hermann, Paris.

Bejczy, A. K. (1980). Smart sensors for smart hands. *Progress in Astronautics and Aeronautics*, 67.

REFERENCES

Ben-Israel, A. and Greville, T. N. E. (1974). *Generalized inverses: theory and applications*. Wiley, New York.
Bidaud, P. and Guinot, J. C. (1987). Application for a manipulator-gripper in an assembly cell. *IEEE conference on robotics and automation, Raleigh*, pp. 2021–32.
Birkhoff, G. and Rota, G. C. (1989). *Ordinary differential equations*. (4th ed). Wiley, New York.
Bogoliubov Jr., N. N. and Mitropolskii, Y. A. (1961). *Asymptotic methods in the theory of non-linear oscillations*. Gordon and Breach, New York.
Boissonnat, J. D., Faverjon, B., and Merlet, J. P. (1988). *Techniques de la robotique: architectures et commande (Tome 1); perception et planification (Tome 2)*. Hermès, Paris.
Borrelly, J. J. and Le Borgne, M. (1988). Implémentation multiprocesseurs d'algorithmes de commande. In *Techniques de la Robotique: architectures et commande (Tome 1)*, pp. 171–93. Hermès, Paris.
Boulic, R. (1986). Conception assistée par Ordinateur de boucles de commande avec capteurs en robotique et en Téléopération. Engineer-Doctor Thesis. University of Rennes.
Boullion, T. L. and Odell, P. L. (1971). *Generalized inverse matrices*. Wiley, New York.
Bremer, H. and Truckenbrodt, A. (1984). Robust control for industrial robots. *Proceedings of the Fifth CISM-IFTOMM symposium on theory and practice of robots and manipulators, Udine*, pp. 195–203. Kogan Page and Hermès.
Brockett, R. W. (1984). Robotic manipulators and the product of exponentials formula. In *Mathematial theory of networks and systems*, pp. 120–29. Springer Verlag Berlin-New York-Paris.
Campbell, S. L. and Meyer, C. D. (1979). *Generalized inverses of linear transformations*. Pitman.
Chevallereau, C. and Khalil, W. (1987). Efficient method for the calculation of the pseudo-inverse kinematics problem. *Proceedings of the IEEE conference on robotics and automation, Raleigh*, pp. 1842–8.
Coiffet, P. (1981). *Les robots (Tome 1) Modélisation et commande*. Hermès, Paris.
Craig, J. J. and Raibert, M. H. (1981). A systematic method of hybrid position/force control of a manipulator. *Journal of Dynamic Systems, Measurement and Control*, **103**, 126–33.
Craig, J. J. (1986). *Introduction to robotics: mechanics and control*. Addison Wesley.
Craig, J. J. (1987). *Adaptive control of mechanical manipulators*. Addison Wesley.
D'Andrea, B. and Levine, J. (1986). CAO for non-linear systems decoupling, perturbation rejection, and feedback linearization with applications to the dynamic control of a robot arm. In *Algebraic and geometric methods in non-linear control theory*, pp. 545–72. Reidel, Dordrecht.
D'Andrea-Novel, B. (1987). Sur la commande d'une classe de systèmes mécaniques. Thesis, Ecole des Mines de Paris.
Desa, S. and Roth, B. (1985). Mechanics: kinematics and dynamics. In *Recent advances in robotics*, pp. 71–130. Wiley, New York.
Desoer, C. A. and Vidyasagar, M. (1975). *Feedback systems: input-output properties*. Academic Press, New York.
Dieudonné, J. and Huet, D. (1968). Eléments d'analyse (Tome 1). Gauthier-Villars.

REFERENCES

Dombre, E. and Khalil, W. (1988). *Modélisation et commande des robots.* Hermès, Paris.

Dubowsky, S. and Desforges, D. T. (1979). The application of model referenced adaptive control to robotic manipulators. *Journal of Dynamic Systems, Measurement and Control*, **101**, 193–200.

Egardt, B. and Samson, C. (1982). Stable adaptive control of non-minimum phase systems. *Systems and Control Letters*, **2**, 137–43.

Eppinger, S. D. and Seering, W. P. (1987). Understanding bandwidth limitations in robot force control. *Proceedings of the IEEE Conference of Robotics and Automation*, Raleigh, pp. 904–9.

Espiau, B. (1984). Note sur l'interprétation de primitives d'actions proximétriques en termes de liaisons cinématiques fictives. *Research report no. 242*, INRIA.

Espiau, B. (1985a). Use of optical reflectance sensors. In *Recent advances in robotics*, pp. 313–58. Wiley, New York.

Espiau, B. and Boulic, R. (1985b). Collision avoidance for redundant robots with proximity sensors. *Proceedings of the third international symposium on robotics research*, Gouvieux, pp. 243–51. MIT Press, Cambridge.

Espiau, B. (1986a). Commande de systèmes redondants et évitement d'obstacles. *Research report no. 495*, INRIA.

Espiau, B. (1986b). An overview of local environment sensing in robotics application. NATO-ARW *Sensors and sensory systems for advanced robotics*, series F, Vol. 43, (ed. P. Dario) pp. 125–51. Springer-Verlag.

Faurre, P. and Robin, M. (1984). *Eléments d'automatique*. Dunod, Paris.

Faverjon, B. (1984). Obstacle avoidance using an octree in the configuration space of a manipulator. *Proceedings of the IEEE conference on robotics and automation*, Atlanta, pp. 504–12.

Fournier, A. (1980). Génération de mouvements en robotique; applications des inverses généralisées et des pseudo-inverses. Thesis. Université de Montpellier.

Freund, E. (1982). Fast non-linear control with arbitrary pole placement for industrial robots and manipulators. *International Journal of Robotics Research*, **1**, 65–79.

Freund, E. and Hoyer, H. (1984). Collision avoidance in multirobot systems. *Second international symposium on robotics research*, pp. 135–46. MIT Press, Cambridge.

Freund, E. and Hoyer, H. (1985). Collision avoidance for obstacles and mobile robots in multirobots systems *Proceedings of the third international symposium on robotics research*, Gouvieux, pp. 253–62. MIT Press, Cambridge.

Gabay, D. and Luenberger, D. G. (1976). Efficiently converging minimization methods based on the reduced gradient. *SIAM Journal of Control and Optimization*, **14**, 42–61.

Gaillet, A. and Reboulet, C. (1983). An isostatic six component force and torque sensor. *Proceedings of the 13th international symposium on industrial robots*, pp. 18–102–11.

Gorla, B. and Renaud, M. (1984). *Modèles des robots manipulateurs, application à leur commande*. Cepadues, Toulouse.

Guinot, J. C., Bidaud, J., and Zeghloul, S. (1985). Modelisation and simulation of a force-position control for a manipulator gripper. *Proceeding of the third international symposium on robotics research*, Gouvieux, pp. 315–19. MIT Press, Cambridge.

REFERENCES

Ha, I. J. and Gilbert, E. G. (1985). Robust tracking in non-linear systems and its applications to robotics. *Proceedings of the 24th conference on decision and control, Fort Lauderdale,* pp. 1009–17.

Hirzinger, G. (1985). Robot learning and teach-in based on sensory feedback. *Proceedings of the third international symposium on robotics research, Gouvieux,* pp. 155–63. MIT Press, Cambridge.

Hogan, N. (1981). Programmable impedance control of industrial manipulators *Proceedings of a conference on CAD/CAM technology,* MIT, Cambridge.

Hogan, N. (1984). Impedance control of industrial robots. *Robotics and Computer-Integrated Manufacturing,* **1,** 97–113.

Hogan, N. (1987). Stable execution of contact tasks using impedance control. *Proceedings of the IEEE Conference on Robotics and Automation,* pp. 1047–54. Raleigh.

Hollerbach, J. M. (1980). A recursive Lagrangian formulation of manipulator dynamics and.a comparative study of dynamics formulation complexity. *IEEE Transactions on Systems, Man and Cybernetics,* **10,** 730–6.

Hollerbach, J. M. and Suh, K. C. (1985). Redundancy resolution of manipulators through torque optimization. *Proceedings of the IEEE conference on robotics and automation, Saint Louis* pp. 1016–21.

Horowitz, R. and Tomizuka M. (1980). An adaptive-control scheme for mechanical manipulators; compensation of non-linearity and decoupling control. Presentation at the winter annual meeting of ASME, paper 80-WA/DSC-6, Dynamics System and Control Division, Chicago.

Inoue, H. (1985). Parallel manipulator, *Proceedings of the third international symposium on robotics research,* pp. 321–7. MIT Press, Cambridge.

Isidori, A. (1985). Control of non-linear systems via dynamic state feedback. In *Algebraic and geometric methods in non-linear control theory,* (ed. M. Fliess and M. Hazewinkel) pp. 121–45. Reidel, Dordrecht.

Isidori, A. (1986). *Non-linear control systems: an introduction.* Springer Verlag.

Kailath, T. (1980). *Linear systems.* Prentice Hall, Englewood Cliffs.

Khalil, W. (1978). Contribution a la commande automatique des manipulateurs avec l'aide d'un modèle mathématique des mecanismes. Thesis. University of Montpellier.

Khalil, W., Liègois, A., and Fournier A. (1979). Commande dynamique de robots. *RAIRO automatique/systems analysis and control,* **13,** 189–201.

Khatib, O. (1980). Commande dynamique dans l'espace operationnel des robots manipulateurs en presence d'obstacles. Thesis. University of Toulouse.

Khatib, O. (1985a). The operational space formulation in the analysis, design and control of robot manipulators. *Proceedings of the third international symposium on robotics research, Gouvieux,* pp. 263–70. MIT Press, Cambridge.

Khatib, O. (1985b). Real time obstacle avoidance for manipulators and mobile robots. *Proceedings of the IEEE conference on robotics and automation, Saint-Louis,* pp. 500–5.

Khatib, O., Slotine, J. J., and Rath, D. (1987). Robust control in operational space for goal-positioned manipulator tasks. *Proceedings of the 1987 international conference on advanced robotics,* pp. 503–12. IFS Publ./Springer Verlag.

Kircanski, M. and Vukobratovic, M. (1984). Trajectory planning for redundant

REFERENCES

manipulators in the presence of obstacles. *Proceedings of the fifth CISM-IFTOMM symposium "Romansy"* pp. 57–63. Kogan Page and Hermes.

Klein, C. A. and Ching-Hsiang, H. (1983). Review of pseudo-inverse control for use with kinematically redundant manipulators. *IEEE Transactions on Systems, Man and Cybernetics*, **SMC-13**.

Klein, C. A. (1984). Use of redundancy in the design of robotic systems. *Proceedings of the second international symposium on robotics research*, pp. 207–14. MIT Press, Cambridge.

Kleinfinger, J. F. (1986). Modélisation dynamique de robots à chaine cinématique simple, arborescente ou fermée, en vue de leur commande. Thesis. Ecole Nationale Supérieure de Mécanique, Nantes.

Klema, V. C. and Laub, A. J. (1980). The singular value decomposition: its computation and some applications. *IEEE Transactions on Automatic Control*, **AC-25**, 164–76.

Koivo, A. J. and Paul, R. P. (1980). Manipulator with self tuning controller. *Proceedings of the IEEE Conference on Cybernetics and Society, Massachussets*, pp. 1085–9.

Krogh, B. H. (1984). A generalized potential field approach to obstacle avoidance control. *SME conference proceedings on robotics research: the next five years and beyond*. Bethlehem, Pennsylvania.

Landau, I. D. (1979). *Adaptive control: the model reference approach*. Marcel Dekker, New York.

Landau, I. D. and Dugard, L. (1986). *Commande adaptative; aspects pratiques et théoriques*. Masson.

De Larminat, P. (1984). On the stabilization condition in indirect adaptive control. *Automatica*, **20**, 793–6.

Lathrop, R. H. (1984). Parallelism in manipulator dynamics. Ph.D. Thesis. MIT, Cambridge.

Le Borgne, M. and Espiau, B. (1984). Modelling and closed-loop control of robots in local operating space. *Proceedings of the IEEE conference on decision and control, Las Vegas*, pp. 1610–15.

Le Borgne, M. (1985). Modélisation des robots manipulateurs rigides. Research report no. 248, IRISA, University of Rennes.

Le Borgne, M. (1987). Quaternions et controle sur l'espace des rotations. Research report no. 377, IRISA, University of Rennes.

Lelic, M. A. and Wellstead, P. E. (1987). Generalized pole-placement self tuning controller; Part 1: Basic algorithm; Part 2: application to robot manipulator control. *International Journal of Control*, **46**, 547–68, 569–602.

Liégeois, A. (1977). Automatic supervisory control of the configuration and behaviour of multibody mechanisms. *IEEE Transactions on Systems, Man and Cybernetics*, **SMC-7**, 868–71.

Lim, K. Y. and Eslami, M. (1987). Robust adaptive controller designs for robot manipulator systems. *IEEE Journal of Robotics and Automation*, **RA-3**, 54–66.

Lilly, K. W. and Orin, D. E. (1986). Multiprocessor implementation of dynamic control schemes for robot manipulators. *ASME computers in engineering conference, Chicago*.

REFERENCES

Liu, M.-H. and Lin, W. (1987). Pole assignment self tuning controller for robotics manipulators. *International Journal of Control*, **46**, 1307–18.

Ljung, L. and Soderstrom, T. (1983). *Theory and practice of recursive identification*. MIT Press, Cambridge.

Ljung, L. (1987). *System identification: theory for the user*. Prentice Hall, Englewood Cliffs.

Loncaric, J. (1987). Normal forms of stiffness and compliance matrices. *IEEE Journal of Robotics and Automation*, **RA-3**, 567–72.

Luh, J. Y. S., Walker, M. W., and Paul, R. (1980a). On-line computational scheme for mechanical manipulators. *ASME Transactions, Journal of Dynamic Systems, Measurement and Control*, **102**, 69–76.

Luh, J. Y. S., Walker, M. W., and Paul, R. P. (1980b). Resolved acceleration control of mechanical manipulators. *IEEE Transactions on Automatic Control*, **AC-25**, 468–73.

Luh, J. Y. S. and Lin, C. S. (1982). Scheduling of parallel computation for a computer controlled manipulator. *IEEE Transactions on Systems, Man and Cybernetics*, **SMC-12**, 214–34.

Maciejewski, A. A. (1984). Obstacle avoidance for kinematically redundant manipulators. M.S. Thesis. The Ohio State University, Columbus.

Marino, R. (1986). Feedback linearization techniques in robotics and power systems. In *Algebraic and geometric methods in non-linear control theory*, pp. 523–43. Reidel, Dordrecht.

Markiewicz, B. R. (1973). Analysis of the computed torque drive method and comparison with conventional position servo for a computer-controlled manipulator. Jet Propulsion Laboratory technical memorandum 33-601. Pasadena.

Mason, M. T. (1981). Compliance and force control for computer controlled manipulators. *IEEE Transactions on Systems, Man and Cybernetics*, **SMC-11**, 418–32.

Mason, M. T. (1982). Compliance; compliance and force control for computer-controlled manipulators. In *Robot motion-planning and control*. (ed. Brady et al.), pp. 305–22 and 373–404. The MIT Press, Cambridge.

Merlet, J. P. (1986). Contribution à la formalisation de la commande par retour d'efforts. Application à la commande de robots parallèles. Thesis. University of Paris.

Merlet, J. P. (1987). C-surface theory applied to the design of an hybrid force-position controller. *Proceedings of the IEEE international conference on robotics and automation, Raleigh*, pp. 1055–9.

Middleton, R. H. and Goodwin, G. C. (1988). Adaptive computed torque control for rigid link manipulators. *System and Control Letters*, No. 10.

Mufti, I. H. (1985). Model reference adaptive control for manipulators: a review. Proceedings of the IFAC symposium on robot control, Barcelona, pp. 27–32.

Nagano, T. (1966). Linear differential systems with singularities and applications to transitive Lie algebras. *Journal of the Mathematical Society of Japan*, No. 18, 398–404.

Nakamura, Y. and Hanafusa, H. (1984). Task priority based redundancy control of robot manipulators. *Proceedings of the second international symposium on robotics research, Kyoto*, pp. 155–62. MIT Press, Cambridge.

Orin, D. E. and Schrader, W. W. (1983). Efficient computation of the Jacobian for robot manipulators. *Proceedings of the first international symposium on robotics research, Bretton Woods*, (eds. M. Brady and R. Paul), pp. 727–34. MIT Press, Cambridge.

Orin, D. E., Olson, K. W., and Chao, H. H. (1986). Systolic architectures for computation of the Jacobian for robots manipulators. In *Specialized computer architectures for robotics and automation*. Gordon and Breach, New York.

Paden, B. and Panja, R. (1988). Globally asymptotically stable 'PD + ' controller for robot manipulators. *International Journal of Control*, 47, 1697–712.

Paul, R. P. C. (1982). *Robot manipulators: mathematics, programming and control*. MIT Press, Cambridge.

Paul, R. P. C. (1987). Problems and research issues associated with the hybrid control of force and displacement. *Proceedings of the IEEE international conference on robotics and automation, Raleigh*, pp. 1966–71.

Praly, L. (1983). Robustness of indirect adaptive control based on pole placement design. *IFAC workshop on adaptive systems on control and signal, San Francisco*.

Raibert, M. H. and Horn, B. K. P. (1978). Manipulator control using the configuration space method. *Industrial robot*, 5, 69–73.

Reboulet, C. and Robert, A. (1985). Hybrid control of a manipulator with an active compliant wrist. *Proceedings of the third international symposium on robotics research, Gouvieux*, pp. 237–41. MIT Press, Cambridge.

Renaud, M. (1981). Geometric and kinematics models of a robot manipulator: calculation of the Jacobian matrix and its inverse. *Proceedings of the 11th international symposium on industrial robots, Tokyo*, pp. 757–64.

Renaud, M. (1983). An efficient iterative analytical procedure for obtaining a robot manipulator dynamic model. *Proceedings of the first international symposium on robotics research, Bretton Woods*, (eds. M. Brady and R. Paul), pp. 749–66. MIT Press, Cambridge.

Salisbury, J. K. (1985). Kinematic and force analysis of articulated hands. In *Recent advances in robotics*, pp. 131–74. Wiley, New York.

Samson, C. (1983a). Commande non-linéaire robuste des robots manipulateurs. Research report no 184, IRISA, University of Rennes.

Samson, C. (1983b). Robust non-linear control of robotic manipulators. *Proceedings of the 22nd IEEE conference on decision and control, San Antonio*, pp. 1211–16.

Samson, C. (1983c). Problèmes en identification et commande de systèmes dynamiques. Thesis. Université of Rennes.

Samson, C. (1987a). An approach for the synthesis and analysis of the robot manipulator's control. Research report no. 356 IRISA, University of Rennes.

Samson, C. (1987b). Robust control of a class of non-linear systems and application to robotics. *International Journal of Adaptive Control and Signal Processing*, 1, 49–68.

Schwartz, J. T. (1969). *Non-linear functional analysis*. Gordon and Breach, New York.

Siestrunck, R. (1973). Quelques aspects nouveaux de la théorie des mécanismes et applications. *Revue Francaise de Mécanique*, No. 45.

Silver, W. (1982). On the equivalence of Lagrangian and Newton–Euler dynamics for manipulators. *International Journal of Robotics Research* 1(2), 60–70.

Slotine, J. J. E. and Sastry, S. S. (1983). Tracking control of non-linear systems using sliding surfaces, with application to robot manipulators. *International Journal of Control*, **38**, 465–92.

Slotine, J. J. E. (1985). The robust control of robot manipulators. *International Journal of Robotics Research*, **4**, 49–64.

Slotine, J. J. E. and Li, W. (1987). On the adaptive control of robot manipulators. *International journal of robotics research*, **6**, 49–59.

Solomon, A. and Davison, E. J. (1983). Design of controllers to solve the robust servo mechanism problem for a class of non-linear systems. *Proceedings of the 22nd IEEE conference on decision and control, San Antonio*, pp. 335–41.

Spivak, M. (1970). A comprehensive introduction to differential geometry. Publish or Perish, Boston.

Spong, M. W., Thorp, J. S., and Kleinwaks, J. W. (1984). The control of robot manipulators with bounded control. Part II: Robustness and disturbance rejection. *Proceedings of the 23rd IEEE conference on decision and control, Las Vegas*, pp. 1047–52.

Spong, M. W. and Vidyasagar, M. (1985). Robust linear control of robot manipulators. *Proceedings of the 24th conference on decision and control, Ft Lauderdale*, pp. 1767–72.

Spong, M. W. and Vidyasagar, M. (1987). Robust linear compensator design for non-linear robotic control. *International Journal of Robotics and Automation*, **3** (4).

Suh, K. C. and Hollerbach, J. M. (1987). Local versus global torque optimization of redundant manipulators. *Proceedings of the IEEE conference on robotics and automation, Raleigh*, pp. 619–24.

Sussman, H. (1973). Orbits of families of vector fields and integrability of distributions. *Transactions of the American Mathematical Society*, No. 180, 171–88.

Takegaki, M. and Arimoto, S. (1981). An adaptive trajectory control of manipulators. *International Journal of Control*, **35**, 219–30.

Townsend, W. T. and Salisbury, J. K. (1987). The effect of Coulomb friction and stiction on force control. *Proceedings of the IEEE conference on robotics and automation, Raleigh*, pp. 887–9.

Utkin, V. (1978). *Sliding modes and their application in variable structure systems*. (Transl. P. Parnakh). MIR, Moscow.

Vukobratovic, M. and Potkonjak, V. (1982a). *Dynamics of manipulation robots: theory and applications*. Springer Verlag.

Vukobratovic, M. and Stokic, D. (1982b). *Control of manipulation robots: theory and applications*. Springer Verlag.

Wampler, C. (1984). Multiprocessor control of a telemanipulator with optical proximity sensors. *International Journal of Robotics Research*, **3**, 40–50.

Wampler II, C. W. (1987). Inverse kinematic functions for redundant manipulators. *Proceedings of the IEEE conference on robotics and automation, Raleigh*, pp. 610–17.

Whitney, D. E. (1982). Quasi-static assembly of compliantly supported rigid parts. *Journal of Dynamic Systems, Measurement and Control*, **4**.

Whitney, D. E. (1985). Historical perspective and stage of the art in robot force control. *Proceedings of the IEEE conference on robotics and automation, St Louis*, pp. 262–8.

REFERENCES

Wonham, W. M. (1985). *Linear multivariable control: a geometric approach.* Springer Verlag.

Yoshikawa, T. (1983). Analysis and control of robots manipulators with redundancy. *Proceedings of the first international symposium on robotics research, Bretton Woods*, pp. 735–48 (eds. M. Brady and R. Paul), MIT Press, Cambridge.

Yoshikawa, T. (1985a). Analysis and design of articulated robot arms from the viewpoint of dynamic manipulability. *Proceedings of the third international symposium on robotics research, Gouvieux*, pp. 273–9. MIT Press, Cambridge.

Yoshikawa, T. (1985b). Manipulability of robotics mechanisms. *International Journal of Robotics Research*, **4**, 3–9.

Young, K. D., Kokotovic, P. V., and Utkin, V. I. (1977). A singular perturbation analysis of high gain feedback systems. *IEEE Transactions on Automatic Control*, **AC-22**, 931–8.

Young, K. K. D. (1978). Controller design for a manipulator using theory of variable structure systems. *IEEE Transactions on Systems, Man and Cybernetics*, **SMC-8** 101–9.

Zhang, H. and Paul, R. P. (1985). Hybrid control of robot manipulators. *Proceedings of the IEEE conference on robotics and automation, St Louis*, pp. 602–7.

INDEX

actuators 169–71, 175, 216, 325, 326
adaptive control 207–15, 216
   synthesis 208–11
admissibility radius 90, 224, 237
aspects 70, 76
assembly tasks 324, 325
atlases 8, 22
attitude errors 26, 33, 35–6, 200–1
attitudes 18, 21–4, 63, 71
avoidability 144

backlash 171, 172, 175
bilinear forms 51, 327, 328, 350

canonical map (π) 30, 55
centrifugal forces 58, 172, 175, 211, 238, 239
chart representations 24–6
charts 8, 9, 24, 26
compact 24
compliance 328, 340
'computed torque method' 198
condition numbers 146
configuration spaces 18, 52
connectivity, single 64
Constant Rank Theorem 12, 15, 117
constraints 48, 303–6
   global 53–4
contacts, virtual 276, 290, 292, 324
contouring tasks 325
controllable systems 5
control schemes
   general 196
   hybrid 131, 134, 301, 347
   linearizing and decoupling 197–202
convexification 187, 195
coordinates, generalized 8, 57, 130–1
coordinate systems, local 8
Coriolis forces 58, 172, 175, 211, 238, 239
cost functions
   distance 150, 151
   energy-based 143, 157–60
   external 143, 149–57
   internal 143–9
   scalar 184
couples 46

C-surfaces 54, 55
current loops 180–2, 196, 203

d.c. motors 159, 169, 170, 202
   conventional control of 176–82
decoupling 183, 184, 194, 197, 216, 346
   ideal 246, 249
delays, measurement 247–8
Denavit–Hartenberg parameters 20–1, 55, 64
diffeomorphisms 9, 11, 24, 85, 258
   global 13
   local 183
differential equations 15–17
differential motion 139–42, 162
differential motion models 71–5, 76
discretization 246–7
dispersion 146
distributions 54
disturbance vectors 241
duality 52
dynamics 61–2

eigenvalues 2, 222, 256–7
eigenvectors 2
elasticity 171, 326, 327, 328, 329
energy, kinetic 158
equilibrium points 16–17
Euclidean norms 2, 4, 87
Euler angles 24, 25–6, 39
exponential map 33
extension 114

feedback 184, 196, 203, 247
fibres 9, 10
forces 18, 153
   contact 327, 337–9
   control of 326, 339–46
   measurement of 326
   virtual 291, 346
force sensors, *see under* sensors
frames 19, 267
   change of 48–9
   choice of 20, 315
   Galilean 57, 173
frequency conditions 336–7, 340

INDEX

friction forces 61, 172, 239, 345
  dry (Coulomb) 171, 238, 328
  viscous 170, 212, 328

gain positivity 201, 222
gains 225
  constant 235–7, 240
  non-linear 227, 237–9, 240
gears 246
geodesics 34–5
geometric models 62, 70
  with homogeneous matrices 63
  with quaternions 63
gradient method 169, 184, 185, 186, 187, 188–9
gradient regulation 187
gravity forces 61, 172, 238, 239
GTC (gradient-like control) 193–5, 196, 250, 252, 253
GVC (gradient and velocity control) 203–4

Hessian matrices 125, 162, 164, 175, 185–6, 195
holomorphic functions 255
homogeneous matrices 63

idempotency 4
identification 208–11, 213, 215
images 1, 38
Implicit Function Theorems 11–15, 79
inertia matrices 57, 158, 172
integral curve 15
interaction matrices, *see under* matrices
interaction screws, *see under* screws
Inverse Function Theorem 11
'inverse problem technique' 198
inverses, generalized 3–4, 162
isomorphisms 3

Jacobian matrices 15, 132–3, 232, 234, 295, 299
  computation of 72–5, 76, 138, 139
  rotational 71
  translational 71
joint coordinates 57
joint limitation avoidance 143, 146–7, 157
joints, rotational 59–60
joints, translational 60–1
joint space, control in 198–9
Jury's criterion 6, 247, 248, 335

Kalman filters 209, 210
kernels 1, 4

kinematics 59–61, 234
  direct 62–3
  inverse 64–71, 76, 106–8

Lagrange equations 57, 58
Lagrange multipliers 122, 161, 163, 164
Lagrangians 122
learning 106, 207
Lie algebra 48
Lie groups 36, 47
linearization 35, 183, 190, 210, 216, 236, 278
linear systems
  invariant 220–4
  stationary 4–8
linkages 53
  dissipative 328
  non-dissipative 328
  virtual 276, 278, 280, 308, 324, 346
    classes of 276, 291, 299, 308
location 18, 20, 25
loops
  closed 168, 236
  inner 171
  open 176–8, 243
Lyapunov method 16, 207

manifolds, differentiable 8–11, 18
manipulability 145, 162
manipulability function, generalized 145, 146
mappings, differentiable 9
  differential of 10
  rank of 11
matrices 1–3
  calibration 329, 330, 332, 341, 347
  compliance 328
  homogeneous 19, 20
  interaction 277, 279, 280, 308, 347
  kinetic energy 199, 223, 232, 237
  non-degenerate 3
  positive 2, 126, 166–7 *see also* positivity conditions
  selection 301, 343
  stiffness 293, 328, 329–30, 332, 342, 343, 350–1
matrix norms, *see* norms
measurement errors 240–5
mechanical impedance 340
minimization
  constrained 121–5, 133, 154, 160, 163–4
  methods, *see* gradient method; Newton method
  secondary cost 121, 132, 133, 251, 253, 297
  transient phase of 160–2

INDEX

models
  dynamic 56–62, 76
    general 57–9
    in joint space 174–5
    in task space 175
  open-loop 176–8
MRAC technique 207

Newton–Euler formalism 57, 59
Newton–gradient method 190
Newton method 169, 184, 185–6, 187, 189–90
Newton–Raphson method 64
noise 240, 243–5
norms 1–2, 4, 87, 259
NTC (Newton-type control) 191–3, 195, 196, 198, 249, 253
null spaces 1, 4, 131
NVC (Newton and velocity control) 204, 206
Nyquist criterion 5

observable systems 5
obstacle avoidance 143, 149–57, 321–4
operational space 199–201, 216
Orin and Schrader algorithm 75
output functions 56, 64, 78, 79 *see also* task functions
output regulation 79, 80

parametrizations 23–4, 36–41, 208, 211–12, 215, 251, 305
  differentials of 39–41
  Euler type 201
paths, differentiable 34
  shortest 34, 35
perturbations, singular 191, 223, 253–4
pitch angles 26, 39
pointing tasks 83
polynomials 180, 181
  characteristic 5, 221
position control 182, 314–17, 326, 340
positivity conditions 126–30, 166–7, 223, 231, 233, 248–53, 342, 346
potential functions 153, 321
precompensation 338, 339
primitives, sensor-based 280, 290, 291, 308, 312
  canonical 280, 281–6
  redundant 280, 287–90, 314
powers, virtual 277, 331, 339
projection operators 4, 165
  orthogonal 123, 163, 297, 343
proportional-derivative, structure 184
pseudoinverses 3–4

quaternions 26–33, 55, 251
  conjugates of 28
  pure 28, 29
  and rotations 29–30, 32–3
  unitary 28, 29, 63, 84

ranges (of matrices) 1
reachable space 70
recursive least squares 209
redundancy, task, *see under* tasks
redundancy rates 111, 117
regularization 101, 121
Renaud algorithm 74–5
resolved acceleration 200–1
resolved motion rate control 139
restrictions 114–15
ρ-admissibility 85, 87
ρ-admissible functions 85, 90
ρ-admissible tasks, *see under* tasks
Riemannian metrics 36
robustness 90, 168, 176, 215, 254
roll angles 26, 39
roll, pitch, yaw representation 26
rotation angles 23, 30–1, 33
rotation axes 23, 30–1, 33
rotation matrices 19, 22, 23, 31, 32–3
rotations
  parametrizations of, *see* parametrizations
  and quaternions 29–30, 32–3
Rouché's theorem, 255–6
Routh criterion 5, 179
Routh tables 6

Schur–Cohn criterion 6
screw products 44, 267–70, 275
screws 42–54, 55, 267, 268
  applications of 51–4
  definition of 42–3
  interaction 270–5, 299, 308, 323
  models of 278
  kinematic 52
  reciprocal 45
  velocity 43, 199, 267
SE_3 20, 47, 268, 327, 342
  differentials of maps on 49–51
  Lie algebra of 48
  trajectory tracking in 104–8
sensors
  binary 266
  contact 266
  external 173
  force 266, 293, 329–32, 346, 347, 348
  internal 173
  non-contact 265, 275, 301, 346

sensors (cont.)
  proximity 266
    optical 266, 271, 348–50
  reflectance 317, 349
  tactile
  thin-field 273–4, 279
Shannon's condition 337
singularities, Jacobian 94, 108, 134, 240, 252
  avoidance of 143, 144, 145
  control problems with 98–101
  geometrical 320
  kinematic 72, 95, 105
    attitude 98
    positioning 96–7
singular points 24, 99, 100, 143
skew symmetric matrix 23
sliders 46, 274, 279, 308
SO_3 22, 23
  distances on 35–6
  geodesics on 34, 35
  local parametrization of 38
Special Euclidean group, see SE_3
spectral radius 2
springs 292
  generalized 327, 328–9, 341, 350–1
    virtual 293
  simple 337–8
stability 16, 168, 218, 234, 296, 332–6, 347
  asymptotic 5, 6, 17, 221
  BIBO 5
  domains of 229
  exponential 17, 221, 222
  global 213–14
  practical tests of 5–8
  uniform 17
Stability Theorem 226–7, 231–5, 241
  proof of 258–64
states
  evolution of 78
  minimal 78
state–space representations 4, 78
stationary linear systems, see linear systems, stationary
stiffness matrices, see under matrices
submanifolds 10, 11
subspaces, reciprocal 275, 276
subsystems, open-loop 169
surface following 311–14

tangent bundles 9
tangent spaces 9, 10, 15, 18
task functions 79
  equivalent 101, 115
  sensor-based 266, 290–3
  simple 134–6

task redundancy, see redundancy, task
tasks 77
  closed-in-time 147, 148
  cyclic 147–9
  feasibility of 81
  global 294, 309, 346
  hybrid 295, 301, 302, 306–8, 341
  periodic 147
  realizable 89
  redundant 112, 253, 294–8, 302
    canonical 117, 294
    geometrically 113
    locally 114–17
    locally canonical 118
    see also redundancy rates
  ρ-admissible 89–92, 118, 121, 218, 220, 251, 252, 320
  weakly-admissible 89, 90
time constants 177
torque coefficients 170
torques, drive 172
  limitation of 143, 159–60
tracking
  perfect 83, 84
  of pointwise target 318–21
tracking error 229–31
trajectories, ideal 88, 89, 94, 103, 143, 195, 295
trajectory tracking 103–8
transmission ratios 171, 174, 204
transmissions 171–2
twists 52, 53, 277

uncertainties 274, 275, 277, 278

valley effects 185, 195
vector fields 15, 16
velocities 18, 267, 270, 305
  angular 37, 59, 267
  desired 139, 140, 202, 203, 205, 206
  frame 41–2, 48, 49, 51–2
  joint 157–9
  quaternion 37
  task 173
velocity control 139, 178–80, 181, 202–7
velocity loops 178, 203

Waldron algorithm 73–4
wrenches 52, 53, 277, 292, 341

yaw angles 26, 39